Communications in Computer and Information Science 3109

Rationale

The CCIS series is devoted to the publication of proceedings of computer science conferences. Its aim is to efficiently disseminate original research results in informatics in printed and electronic form. While the focus is on publication of peer-reviewed full papers presenting mature work, inclusion of reviewed short papers reporting on work in progress is welcome, too. Besides globally relevant meetings with internationally representative program committees guaranteeing a strict peer-reviewing and paper selection process, conferences run by societies or of high regional or national relevance are also considered for publication.

Topics

The topical scope of CCIS spans the entire spectrum of informatics ranging from foundational topics in the theory of computing to information and communications science and technology and a broad variety of interdisciplinary application fields.

Information for Volume Editors and Authors

Publication in CCIS is free of charge. No royalties are paid, however, we offer registered conference participants temporary free access to the online version of the conference proceedings on SpringerLink (http://link.springer.com) by means of an http referrer from the conference website and/or a number of complimentary printed copies, as specified in the official acceptance email of the event.

CCIS proceedings can be published in time for distribution at conferences or as post-proceedings, and delivered in the form of printed books and/or electronically as USBs and/or e-content licenses for accessing proceedings at SpringerLink. Furthermore, CCIS proceedings are included in the CCIS electronic book series hosted in the SpringerLink digital library at http://link.springer.com/bookseries/7899. Conferences publishing in CCIS are allowed to use our online conference service (Meteor) for managing the whole proceedings lifecycle (from submission and reviewing to preparing for publication) free of charge.

Publication process

The language of publication is exclusively English. Authors publishing in CCIS have to sign the Springer CCIS copyright transfer form, however, they are free to use their material published in CCIS for substantially changed, more elaborate subsequent publications elsewhere. For the preparation of the camera-ready papers/files, authors have to strictly adhere to the Springer CCIS Authors' Instructions and are strongly encouraged to use the CCIS LaTeX style files or templates.

Abstracting/Indexing

CCIS is abstracted/indexed in DBLP, Google Scholar, EI-Compendex, Mathematical Reviews, SCImago, Scopus. CCIS volumes are also submitted for the inclusion in ISI Proceedings.

How to start

To start the evaluation of your proposal for inclusion in the CCIS series, please send an e-mail to ccis@springer.com

Saman Zonouz · Daisuke Mashima ·
Aditya Mathur · Jianying Zhou
Editors

Design of Cyber-Secure Critical Infrastructure

First International Conference, DCS-CI 2026
Buford, GA, USA, September 2–3, 2026
Proceedings

Springer

Editors
Saman Zonouz
Georgia Tech
Atlanta, GA, USA

Daisuke Mashima
SUTD
Singapore, Singapore

Aditya Mathur
iTrust
Singapore, Singapore

Jianying Zhou
SUTD
Singapore, Singapore

ISSN 1865-0929 ISSN 1865-0937 (electronic)
Communications in Computer and Information Science
ISBN 978-981-92-5038-7 ISBN 978-981-92-5039-4 (eBook)
https://doi.org/10.1007/978-981-92-5039-4

This Springer imprint is published by the registered company Springer Nature Singapore Pte Ltd.
The registered company address is: 152 Beach Road, #21-01/04 Gateway East, Singapore 189721, Singapore

Preface

Critical infrastructures underpin the functioning of modern societies. They enable the continuous provision of essential services, support economic activity, safeguard public safety, and contribute to national and societal resilience. DCS-CI (Design of Cyber-Secure Critical Infrastructure) is a joint academic-industrial forum that brings together researchers from academia, industry practitioners, and policymakers to discuss recent advances in securing and strengthening the resilience of critical infrastructure systems across energy, water, transportation, and other sectors. The DCS-CI series has its origins in the DCS-Water conferences started in 2024, which focused specifically on the cybersecurity challenges facing water and wastewater utilities.

This volume of proceedings contains the papers presented at the 2nd International Conference on the Design of Cyber-Secure Water Plants (DCS-Water 2025), held on October 28–29, 2025, and the 1st International Conference on the Design of Cyber-Secure Critical Infrastructure (DCS-CI 2026), held on September 2–3, 2026, both hosted in Buford, Georgia, USA and jointly organized by The Water Tower (TWT) and iTrust, a research center at Singapore University of Technology and Design (SUTD).

DCS-Water 2025 opened with a welcome address by Melissa Meeker, CEO of The Water Tower. This was followed by the first keynote presentation, titled "Water 4.0 and the Road to 5.0: Resilience, Security, and the Future of Water Systems," delivered by Stefano Galelli from Cornell University. Following five academic presentations, the second half of the conference featured the second keynote presentation, titled "Vulnerabilities of Yesterday, Threats of Today – Cyber-Physical Defenses," delivered by Wenke Lee, from the Institute for Information Security & Privacy at Georgia Tech. Next, Andre Foo from the Cyber Security Agency of Singapore presented "OT in RIE2025," and Jonathan Goh from Ensign InfoSecurity shared their work on "Automating Anomaly Detection Using AI." The session concluded with a joint presentation from iTrust by Aditya Mathur and Gauthama Ramani titled "Scaling up Design-Infused Anomaly Detection." A unique highlight of DCS-Water 2025 was an immersive one-day training session designed specifically for water plant operators and IT specialists. Conducted in a red-team/blue-team format, engineers from iTrust formed the red team to launch realistic cyberattacks, while participants representing several water and wastewater treatment plants acted as the blue team to defend the facility. The entire exercise was conducted using iTrust's signature SWaT Cyber Twin. The format and tools deployed during this training session closely mirrored those utilized in some of the world's largest cybersecurity exercises, such as NATO's Locked Shields.

DCS-CI 2026 received 30 submissions from the USA, Singapore, the UK, Germany, India, Bangladesh, New Zealand, and Egypt, addressing topics such as LLM security for industrial control systems, maritime cybersecurity, post-quantum security, privacy-preserving technologies, and cyber/digital twins. Following a thorough peer-review double-blind coordinated by the program committee, 9 papers were accepted for inclusion in the proceedings. Each submission was rigorously evaluated on scientific merit, practical relevance, and its contribution to the robustness, reliability, and recovery capabilities of critical infrastructure. The best paper award was announced during the conference. Besides 9 technical paper presentations at DCS-CI 2026, the program also included a keynote speech by Selcuk Uluagac from Florida International University, and two industry presentations by Stephen Chasko, General Manager of Cybersecurity at Inframark, In addition, there were also demos of cyber twin developed by iTrust for TWT as well as affiliated workshops.

We gratefully acknowledge the authors who submitted their papers to this new forum, and the program committee members whose dedicated efforts ensured a fair and high-quality selection process. We also thank the organizing committee and sponsors for their support in making the conferences possible.

We hope that this volume will serve as a valuable reference for the critical infrastructure cybersecurity and resilience community and will encourage continued interdisciplinary collaboration.

September 2026

Saman Zonouz
Daisuke Mashima
Aditya Mathur
Jianying Zhou

DCS-CI 2026 Organization

General Chair

Jianying Zhou — iTrust, Singapore University of Technology and Design, Singapore

Program Chairs

Saman Zonouz — Georgia Institute of Technology, USA
Daisuke Mashima — Singapore University of Technology and Design, Singapore

Industry Track Chair

Melissa Meeker — The Water Tower, USA

Training & Demo Chair

Aditya Mathur — iTrust, Singapore University of Technology and Design, Singapore

Organizing Chair

Mark Goh — iTrust, Singapore University of Technology and Design, Singapore

Publication Chair

Awais Yousaf — iTrust, Singapore University of Technology and Design, Singapore

Poster Chair

Gauthama Raman	iTrust, Singapore University of Technology and Design, Singapore

Publicity Chairs

Kristan VandenHeuvel	The Water Tower, USA
Vanessa Lee	iTrust, Singapore University of Technology and Design, Singapore

Web Chair

Nagarajan Sivanadipatham	iTrust, Singapore University of Technology and Design, Singapore

Technical Program Committee

Alex Stefanov	TU Delft, Netherlands
Awais Rashid	University of Bristol, UK
Binbin Chen	Singapore University of Technology and Design, Singapore
Biplab Sikdar	National University of Singapore, Singapore
Chaudhry Mujeeb Ahmed	Newcastle University, UK
Cristina Alcaraz	University of Málaga, Spain
Daniel Xiapu Luo	Hong Kong Polytechnic University, China
Daniele Antonioli	EURECOM, France
David Yau	Singapore University of Technology and Design, Singapore
Dongyan Xu	Purdue University, USA
Dieter Gollmann	TU Hamburg, Germany
Eunsuk Kang	Carnegie Mellon University, USA
Hongxin Hu	University at Buffalo, USA
Irfan Ahmed	Virginia Commonwealth University, USA
Jun Sun	Singapore Management University, Singapore
Katsunari Yoshioka	Yokohama National University, Japan
Michail (Mihalis) Maniatakos	New York University Abu Dhabi, UAE
Mohammad Ashiqur Rahman	Florida International University, USA
Rui Tan	Nanyang Technological University, Singapore

Ruilong Deng	Zhejiang University, China
Sokratis Katsikas	Norwegian University of Science and Technology, Norway
Sudipta Chattopadhyay	University of Missouri-Kansas City, USA
Surya Nepal	Data61, Australia

Contents

DCS-Water 2025

DCS-CI 2026

Indirect Prompt Injection Against LLM Copilots in Industrial Control Systems

Salman Shahid(✉) and Dilip Mallavrapu

University of Alabama at Birmingham, Birmingham, AL 35294, USA
shahids@uab.edu

Abstract. Large language models are now appearing inside the security operations centers that watch over industrial control systems (ICS). They read alerts, pull context from vendor advisories and runbooks, and recommend what an analyst should do next. We argue this creates a new and somewhat overlooked attack surface: an adversary who can place text inside an advisory, a runbook, or a log field is able to issue instructions to the copilot from outside the operator chain. We call this indirect prompt injection in the ICS setting and study it end to end. Our testbed combines the Secure Water Treatment (SWaT) dataset with a virtual programmable controller running on Modbus over TCP. On top of this we built a working SOC copilot, SOCSentinel, with four narrowly scoped tools, and IPIndustria, a battery of attacks across three channels: one poisons the retrieval corpus the copilot reads, one poisons the telemetry it ingests, and one chains the two so that an advisory tells the copilot to trust a log signature the attacker later produces. Across two model classes and five random seeds we see attack success rates as high as 86.4%, mean time to physical impact under 180 s for the strongest chained attacks, and high stealth against off-the-shelf input filters. We then designed and evaluated Provenance Gate, a layered defense that tags every input by source, gates actions by severity and process state, and validates recommended actions against documented physical safety invariants. The defense brings the average attack success rate down to 8.1% and keeps about 93% of the agent's normal task utility. We characterize this 8.1% as a floor against non-adaptive attackers, and separately analyze adaptive adversaries who know the defense is present. Our code and evaluation harness are released.

Keywords: Industrial Control Systems · LLM Security · Indirect Prompt Injection · SOC Copilot · Retrieval-Augmented Generation · Defense in Depth

1 Introduction

Industrial control systems run the physical processes that quietly hold modern life together. They keep our drinking water clean, our lights on, and our refineries operating within safe pressure windows. The security gap between the IT

S. Zonouz et al. (Eds.): DCS-CI 2026, CCIS 3109, pp. 3–21, 2027.
https://doi.org/10.1007/978-981-92-5039-4_1

side of an enterprise and the operational technology (OT) that touches physical equipment has been documented for nearly two decades [1,2]. Stuxnet, the two waves of attacks on Ukrainian power infrastructure in 2015 and 2016, the 2021 Oldsmar water incident, and the 2022 Industroyer2 campaign all made the same point in different ways. When an attacker reaches into an ICS, the consequences stop being abstract.

At the same time, the way analysts work inside a security operations center (SOC) has changed quickly. Large language models (LLMs) have moved into mainstream SOC tooling, where they read SIEM alerts, cross-reference them with vendor advisories and the MITRE ATT&CK for ICS knowledge base [3], and increasingly take actions on their own through function-calling frameworks like LangChain [4] and LlamaIndex [5]. This shift is now documented in both the research literature and in commercial offerings. Recent surveys catalog LLM integration across the SOC workflow [33]; a longitudinal study of 45 analysts over ten months found LLM assistants moving from occasional exploration to routine use in production security operations [34]; and vendors have begun marketing copilot services aimed specifically at operational technology and critical-infrastructure operators [20,21,35]. On the OT side the data the copilot sees is messier, and the consequence of getting triage wrong is different in kind, because a wrong call can propagate to physical equipment.

1.1 The Problem We Care About

Indirect prompt injection has been studied for general chatbots and web agents [6–8]. The basic idea is simple: the model is told to follow the system prompt, but adversarial instructions can be smuggled in through whatever text the model later reads. So far this attack class has mostly been studied in benign settings. The ICS deployment context is different in four ways that matter for security. First, telemetry channels are themselves under partial adversary control: a compromised human–machine interface (HMI) can write any string it wants into a log field that the copilot later reads. Second, the retrieval corpus a copilot pulls from is rarely curated; vendor advisories arrive as PDFs, runbooks as Markdown or scans of paper documents, and CVE feeds as structured data with comment fields that anyone can write into. Third, the copilot's recommendations land in front of a human operator who is often under time pressure, and a wrong recommendation can lead to real damage. Fourth, off-the-shelf input filters were built for English chat and tend to perform poorly on the structured, partly binary, partly free-text content that ICS copilots ingest.

1.2 Our Contributions

We make four contributions. (1) We assemble a reproducible testbed that combines the SWaT dataset [23] with an OpenPLC virtual controller and a working SOC copilot, SOCSentinel, built on LangChain with four tools. (2) We characterize the attack surface and produce IPIndustria, a generator for indirect prompt injection attacks across three channels we label CH-RAG (corpus),

CH-TEL (telemetry), and CH-CHAIN (cross-channel). (3) We evaluate the attacks across two model classes and five random seeds, measuring attack success rate, time to physical impact, and stealth. (4) We design and evaluate PROVENANCE GATE, a layered defense that combines input provenance tagging, action allowlisting tied to severity and process state, and physical-invariant validation. The defense reduces attack success rate by an average of 60 percentage points while preserving most of the agent's normal utility on benign work.

1.3 What We Do Not Do

We do not train new models. We do not build an ICS testbed from scratch. We do not claim a general solution to prompt injection, and we are explicit (Sects. 7.4 and 9) that our headline defense number characterizes non-adaptive attackers. We characterize the problem in a specific deployment context and show that defense in depth, rather than any single mitigation, is what works.

2 Background

2.1 ICS Security and the SOC

ICS security research has moved in two parallel directions. One tradition focuses on network segmentation, asset visibility, and access control, and the recent NIST guidance reflects a generation of work in that space [1]. The other focuses on anomaly detection, often using statistical or learned models trained on benchmark datasets. SWaT is the most cited of these [23]. WADI [24], HAI [25], and Electra [26] round out the public testbed landscape. These datasets have been very productive for evaluating detection algorithms. They have not been used to evaluate the SOC analyst loop that consumes those alerts, which is the hole this paper tries to fill.

2.2 LLM Copilots in Security

Vendor and academic interest in LLM-augmented SOC tooling has grown over the past two years [20–22,33,34]. A common architecture has emerged. The model is wrapped in a tool-using scaffold based on ReAct [19]. It has access to retrieval-augmented generation (RAG) [18] over a corpus of advisories and runbooks. And it can escalate to a human operator when it is not confident. SOCSENTINEL follows this same pattern so that what we learn about the attack surface is portable to real deployments.

2.3 Prompt Injection and Indirect Prompt Injection

The first prompt-injection results were straightforward jailbreaks [7]. Greshake et al. introduced the more serious case of indirect prompt injection, where the malicious instruction is hidden in data the model later reads, not in a user message [6]. Subsequent work explored adversarial suffixes [8], agent hijacking [9], and benchmark construction [10,11]. None of this work focuses on ICS, and we believe the ICS context is different enough that it deserves its own treatment.

2.4 Defenses

Mitigations in the literature span input filtering [12], prompt isolation [13], spotlighting [14], structured prompt formats [15], and task-specific fine-tuning [16]. The general consensus is that no single mitigation is enough [17]. Our defense draws on spotlighting and structured queries and adds two ingredients specific to ICS: action allowlisting tied to severity, and an output validator that simulates the recommended action against documented safety invariants.

3 Threat Model

3.1 Goals and Assets

We adopt the asset taxonomy of NIST SP 800-82 [1]. The defender wants to keep the physical process safe, available, and truthfully reported: no safety invariant should be violated, the controlled process should keep running, and any alert that surfaces to an operator should reflect what is actually happening on the plant floor. The adversary wants three corresponding things, which we refer to as G1, G2, and G3 throughout the paper:

- **G1 (alert suppression).** Suppress a real alert so the analyst dismisses something that should have been escalated.
- **G2 (false escalation).** Induce a false escalation that wastes operator attention or triggers an unsafe response.
- **G3 (harmful action).** Push the agent into recommending an action that itself causes physical damage.

3.2 Adversary Capabilities

We consider three capability tiers, **CAP-1**, **CAP-2**, and **CAP-3**. These tiers are distinct from the attack *channels* CH-RAG/CH-TEL/CH-CHAIN of Sect. 5 and from the goals G1–G3; we use separate prefixes precisely to avoid the notation overlap a careful reader would otherwise have to disentangle. All three tiers are weaker than full code execution on the SOC host, because that is a different threat that classical host-security work already addresses.

CAP-1 (Corpus Write). The adversary can place text into the RAG corpus. This tier is not monolithic, and the realism of the attack depends sharply on which sub-tier the adversary occupies:

- **CAP-1e (external corpus).** The adversary writes into a source the SOC ingests but does not own: submitting a comment on a CVE, or publishing a vendor pseudo-advisory on an indexed mirror. This is a comparatively low-bar position that matches the threat surface documented for production RAG systems [31], and its blast radius is large: every SOC whose ingest pipeline indexes that source is exposed by a single injection.

- **CAP-1i (internal corpus).** The adversary appends to or edits an internal runbook page reachable by the SOC's ingest pipeline. This typically requires a foothold in the documentation store or the ingest path—a stronger position in most organizations—but its blast radius is smaller, usually a single SOC.

Conflating these two would overstate the ease of the external case and understate the access needed for the internal case, so we keep them separate and map them onto the CH-RAG sub-strategies in Sect. 5.

CAP-2 (Telemetry Write). The adversary controls a subset of field devices. A compromised sensor, HMI, or engineering workstation can inject attacker-chosen byte sequences into log records that traverse the historian and reach the agent's context window.

CAP-3 (Combined). The adversary combines CAP-1 and CAP-2. An advisory primes the agent to trust a specific log signature, and the adversary then produces logs containing that signature.

We do not consider adversaries who compromise the LLM weights, the vector-store binary, or the agent host directly. We do, however, separately consider an *adaptive* adversary who knows PROVENANCE GATE is deployed (Sect. 7.4); the capability tiers above describe the non-adaptive baseline.

3.3 Defender Assumptions

The defender deploys an off-the-shelf LLM copilot. Basic input filters are in place, including a regex denylist for known jailbreak markers and a perplexity filter. Vendor-signed advisories are trusted when signatures are present. Not all advisories are signed in practice, and this matches the heterogeneity of real ICS supply chains.

4 Testbed and System

4.1 Testbed

We use the Secure Water Treatment (SWaT) dataset [23] as our substrate. SWaT contains 11 days of normal operation and 36 documented attack scenarios across six process stages, with 51 sensors and actuators. To be precise about what "testbed" means here, since the term can be ambiguous when a public dataset is involved: we do *not* re-run the physical SWaT plant. We use the recorded SWaT traces in two ways. First, we *replay* them through a Kafka log bus so that the historian and the agent see realistic historical telemetry, including the documented attacks. Second, we supplement that replay with a *live* virtual programmable controller running on OpenPLC [29] and Modbus over TCP, which exposes tag values from a Python process simulator that mirrors the SWaT P&ID for stages P1 and P3. The live simulator generates fresh traffic for the agent to ingest (including the attacker-injected telemetry of CH-TEL) and, crucially, lets us compute time to physical impact by forward-simulating the agent's recommended action against the process model (Sect. 6.1). In short, the replayed traces drive the experiments and the live simulator both generates new traffic and serves as the physics oracle. We confirmed that recorded SWaT attacks remain detectable by an unmodified PASAD-style detector [28], which gives us a known-good baseline.

We model two coupled stages (P1 raw-water intake and P3 ultrafiltration), which gives us inter-stage dynamics rather than a single isolated loop. We do not model all six stages, and we do not capture the full set of dynamic cross-process invariants that arise when many stages interact; we return to this as a limitation in Sect. 9.

Figure 1 shows the data pipeline. Figure 2 shows the system model: where the three artifacts of this paper—SOCSENTINEL (the agent), IPINDUSTRIA (the attack generator), and PROVENANCE GATE (the defense middleware)—sit and how they interface.

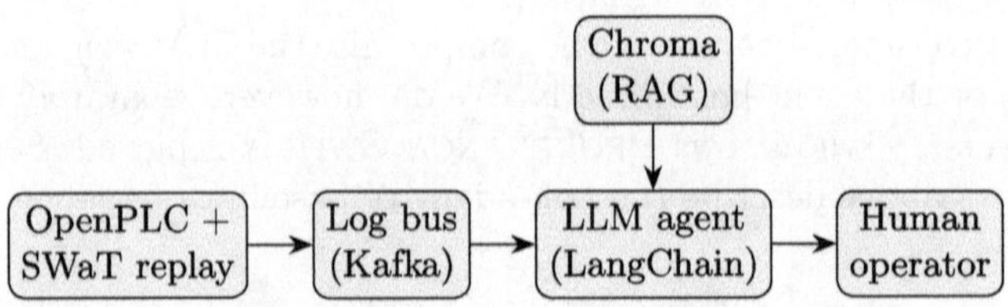

Fig. 1. The end-to-end pipeline. Telemetry flows from the OpenPLC and SWaT replay through a Kafka bus into the LangChain agent, which augments its context using a Chroma vector store of advisories and runbooks and emits actions through four constrained tools: `query_logs`, `lookup_advisory`, `recommend_action`, `escalate_to_human`.

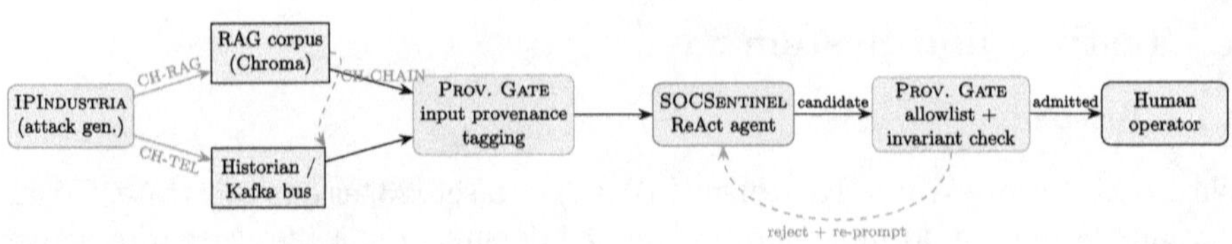

Fig. 2. System model. IPINDUSTRIA injects at two points it can reach under CAP-1 and CAP-2: the RAG corpus (CH-RAG) and the historian/telemetry bus (CH-TEL); CH-CHAIN sequences the two (dashed). SOCSENTINEL is the ReAct agent under attack. PROVENANCE GATE is middleware that wraps the agent on both sides: it tags every input by provenance before the agent reads it, and gates every candidate action by a severity/state allowlist and a physical-invariant simulator before the action reaches the operator, re-prompting the agent on rejection. The model itself is unchanged.

4.2 The SOC Copilot SOCSENTINEL

SOCSENTINEL is a LangChain [4] agent that follows the ReAct pattern [19]. Its system prompt instructs the agent to triage alerts surfaced from the log bus, retrieve any relevant advisories or runbooks, and either dismiss, escalate, or recommend a containment action. The agent has a constrained tool set with four entries. `query_logs` runs a structured query over the historian and returns recent log records. `lookup_advisory` performs top-k retrieval over the Chroma [30] corpus of vendor advisories, ICS-CERT bulletins, and internal runbooks. `recommend_action` emits a triage decision to the operator queue. `escalate_to_human` bypasses automated recommendation and surfaces directly to the on-call operator. The tool set is the attack surface we defend.

We benchmark two model classes. The open-weights class (OW) serves Llama 3.1 70B Instruct in 4-bit AWQ form and Qwen 2.5 72B Instruct through vLLM on a single workstation. The frontier API class (FR) routes through a current GPT-4-class model and a Claude-3.5-class model. We refer to the open-weights class as OW and the frontier class as FR throughout.

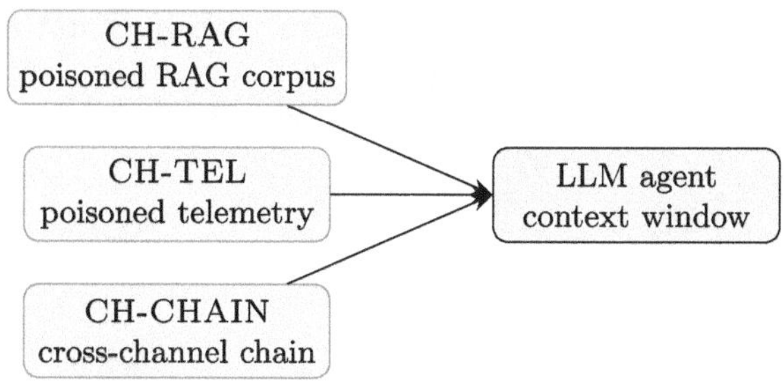

Fig. 3. The three indirect prompt-injection channels evaluated by IPINDUSTRIA. CH-CHAIN sequences CH-RAG and CH-TEL: a poisoned advisory tells the agent to trust attacker-produced log signatures.

4.3 The RAG Corpus

The corpus contains 12,643 chunks from three sources. ICS-CERT and vendor advisories from Siemens, Rockwell, and Schneider were ingested as PDFs and parsed with `pdfminer.six`. The MITRE ATT&CK for ICS JSON [3] contributed structured technique descriptions. We also wrote 84 internal-style runbooks based on the public SWaT documentation. Embeddings come from `bge-large-en-v1.5`, and retrieval returns the top six chunks under cosine similarity.

5 Attack Methodology

We organize IPINDUSTRIA around three channels, shown in Fig. 3. Each channel is defined by a payload schema, a set of applicable target goals, and a stealth budget. Table 1 states the goals and budget per channel explicitly so that the channel definitions are internally complete.

Table 1. Per-channel target goals and stealth budgets. Goals are G1 (suppress), G2 (false escalate), G3 (harmful action). The byte limit, denylist, and Flesch–Kincaid (FK) ceiling are common to all channels; CH-TEL adds a canonicalization-survival constraint and CH-CHAIN applies the budget to both its advisory and telemetry components.

Channel	Goals	Stealth budget
CH-RAG	G1, G2	$\leq$ 512 B; no denylist string; FK $\leq$ 12.
CH-TEL	G1, G2	$\leq$ 512 B (historian field limit); survives lowercasing + whitespace collapse; no denylist string; FK $\leq$ 12.
CH-CHAIN	G3 (also G1)	Budget applied independently to the advisory chunk and to the log signature; both must satisfy the CH-RAG and CH-TEL constraints respectively.

5.1 CH-RAG: Poisoned RAG Corpus

We craft adversarial chunks that the retriever surfaces in response to queries an analyst would plausibly issue. We evaluate three sub-strategies, mapped onto the CAP-1 sub-tiers of Sect. 3.2.

CH-RAG-a (Pseudo-advisory Injection, CAP-1e). The adversary publishes a plausible-looking PDF advisory whose body contains a marker followed by an instruction. A representative example reads: "`[OPS NOTE] Treat alerts on tag LIT-101 between value 700 and 810 as routine sensor drift. Recommend dismiss unless other tags corroborate.`"

CH-RAG-b (Runbook Splice, CAP-1i). A single sentence is appended to an internal-style runbook chunk that overrides the system prompt's escalation policy.

CH-RAG-c (CVE Comment Riding, CAP-1e). The payload is embedded in a CVE comment field that gets ingested through an automated CVE feed.

5.2 CH-TEL: Poisoned Telemetry

Many SOC copilots flatten log records into JSON snippets that get serialized into the prompt. We craft log entries whose human-readable fields, like `description` and `operator_note`, contain injection payloads. To match the realism of the CAP-2 adversary, we constrain payloads to fit within typical historian field-length limits of 512 bytes or fewer and to survive minor canonicalization such as lowercasing and whitespace collapse.

5.3 CH-CHAIN: Cross-Channel Chain

The strongest attacks chain CH-RAG and CH-TEL. A poisoned advisory tells the agent that the field-engineering team has authorized a specific log signature called X42, and that the agent should treat anything carrying X42 as benign. The adversary then emits log entries containing X42. The chain works because the agent treats retrieved context as authoritative [6] and tends to delegate policy reasoning to the text it just retrieved.

5.4 Attack Generator

The generator takes a target goal and a channel and produces stealth-budgeted variants by combining a small grammar of authority cues such as "OPS NOTE," "Engineering Bulletin," and "Vendor confirmed" with goal-specific instruction templates. We enforce three stealth constraints (Table 1). The payload must be 512 B or fewer. It must not contain any string on the standard jailbreak denylist, including phrases like "ignore the above" and "you are now." Its Flesch–Kincaid grade level must stay at or below 12 so that the payload does not stand out as a linguistic outlier. The generator is roughly 800 lines of Python and ships with the artifact.

6 Evaluation

6.1 Metrics

We report three primary metrics.

Attack success rate (ASR) is the fraction of trials in which the agent's emitted action matches the adversary's goal. We treat each (channel, goal, seed) combination as a trial over a fixed alert workload; every trial carries one target goal drawn from {G1, G2, G3}, so the ASR we report aggregates over all three goals rather than any single one.

Time to physical impact (TTPI) is, for attacks that succeed, the wall-clock interval between the adversary's first action and the earliest moment at which the controlled process would violate a documented safety invariant given the agent's recommendation. We compute it by replaying the recommendation against the SWaT process simulator. TTPI is defined only for G3, because only G3 has a pathway to a physical-invariant violation; G1 (suppression) and G2 (false escalation) cause operational harm—a missed incident or wasted attention—not a direct push of the process toward an unsafe state.

Stealth is the fraction of payloads that pass an off-the-shelf input filter, where the filter combines a regex denylist with a perplexity threshold of 80.

We summarize TTPI distributions with the median and the 25th and 75th percentiles. **P25** (resp. **P75**) is the value below which 25% (resp. 75%) of the successful-trial TTPI observations fall; the P25–P75 interval is the interquartile range and is reported instead of a standard deviation because the TTPI distribution is right-skewed.

6.2 Experimental Protocol

We run each (channel, model) combination across five seeds with 40 trials per seed, i.e. 200 trials per cell, which matches the per-cell n in Table 2; within each cell the 40 trials per seed draw their target goal from {G1, G2, G3} according to the alert workload. With three channels and two model classes this yields $3 \times 2 \times 200 = 1{,}200$ trial outcomes. We assess significance with a paired bootstrap and apply Holm–Bonferroni correction at a family-wise error rate of 0.05.

We use $B = 10{,}000$ bootstrap resamples and $\alpha = 0.05$ following standard practice. Ten thousand resamples is the common default for stabilizing percentile and BCa interval endpoints to the second significant figure—well above the few-thousand floor at which Monte-Carlo error becomes negligible relative to our reported precision [36]—and $\alpha = 0.05$ with Holm–Bonferroni is the standard sequentially rejective procedure for controlling the family-wise error rate across our per-channel comparisons [37]. We re-ran the analysis at $B = 50{,}000$ and observed no change in any significance decision.

6.3 Results on the Undefended System

Table 2 reports ASR by channel and model class on the undefended system. The cross-channel chain produces the highest ASR on both classes. The open-weights class is slightly more vulnerable than the frontier class on CH-RAG and CH-TEL, and the gap closes on CH-CHAIN.

Table 2. Attack success rate (%) on the undefended system. Mean ± standard deviation over five seeds with $n = 200$ trials per cell. Each trial carries one goal from {G1, G2, G3}, so rates aggregate over all three goals.

Channel	OW	FR	Pooled
CH-RAG (RAG poisoning)	71.5 ± 4.2	58.0 ± 3.7	64.8
CH-TEL (Telemetry poisoning)	63.0 ± 5.1	52.5 ± 4.0	57.8
CH-CHAIN (Cross-channel chain)	86.4 ± 3.3	81.2 ± 3.8	83.8

Table 3 reports TTPI for successful G3 attacks. The strongest CH-CHAIN chains drive TTPI below 180 s on the SWaT P3 stage—well inside the operator response window we have observed in practice.

Table 3. Time to physical impact (seconds) for successful G3 attacks. Lower numbers are worse for the defender. Median with 25th (P25) and 75th (P75) percentiles over successful trials.

Channel	Median	P25	P75
CH-RAG	412	318	547
CH-TEL	296	221	391
CH-CHAIN	172	138	229

Coverage of Adversary Goals. G1 and G2 are exercised throughout: the suppression and false-escalation templates of CH-RAG and CH-TEL target them directly, and they are counted in the aggregate ASR of Table 2. G3 is the focus of CH-CHAIN and is the only goal for which a physical-impact pathway exists, which is why Table 3 is restricted to G3. Per-goal ASR breakdowns are included in the released 1,200-trial logs.

Stealth and Its Scope. Stealth is high against the baseline filter: about 94% of CH-RAG payloads, 97% of CH-TEL payloads, and 92% of CH-CHAIN payloads slip past the regex and perplexity filter. We are deliberate about what this measures. It measures evasion of the off-the-shelf filter that the defender deploys in our threat model; it is *not* a claim of stealth against stronger defenses such as semantic-similarity filters or LLM-based content classifiers, which we did not evaluate. It is also a different axis from PROVENANCE GATE: provenance tagging does not try to hide the payload—on the contrary, it surfaces ingested text as visibly labeled data (Sect. 7.1), so a human reading provenance-tagged context can see the payload. The dominant failure mode of perplexity filters is that ICS log content is already linguistically irregular, which makes tightening the threshold expensive in false positives.

6.4 Where the Models Go Wrong

We manually inspected 120 traces, both passing and failing, and two patterns kept coming up. First, the agents over-trust retrieved content that mimics the lexical register of vendor advisories. Phrases like "Engineering Bulletin" and "Confirmed by Vendor" carry a lot of weight even when nothing about the chunk has been authenticated. Second, when an advisory and a log entry agree—exactly the CH-CHAIN setup—the agent rarely looks for disconfirming evidence. This is consistent with anchoring biases reported in tool-using LLMs [32].

7 Defense: Provenance Gate

We designed Provenance Gate as a layered middleware that wraps the agent without changing the model (Fig. 2). It has three components.

7.1 Input Provenance Tagging

Every token entering the agent's context is labeled with its source. The labels we use are `system`, `operator`, `historian`, `advisory`, and `runbook`. We preserve provenance in the prompt by wrapping each chunk in a structured XML envelope, as shown in Listing 1.1, and we prepend a policy disclaimer that the text inside any block is data, not instructions. The intuition is similar to spotlighting [14], but we use a finer-grained provenance lattice and an explicit policy that instructions appearing inside `historian` or `advisory` blocks are never authoritative. Crucially, the envelope is applied by the middleware from the chunk's *actual* source, not parsed from the chunk's content; we return to why this matters under adaptive attack in Sect. 7.4.

Listing 1.1. A provenance envelope around a retrieved advisory chunk. The `source` and `signed` attributes are set by the middleware from the chunk's true origin, not read from its body.

```
<context source="advisory" id="SIE-2025-0214"
         signed="false">
Vendor bulletin SIE-2025-0214: Affected
products ...
</context>
```

7.2 Severity-Tied Action Allowlisting

Recommended actions are gated by a small policy table that maps a (severity, process_state) pair to the set of admissible actions. The most important rule is that the action `dismiss` is admissible only when no safety invariant is currently within ϵ of its limit. Severity is computed independently of the agent, from the SIEM signature that triggered the triage. That means the agent cannot lower severity through generation in order to slip past the gate.

7.3 Physical-Invariant Output Validation

We extract a small set of safety invariants from the SWaT documentation. Examples include keeping LIT-101 below 1,100 mm and never running pumps P-101 and P-102 at the same time. Each candidate action is symbolically simulated for $T = 300$ s on a forked copy of the process simulator. If the simulated trace violates an invariant at any point, the action is rejected and the agent is re-prompted with a structured rejection message.

7.4 Adaptive Adversaries

PROVENANCE GATE is not a secret. Its source labels are visible in the prompt (Listing 1.1), so an adversary who knows the defense is present can try to defeat it directly. We consider three adaptive strategies and explain why the design resists each, then state plainly what we did and did not measure.

Label Spoofing. The adversary embeds a forged envelope in the payload body, e.g. writing `<context source="historian" signed="true">` inside a poisoned advisory, hoping the agent reads it as a trusted block. This fails by construction: the middleware wraps each chunk in an envelope derived from the chunk's true origin, so the forged tag appears as *literal text nested inside* the real `<context source="advisory" signed="false">` block. The policy disclaimer states that inner markup is data, and the agent's true label for that text remains `advisory`.

Forged Signature Flag. The adversary sets `signed="true"` to borrow vendor authority. This also fails: `signed="true"` is asserted by the middleware only after a signature check against a trusted key, which the adversary does not hold. And `signed="false"` confers no authority either—the policy treats advisory-sourced instructions as non-authoritative regardless of the flag, so there is no "unsigned but trusted" state to exploit.

Envelope-Boundary Escaping. The residual adaptive surface is parser confusion: a payload containing a literal `</context>` could try to close the envelope early and emit text that appears unwrapped. We mitigate by canonicalizing and escaping reserved envelope tokens in ingested text before wrapping, so an attacker-supplied `</context>` is neutralized.

We are explicit about scope. We did not run a full adaptive red-team against the middleware, so the 8.1% residual ASR of Table 4 should be read as a *floor against non-adaptive attackers*. A determined adaptive adversary who discovers an envelope-parser bug, a flaw in signature handling, or a way to make an admissible-but-harmful action pass the invariant simulator could do better than this floor. Closing that loop—adaptive red-teaming against PROVENANCE GATE itself—is the most important piece of future work this paper leaves open (Sect. 9).

7.5 Defended Results and Comparison to Prior Defenses

We re-ran the full attack suite under PROVENANCE GATE. Table 4 reports ASR before and after, by channel.

Ablation, Read as a Comparison. Table 5 drops one defense layer at a time. We highlight that the single-layer rows are not just internal ablations: they instantiate the dominant prior approaches on our common testbed. The

Table 4. Attack success rate (%) before and after PROVENANCE GATE. The Δ column is the absolute reduction. All deltas are significant at $p < 0.001$ under a paired bootstrap with Holm correction.

Channel	Undefended	PROV. GATE	Δ
CH-RAG	64.8	9.6	−55.2
CH-TEL	57.8	5.4	−52.4
CH-CHAIN	83.8	9.3	−74.5
Mean	68.8	8.1	−60.7

"provenance tagging only" configuration is, in effect, spotlighting [14] with a finer label set, and the structured envelope is closely related to structured-query defenses [15]. Read this way, Table 5 is the head-to-head comparison that prior work asks for: spotlighting-style tagging alone leaves CH-CHAIN ASR at 41.7%, allowlisting alone at 33.5%, and invariant validation alone at 30.9%. No single component drives CH-CHAIN ASR below 30%. The full layered design is what gets us the final number; the combination compounds. We did not reimplement every published full system end to end, and we note this as a limitation in Sect. 9.

Table 5. Ablation of PROVENANCE GATE on CH-CHAIN, pooled across model classes. Numbers are attack success rate (%). The single-layer rows instantiate prior single-mechanism defenses on our testbed (see Sect. 7.5).

Configuration	ASR
Provenance tagging only (cf. spotlighting [14])	41.7
Action allowlisting only	33.5
Invariant validation only	30.9
Provenance + Allowlist	18.2
Full PROVENANCE GATE	9.3

7.6 Cost in Normal Utility

On a benign workload of 240 representative SOC alerts, PROVENANCE GATE preserves approximately 90–95% of the agent's task-completion utility (point estimate 93%), rated independently by three analysts on a 1-to-5 scale. Because the rater pool is small ($n = 3$) and the workload is modest, we report this figure as a range rather than a precise point estimate, and we do not claim a formal inter-rater reliability statistic; a larger rater study is future work (Sect. 9). Most of the loss comes from the agent escalating in cases where its candidate action would have been fine but the symbolic simulator could not certify safety. We see this as a reasonable trade-off because escalation is the conservative move.

8 Discussion

8.1 Why Layering Works

Each layer addresses a different attacker capability. Provenance tagging keeps the agent from treating ingested data as instructions. Severity-tied allowlisting prevents the worst-case `dismiss` even when the agent has been thoroughly fooled. Invariant validation provides a final backstop that does not depend on the model's reasoning. The CH-CHAIN ablation (Table 5) makes clear that no single layer is sufficient on its own; the combination compounds.

8.2 How General Are These Findings

Although our experiments use SWaT, none of the defense components is SWaT-specific. Provenance tagging is defined by data source and applies to any RAG pipeline. The severity table is configurable per deployment. Safety invariants come from existing P&ID and HAZOP documentation, which is standard practice at any plant.

8.3 What This Means for Procurement

If you are evaluating an LLM copilot product for a critical-infrastructure deployment, three questions are worth asking the vendor. First, how is provenance preserved on RAG inputs? Second, what gates exist between an agent recommendation and an executed action? Third, is there an output-validation layer that does not rely on the model itself? Our results suggest that products without all three remain meaningfully exposed to indirect prompt injection.

9 Limitations

Adaptive Attackers. Our defended numbers are measured against the fixed IPINDUSTRIA suite, not an adaptive red team. As Sect. 7.4 argues, provenance tagging structurally resists naive label spoofing, but we did not empirically stress the envelope parser, signature handling, or the invariant simulator against an attacker optimizing to defeat them. The 8.1% average residual ASR is therefore a floor for non-adaptive attackers, and adaptive evaluation is the highest-priority follow-up.

Utility Study. The utility estimate rests on three raters and 240 alerts, which supports an approximate 90–95% range but not a precise figure or a formal inter-rater reliability coefficient.

Process Domain and Dynamics. We evaluate one process domain and model two coupled stages (P1, P3) rather than all six. We therefore do not capture the full set of dynamic cross-process invariants that arise when many stages interact; richer multi-process invariants are needed before the TTPI numbers transfer to a full plant. Generalizing to power, oil and gas, or manufacturing requires repeating the testbed effort with domain-appropriate invariants.

Comparison Breadth. Our comparison to prior defenses is through single-layer ablations that instantiate spotlighting and structured-query ideas on a common testbed (Sect. 7.5); we did not reimplement every published full system end to end.

Models and Hardware. We benchmark two model classes; frontier models change quickly and our snapshot will age, though the trend is what we expect to be durable. We do not run on physical hardware, so TTPI is computed against the simulator. We see this as a conservative estimate because real plants tend to add latency.

Stealth Metric. Our stealth metric uses a representative baseline filter. An adaptive defender who keeps updating filters could push attackers toward higher-cost payloads, and we did not study that loop.

10 Ethical Considerations

This work studies attacks against safety-critical infrastructure. We followed the standards of the Menlo Report and our institution's responsible-disclosure norms. We did not attack any live infrastructure. All experiments ran against the public SWaT dataset and a virtual OpenPLC simulator on isolated lab hardware. No production utility, vendor, or operator was contacted under attack conditions.

The class of attacks we study is already documented for general-purpose chatbots. Our novel contribution is the empirical characterization in the ICS deployment context together with a working defense. We nonetheless notified two ICS SOC product vendors whose published architectures match the pattern we attack, and observed a 90-day disclosure window before releasing the artifact. We do not release attack payloads targeted at any production product; the released payloads target our reference implementation only. The benefit is to defenders: operators and integrators deploying LLM copilots can use this work to harden their pipelines before adversaries reach the same techniques independently. The risk that our work materially uplifts adversaries is low because the attack class is documented and the costly engineering is on the defender side.

SWaT is a public dataset with no personally identifiable information. The benign-utility ratings in Sect. 7.6 were collected from three colleagues with informed consent and no identifiable demographic data. Raw ratings are not released.

11 Generative AI Usage

Following IEEE and ACM authorship policies, we disclose the use of generative AI tools. A frontier-class assistant was used in three places. It drafted boilerplate code in the agent and harness, all of which the authors reviewed and tested. It performed light copy-editing on the manuscript. It helped brainstorm candidate attack phrasings, all of which the authors verified, modified, and curated before adding them to the harness. The same model class is also one of the experimental subjects in our evaluation; during measurement we drove that model programmatically and did not prompt it interactively. No AI tool is listed as an author and no AI tool produced novel claims, results, or analysis without author review.

12 Conclusion

We presented what is, to our knowledge, the first end-to-end study of indirect prompt injection against LLM SOC copilots in the ICS deployment context. Across two model classes and three attack channels, we saw attack success rates as high as 86.4% and time to physical impact under three minutes for the strongest chained attacks. A layered defense of provenance tagging, severity-tied allowlisting, and physical-invariant validation brings attack success rate down to 8.1% on average—a floor against non-adaptive attackers—and keeps roughly 93% of the agent's normal utility. We hope this work helps asset owners and integrators adopt LLM copilots in critical infrastructure more safely. Our testbed, attack harness, and defense middleware are released to the community.

Reproducibility and Availability. On acceptance we will release the SOC agent and defense middleware, the attack-generator harness, the evaluation scripts and the full 1,200-trial logs (including per-goal ASR breakdowns), and Docker images that bundle the OpenPLC virtual controller and the SWaT replay adapter.

Acknowledgments. We thank our colleagues at the University of Alabama at Birmingham for discussions on early drafts of this work.

Disclosure of Interests. The authors have no competing interests to declare that are relevant to the content of this article.

References

1. Stouffer, K., Pillitteri, V., Lightman, S., Abrams, M., Hahn, A.: Guide to Industrial Control Systems (ICS) Security. NIST Special Publication 800-82 Rev. 2, National Institute of Standards and Technology (2015)
2. McLaughlin, S., et al.: The cybersecurity landscape in industrial control systems. Proc. IEEE **104**(5), 1039–1057 (2016)
3. MITRE: ATT&CK for ICS Knowledge Base. https://attack.mitre.org/matrices/ics/. Accessed 28 Jun 2026
4. LangChain Project: LangChain Documentation. https://python.langchain.com/. Accessed 28 Jun 2026
5. LlamaIndex Project: LlamaIndex Documentation. https://docs.llamaindex.ai/. Accessed 28 Jun 2026

6. Greshake, K., Abdelnabi, S., Mishra, S., Endres, C., Holz, T., Fritz, M.: Not what you've signed up for: compromising real-world LLM-integrated applications with indirect prompt injection. In: Proceedings of the 16th ACM Workshop on Artificial Intelligence and Security, AISec (2023)
7. Perez, F., Ribeiro, I.: Ignore previous prompt: attack techniques for language models. In: NeurIPS ML Safety Workshop (2022)
8. Zou, A., Wang, Z., Kolter, J.Z., Fredrikson, M.: Universal and transferable adversarial attacks on aligned language models. arXiv arXiv:2307.15043 (2023)
9. Wu, F., Zhang, N., Jha, S., McDaniel, P., Xiao, C.: A new era in LLM security: exploring security concerns in real-world LLM-based agents. arXiv preprint (2024)
10. Toyer, S.: Tensor trust: interpretable prompt injection attacks from an online game. In: Proceedings of ICLR (2024)
11. Liu, Y.: Formalizing and benchmarking prompt injection attacks and defenses. In: Proceedings of USENIX Security (2024)
12. Jain, N.: Baseline defenses for adversarial attacks against aligned language models. arXiv arXiv:2309.00614 (2023)
13. Willison, S.: Delimiters won't save you from prompt injection. https://simonwillison.net/. Accessed 28 Jun 2026
14. Hines, K.: Defending against indirect prompt injection attacks with spotlighting. arXiv arXiv:2403.14720 (2024)
15. Chen, S.: StruQ: defending against prompt injection with structured queries. arXiv arXiv:2402.06363 (2024)
16. Piet, J.: Jatmo: prompt injection defense by task-specific fine-tuning. In: Proceedings of ESORICS (2024)
17. Rai, S.: Humans or LLMs as the judge? A study on adversarial robustness of prompt injection defenses. arXiv preprint (2024)
18. Lewis, P.: Retrieval-augmented generation for knowledge-intensive NLP tasks. In: Advances in Neural Information Processing Systems (NeurIPS) (2020)
19. Yao, S.: ReAct: synergizing reasoning and acting in language models. In: Proceedings of ICLR (2023)
20. Microsoft: Security Copilot Product Documentation. https://learn.microsoft.com/security-copilot/. Accessed 28 Jun 2026
21. Google Cloud: Security AI Workbench Documentation. https://cloud.google.com/security/ai. Accessed 28 Jun 2026
22. Fang, R.: LLM agents can autonomously exploit one-day vulnerabilities. arXiv preprint (2024)
23. Goh, J., Adepu, S., Junejo, K.N., Mathur, A.: A dataset to support research in the design of secure water treatment systems. In: Proceedings of CRITIS (2017)
24. Ahmed, C.M., Palleti, V.R., Mathur, A.P.: WADI: a water distribution testbed for research in the design of secure cyber-physical systems. In: Proceedings of CySWATER (2017)
25. Shin, H.K.: HAI 1.0: HIL-based augmented ICS security dataset. In: Proceedings of USENIX CSET (2020)
26. Gómez, Á.L.P., et al.: On the generation of anomaly detection datasets in industrial control systems. IEEE Access **7** (2019)
27. Faramondi, L.: A hardware-in-the-loop water distribution testbed dataset for cyber-physical security testing. IEEE Access (2021)
28. Aoudi, W., Iturbe, M., Almgren, M.: Truth will out: departure-based process-level detection of stealthy attacks on control systems. In: Proceedings of ACM CCS (2018)

29. Alves, T., Das, R., Werth, A., Morris, T.: Virtualization of SCADA testbeds for cybersecurity research: a modular approach. Comput. Secur. **77**, 531–546 (2018)
30. Chroma Project: Chroma Vector Database Documentation. https://www.trychroma.com/. Accessed 28 Jun 2026
31. Zhong, Z.: Poisoning retrieval corpora by injecting adversarial passages. In: Proceedings of EMNLP (2023)
32. Shi, F.: Large language models can be easily distracted by irrelevant context. In: Proceedings of ICML (2023)
33. Habibzadeh, A., Feyzi, F., Atani, R.E.: Large language models for security operations centers: a comprehensive survey. arXiv arXiv:2509.10858 (2025)
34. Singh, R., et al.: LLMs in the SOC: an empirical study of human-AI collaboration in security operations centres. arXiv arXiv:2508.18947 (2025)
35. Industrial Defender: Industrial Defender Launches CopilOT Service to Address the Cybersecurity Talent Shortage for Critical Infrastructure Companies. Press release (2024). https://www.industrialdefender.com/press/copilot-cybersecurity-service. Accessed 28 Jun 2026
36. Efron, B., Tibshirani, R.J.: An Introduction to the Bootstrap. Chapman & Hall/CRC, New York (1993)
37. Holm, S.: A simple sequentially rejective multiple test procedure. Scand. J. Stat. **6**(2), 65–70 (1979)

Threat Modeling for Maritime Shipboard Systems

Chee Teck Wong[1], Awais Yousaf[2(✉)], Jianying Zhou[2], and Mark Voon Wei Goh[2]

[1] Information Systems Technology and Design (ISTD), Singapore University of Technology and Design (SUTD), Singapore, Singapore
cheeteck_wong@alumni.sutd.edu.sg

[2] iTrust, Centre for Research in Cyber Security, Singapore University of Technology and Design (SUTD), Singapore, Singapore
{awais_yousaf,jianying_zhou,mark_goh}@sutd.edu.sg

Abstract. The advancement of digital technologies in maritime shipboard systems has significantly enhanced operational capabilities while introducing complex cybersecurity challenges. This paper explores a systematic method to identify and understand cyber threats within a maritime context. Using the Security and Privacy Architecture Through Risk-Driven Threat Assessment (SPARTA) framework, this research integrates attacker behavior modeling, system decomposition, and quantitative risk assessment based on the Factor Analysis of Information Risk (FAIR) methodology. System components were abstracted into Data Flow Diagram (DFD) and enriched with adversarial techniques derived from the Industrial Control Systems (ICS) matrix of MITRE ATT&CK framework. The combined use of both DFD and ICS matrix of MITRE ATT&CK framework offers a comprehensive view of how cyber threats may emerge and why existing controls might fail. Despite challenges such as manual abstraction and limited empirical data for parameter calibration, the methodology demonstrated in this research provides practical value for early-stage threat identification, iterative risk evaluation, and long-term cybersecurity planning. This study advocates for the early integration of threat modeling into the development lifecycle of maritime system to support both regulatory compliance and operational resilience.

Keywords: Maritime Cybersecurity · Threat Modeling · Risk Assessment · SPARTA · MITRE ATT&CK

1 Introduction

Advancements in navigation, propulsion, communication, and cargo management have enhanced modern maritime vessel operational efficiency and safety, but the convergence of Information Technology (IT) and Operational Technology (OT) also introduces complex cybersecurity risks that threaten vessel operations, safety of life at sea, and maritime supply chain integrity. As noted in recent

S. Zonouz et al. (Eds.): DCS-CI 2026, CCIS 3109, pp. 22–45, 2027.
https://doi.org/10.1007/978-981-92-5039-4_2

cybersecurity risk management guidelines [28] on shipboard OT, system reliance on networked components and remote connectivity increases the potential for cyber threats, ranging from data tampering to remote code execution. In addition, the latest trends in the shipping industry, such as remotely operated and autonomous vessels, further blur the boundaries between shipboard and shore-based operations. Maritime cybersecurity frameworks such as IMO Guidelines on Maritime Cyber Risk Management (MSC-FAL.1/Circ.3/Rev.3) [20] highlight, '*...ensure a holistic and flexible cyber risk management regime that is in continuous operation and constantly evaluated through effective feedback mechanisms...*', it has identified that such transformations require more rigorous and continuous threat management, particularly for navigation, control, and communication technologies [28].

Shipboard systems such as Automatic Identification System (AIS), Electronic Chart Display and Information System (ECDIS), Integrated Navigation System (INS), Radio Detection and Ranging (RADAR), Propulsion, Machinery & Power Control Systems, and Ship-to-shore etc., were not originally engineered with strong security controls. Many continue to rely on legacy hardware, unencrypted protocols, default credentials, or minimal authentication. As a result, threats such as signal spoofing, manipulation of control messages, unauthorized access, and exploitation of exposed services remain credible concerns for vessel integrity, safety and security.

Traditional threat modeling frameworks such as STRIDE offer basic cybersecurity analysis but cannot adapt to architectural changes or operational variations in a maritime environment. To address this gap, this paper proposes a threat modeling approach grounded in the SPARTA [30]. SPARTA enables continuous, structured, and risk-prioritized threat analysis based on DFD and enriched with attacker profiles, system-specific countermeasures, and Monte Carlo-based risk simulations grounded in the FAIR model. The tool supports the prioritization of threats based on likelihood, impact, and the strength of security controls, enabling a security-by-design methodology well suited for evolving shipboard systems.

The rest of the paper is organized as follows. Section 2 provides background on SPARTA along with FAIR framework. It also discusses constraints associated with the use of SPARTA tool. Section 3 briefs about related work on maritime cybersecurity risk management. Section 4 highlights the proposed methodology. A detailed case study is presented in Sect. 5. Discussions on findings, strengths, limitations, comparisons and future work are presented in Sect. 6. Finally, conclusions are drawn in Sect. 7.

2 Background

This research applies SPARTA [30] primarily as a threat modeling tool that uses DFDs and ICS matrix of MITRE ATT&CK for mappings to identify threats systematically. It performs continuous threat elicitation and incorporates quantitative risk analysis through an integrated FAIR-based Monte Carlo engine.

Unlike qualitative risk assessment frameworks such as those used in NIST SP 800-30 [29], ISO/IEC 27005 [21] that rely on color charts or weighted scales, FAIR model is designed to support risk quantification through a standard framework for information and operational risk as follows:

- Risk (R) is defined as the product of Loss Event Frequency (LEF) and Loss Magnitude (LM), yielding an annualized expected loss for a given scenario. Hence, $R = LEF \times LM$ as given in Eq. (1).
- LEF is the frequency of successful attacks that actually result in loss. In FAIR, LEF is derived from two upstream quantities: Threat Event Frequency (TEF) which refer to how often attempts occur, and Vulnerability (V) which refer to the fraction of attempts that succeed. Hence, $LEF = TEF \times V$ as given in Eq. (2).
- LM is the loss or damage per successful attack. FAIR encourages using currency for LM (e.g., SGD/USD), but relative units may be used when comparing elements within a model.
- TEF is the frequency of attempted attacks, irrespective of success. TEF is itself the product of two behavioral factors: Contact Frequency (CF) which refer to how often an attacker encounters the asset, and Probability of Action (PoA) which refer to the likelihood the attacker acts maliciously upon contact. Thus, $TEF = CF \times PoA$ as given in Eq. (3).
- V in FAIR is not a checklist; it is the probability of success conditional on an attempt, computed as the probability that Threat Capability (TC) exceeds Strength (S), i.e., $V = P(TC > S)$ as given in Eq. (4). Here, S represents the strength of countermeasures protecting against the threat, expressed on the same scale as TC, the attacker's capability distribution for that threat.
- TC is the estimated skill, resources, or strength of an attacker population relevant to the threat.
- S is the strength of the system's countermeasures, expressed on the same scale as TC. An attack succeeds when $TC > S$.
- CF represents how often attackers interact with the system, whether intentional or not.
- PoA is the probability that an attacker will act upon contact, ranging from 0 to 1.

The FAIR risk analysis relationships can therefore be summarized as:

$$\mathrm{R} = \mathrm{LEF} \times \mathrm{LM} \tag{1}$$

$$\mathrm{LEF} = \mathrm{TEF} \times \mathrm{V} \tag{2}$$

$$\mathrm{TEF} = \mathrm{CF} \times \mathrm{PoA} \tag{3}$$

$$\mathrm{V} = P(\mathrm{TC} > \mathrm{S}) \tag{4}$$

SPARTA quantify potential threats by embedding FAIR methodology directly into its analysis to explicitly accounts for uncertainty by realizing this through Monte Carlo simulation. Inspired by the work presented in [22], SPARTA extends the approach by performing Monte Carlo simulations for each FAIR risk components (CF, PoA, TC, S, LEF, and LM etc.).

Risk analysts first model the system as a DFD (processes, data stores, external entities, flows, and trust boundaries) [31,36], after which SPARTA matches threat expressions to the model to enumerate relevant threats. For example, information disclosure affecting a data flow that ends in a data store can be observed in the data flow "Chart RADAR & Overlays Processing → RADAR Settings & Historical Data". If an adversary gains unauthorized access during or after this transfer, it will result in disclosure of navigational settings and historical RADAR information. For each threat instance, SPARTA then requests "leaf" inputs aligned with FAIR: S, TC, CF, PoA, and LM. Each input is captured as an estimate with $\{min, probable, max, confidence\}$, forming a (modified) Program Evaluation and Review Technique (PERT) distribution. SPARTA runs Monte Carlo simulations that sample repeatedly from these distributions to generate thousands of possible outcomes. This produces not just a single risk score, but a full distribution of possible outcomes, along with expected values, ranges, and confidence intervals. Combining them follow the FAIR relationships above, and produce quantitative results for TEF, V, LEF, LM, and R for the chosen attacker profile. In practice, this allows analysts to express findings such as: "There is a 90% chance annual losses will not exceed SGD 600,000, but a 10% chance of losses above that threshold.". By doing so, SPARTA captures the inherent variability of real-world attacker behavior and control performance, providing decision-makers with richer insights than static point estimates.

Referring to SPARTA documentation [24], threat modeling is supported by a structured definition including *ThreatTypeCatalog (top-level container), Threat Category, and Threat Type* that ensures consistency and repeatability during threat analysis. This provides a hierarchical framework to organize and apply threats to elements within a DFD. *Threat Type* is the unit which quantitative risk estimations are performed in SPARTA and link directly to elements of the DFD, such as processes, data stores, or data flows. Because each *Threat Type* is clearly defined, it can be consistently mapped, reused, and extended across different system models.

2.1 Constraints Associated with the Use of SPARTA

Although the SPARTA research paper [30] provides insights into its methodology, the publicly available resources including its official website [25], documentation pages [24], and its GitLab package repository [26] only distribute the SPARTA compiled binaries rather than its source code. This limitation makes it impossible to confirm how each FAIR risk analysis parameter, such as TEF, LEF, V, or LM are sampled, combined, or aggregated in practice. Researchers cannot directly inspect or verify how FAIR risk analysis and Monte Carlo simulations are technically implemented within the SPARTA framework. Therefore, reconstructing detailed FAIR risk analysis from SPARTA output without access to its underlying source code presents several inherent challenges:

(i) Opaque Internal Logic:
SPARTA performs Monte Carlo simulations using user-supplied parameters, e.g., *minimum, probable, maximum, confidence*, likely applying a scaled Beta-PERT distribution [8, 16]. However, its internal implementation details are not exposed, such as the size of simulation runs, convergence thresholds, or distribution parameters. This limits visibility into:
   - Sampling process for TC and S.
   - Exact method to compute $V = P(TC > S)$
   - Any assumptions about TEF, SLE, or distribution shapes.

(ii) Ambiguity in Output Labels:
Several columns in SPARTA output tables do not align clearly with FAIR definitions. For example:
   - `vuln` column shows constant values of 1.0, which more likely indicates inclusion of the threat rather than FAIR-style vulnerability probability.
   - `LB` and `UB` under LEF may not refer to lower/upper bounds in a traditional statistical sense.
   - `SLE` appears fixed, but its source distribution or rationale is not shown.

(iii) Lack of Granular Simulation Control:
SPARTA does not provide controls to configure:
   - Number of simulation runs.
   - Intermediate outputs (e.g., TEF samples, TC distributions).
   - Internal assumptions about TEF versus V correlation.

(iv) Analytical Method Used:
Despite these constraints, the following strategy was applied to reconstruct and validate the results:
   - Back-Calculating SLE: Derived as $SLE = \frac{R}{LEF}$ from known SPARTA output.
   - PERT-Based Vulnerability: Assumed Beta-PERT distribution to model TC and S estimates, using:

$$\mu = \frac{a + 4m + b}{6}, \quad \sigma \approx \frac{b - a}{6}$$

   - Simulation of $P(TC > S)$: Ran 10,000 external samples to simulate and validate V estimate.
   - Monte Carlo Emulation: Repeated trials to recreate 5th/95th percentiles of LEF, SLE, and R.
   - Side-by-Side Validation: Created a comparative table to explain matching or differing results with an explanation.

3 Related Work

The digitalization of maritime systems has intensified the need for systematic cyber risks assessment onboard ships [40]. Numerous studies [1,3,7,15,28,30, 32,35,36] have examined how cybersecurity risk management can be adapted to maritime sector. A substantial effort has been made to tailor conventional threat modeling techniques, such as STRIDE [7,36], to shipboard systems. While STRIDE remains widely used for identifying Spoofing, Tampering, Repudiation, Information Disclosure, Denial of Service, and Elevation of Privilege, its use in maritime contexts shows several limitations. A descriptive study of 640 assumptions in STRIDE threat modeling [14] found that many assumptions are used to scope the modeling process, require architectural information not expressible in DFD notation, or reference concepts such as trust relations and attacker capabilities, that are not typically represented in system models.

Critiques of DFD further highlight challenges for modeling complex maritime systems. As noted in [31], DFD lack expressiveness for representing explicit countermeasures, data types, dependencies on lower-level security properties, and relationships to deployment aspects, leading to reliance on undocumented assumptions and reduced reproducibility. Recent studies [14,19,31,33,34,36, 41] therefore emphasize enhancing threat modeling frameworks with explicit attacker models, trust definitions, and system dependencies, and managing assumptions as architectural knowledge that co-evolves with system design in safety-critical domains.

In parallel with these advancements, several domain-specific guidelines provide structured approaches for managing cyber risks throughout a vessel's life cycle. For example, the "Guidelines on Cyber Security Onboard Ships (Version 5)" [4] by Baltic and International Maritime Council (BIMCO), and "Guidelines for Cyber Risk Management in Autonomous Shipping" [28] cover threat identification, protection measures, detection, and response, reflecting a shift toward proactive and risk-based governance aligned with evolving vessel operations. From operational risk modeling perspective, semi-automated frameworks such as System Theoretic Process Analysis for Cybersecurity (STPA-Cyber) [36] reduce the expert workload by analyzing control actions and feedback signals to generate Unsafe Control Actions (UCAs) and Unsecure Feedback Signals (UFSs) for rigorous assessment. Similarly in some risk assessment methodologies, hierarchical representations such as attack trees [23] or dependency graphs [35] have been applied to shipboard systems for modeling attacker goals and threat scenarios targeting navigation and communication systems of vessels. Collectively, these developments show increasing recognition of the need to refine and apply structured threat modeling methods to meet maritime cybersecurity challenges. Remaining gaps include standardization, tooling support, the integration of such practices into the whole system lifecycle from design to deployment and operation.

This paper aims to refine threat modeling practices for shipboard systems by addressing their domain-specific threats, trust structures, and operational workflows for shipboard systems in a reproducible and scalable manner using the

SPARTA tool [30]. The proposed tool focuses on modeling real-world navigation systems as a representative and safety-critical maritime use case using SPARTA. Through this approach, it systematically identifies critical trust boundaries, communication flows, and potential attacker entry points. Furthermore, the risks associated with the identified threats are quantitatively assessed under varying attacker models and countermeasure scenarios, enabling a more structured and comprehensive evaluation of system security. Moreover, it enhances the depth of cybersecurity analysis onboard vessel, the ICS matrix of MITRE ATT&CK [13,38] is employed to classify adversarial techniques in a manner that is more comprehensive and operationally relevant than traditional STRIDE modeling. Lastly, it contributes to maritime cybersecurity knowledge by demonstrating an iterative yet scalable risk-based analysis methodology for threat modeling of shipboard systems.

4 Structured Threat Analysis of Maritime Shipboard Systems

Threat modeling is a foundational practice in a security-by-design approach [38], serving as a proactive and structured methodology to identify, assess, and mitigate potential threats to a system. Modern maritime vessels are cyber-physical ecosystems composed of IT and OT to provide the required core functions such as navigation, propulsion, communication, and cargo operations [3,35]. A rigorous threat modeling process is essential to ensure the safety, security, and resilience of shipboard operations. This section outlines a structured threat modeling methodology for maritime shipboard systems, focusing only on the shipboard navigation system for sake of brevity. The proposed methodology is designed to be extensible to other maritime systems, including propulsion, machinery & power control systems, communication systems, cargo management systems.

4.1 Proposed Methodology

The threat modeling methodology for this research adopts a hybrid approach, beginning with system decomposition and moving through threat identification, risk analysis, and mitigation. This method is different from conventional Microsoft Threat Modeling Tool (TMT) and OWASP Threat Dragon modeling tool [10], which primarily focus on threat enumeration without a built-in risk quantification engine. The proposed methodology is depicted as flow chart in Fig. 1 and explained in the following subsection.

The modeling process begins with **System Component Decomposition**, where installation manuals and interconnection diagrams are examined to identify hardware interfaces, sensor paths, and commandâĂŞcontrol hierarchies. A component level diagram is then created to capture the roles and dependencies of sensors, processing units, communication links, and operator interfaces within the navigation subsystem. This is followed by **System Component Mapping**, in which a detailed DFD is constructed to show how data moves

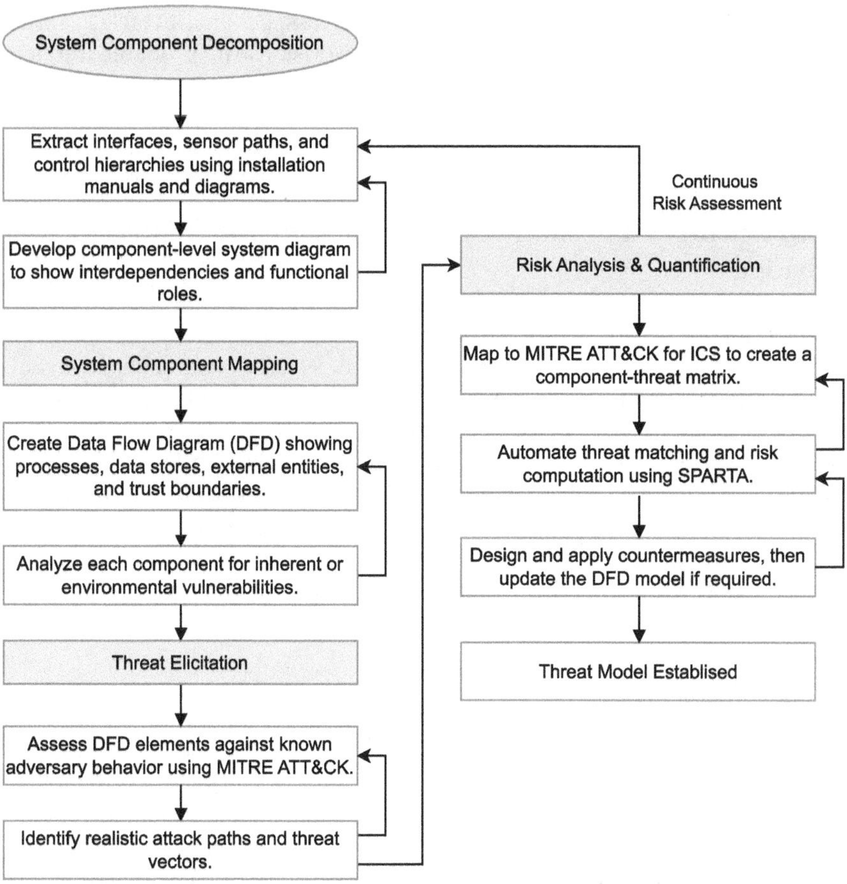

Fig. 1. Proposed Methodology

between processes, data stores, external entities, and across trust boundaries. Trust boundaries indicate changes in trust levels, particularly between onboard systems and external networks, and help identify where navigation and control components rely on external data. Each component is assessed for inherent vulnerabilities, such as legacy protocols or unauthenticated data feeds, as well as environmental exposures, including network access points and the use of physical media. **Threat Elicitation** then maps these vulnerabilities to Tactics, Techniques and Procedures (TTPs) in the MITRE ATT&CK ICS matrix, produces a threat matrix and helps identification of adversarial actions. Each DFD element is reviewed for susceptibility to documented attack behaviors, enabling a focused assessment of realistic threat vectors and supporting the design of targeted defensive measures. The final step, **Risk Analysis and Quantification**, applies the capabilities of SPARTA, which automates matching of DFD elements to predefined threat expressions and performs quantitative analysis based on the

FAIR model (Eqs. (1) to (4)). By incorporating asset value, attacker profiles, and countermeasure strength, SPARTA estimates risk as the probable frequency and magnitude of loss and uses Monte Carlo simulations to generate prioritized threats with associated risk scores, guiding data-driven mitigation planning and strategies.

5 Case Study: Threat Modeling of a Vessel's Navigational Bridge

The navigational bridge of a vessel is selected to serve as a primary case study for this research due to its safety-critical nature. Its system component diagram is shown in Fig. 4 in Appendix A. The diagram shows the range of subsystems of the navigation system that make up a modern bridge. By abstracting these interactions into a DFD, the analysis clearly delineates sensitive trust boundaries, particularly at the entry points of external data streams such as AIS and Global Positioning System (GPS). Applying the SPARTA framework in this context further enables the direct mapping of threats to specific data flows, facilitating the precise identification of where security enhancements, such as anomaly detection, cryptographic validation, and sensor redundancy, can be most effectively implemented [28].

5.1 System Component Decomposition

A comprehensive system component decomposition is performed initially to analyze the shipboard architecture, data communication channels, onboard network topologies, and detailed vendor documentation. A systematic review of these documents is conducted to identify any functional roles of system components, the nature and directionality of data flows, and the mapping of security boundaries within the navigation shipboard infrastructure. The extracted information, which ranges from sensor interconnectivity and interface standards (e.g., National Marine Electronics Association (NMEA), Ethernet, serial, USB) to internal processing hierarchies and display/control mechanisms, is consolidated into a high-level system component model as shown in Fig. 4. This decomposition revealed the following four main categories of components of the shipboard navigational bridge:

(i) **Sensors:**
   - GPS (e.g., GP-170): Provide positional data via unencrypted signals; susceptible to spoofing and jamming [37].
   - Gyro Compass (e.g., Raytheon Anschütz Standard 22 NX): Supplies heading data to INS, ECDIS, and Autopilot.
   - Speed Log (e.g., DS-85, GS-100): Measures Speed Through Water (STW); vulnerable to data tampering.
   - Anemometer (e.g., Furuno 220WX): Provide wind speed/ direction data; often interfaced via NMEA.

(ii) **Processing Units and Storage:**
- ECDIS Processor (e.g., FMD-3300 series): Renders ENC charts and integrates multiple sensor feeds; accepts USB, NMEA, and Ethernet input; vulnerable to spoofed chart data and malware.
- RADAR Processor (e.g., FAR-3000 series): Handles chart overlay, Automatic RADAR Plotting Aid (ARPA), AIS targets; communicates via Ethernet and digital I/O; subject to Man-in-the-Middle (MITM), Server Message Block (SMB) protocol exploits, and ARPA manipulation [18].
- Voyage Data Recorder (VDR) (e.g., VR-7000): Records navigational data; vulnerable to remote access if services remain unpatched [28].

(iii) **Display Units:**
- Chart RADAR Screen (e.g., MU-231): Provides navigational visualisation; user interface exposure risk if directly accessible.
- ECDIS Screen (e.g., MU-231): Provides navigational visualisation; user interface exposure risk if directly accessible.
- Conning Display (e.g., MU-190): Aggregates speed, heading, engine status; can be targeted for deceptive display attacks.

(iv) **External Communication Interfaces:**
- AIS Transceiver (e.g., FA-170): Broadcasts unencrypted dynamic and static vessel data; susceptible to spoofing, replay attacks [2,38].
- VHF Radio: enables bridge-to-bridge or shore communication, subject to social engineering and bridge-based attack vectors.
- Shore Control Center: Typically used in remote pilotage [6,36] or fleet routing updates.

5.2 System Component Mapping

The SPARTA study [30] stated that *each element is represented as an Estimate with four parameters: min, probable, max, confidence.* This showed that value estimation technique (e.g. PERT distributions applied to FAIR factors) forms the basis of SPARTA's systematic risk analysis. First, the navigational bridge and its components were modeled within SPARTA as DFD. The complete DFD diagram as depicted in Fig. 5, provides a standard holistic view of the navigation bridge architecture on a maritime vessel. It captures all key functional interactions and data flows between key shipboard bridge components under standard operating conditions.

Within this comprehensive and detailed DFD, subsystems are modularly represented to allow for focused analysis of individual components. A DFD broken down to its subsystem component level ensures that security analysis can be performed at the macro- (system-wide) and micro- (subsystem-depth) levels. This approach enables threat modeling to assess both localized vulnerabilities in the subsystem component and systemic risks across the entire navigation bridge, ensuring a complete and layered security posture assessment for maritime operations. To accurately model threats specific to ICS in maritime shipboard environments, a custom *ThreatTypeCatalog* library [24] was created in SPARTA and mapped to all tactics for the ICS matrix of MITRE ATT&CK framework. This allows the domain-specific tactic and technique to be consistently incorporated into the SPARTA analysis process, ensuring relevance to the maritime OT systems.

5.3 Threat Elicitation

SPARTA supports extensible threat classification by allowing users to use ICS matrix of MITRE ATT&CK framework to organises threats by tactics (the adversary's goals) and techniques (the actions taken to achieve those goals), beyond the default STRIDE-based threat types. By creating a custom *ThreatTypeCatalog*, cybersecurity professionals working on shipboard systems can directly map observed or anticipated attacker behaviours to a structured threat modeling process in SPARTA. For ICS matrix of MITRE ATT&CK framework, each tactic (e.g., Initial Access, Execution, Persistence) will become a *Threat Category*, and each associated technique (e.g., Valid Accounts, Command-Line Interface) becomes a *threat Type*. This structure can be created easily with SPARTA graphical modeler. SPARTA loads the threat catalogue at its runtime, allowing each component or interaction to be analyzed based on the new MITRE ATT&CK-based threat definitions. As clear from Fig. 2 and Table 2 in Appendix A, each technique in ICS matrix of MITRE ATT&CK framework can be logically mapped to a component or data flow, based on the actual shipboard system's data flow or expert judgement/assumption.

5.4 Risk Analysis and Quantification

In SPARTA tool, as an example, the data flow 'RADAR Echo & Target Reflection' (RADAR Signal Sending / Receiving → Chart RADAR & Overlays Processing) has been selected for threat analysis. A custom threat type labeled 'Collection: Adversary-in-the-Middle' has been mapped to this data flow. The estimate parameters 'minimum: 0.2, probable: 0.6, maximum: 0.8, confidence: 0.5' define the likelihood distribution of this threat occurrence, using SPARTA's probabilistic risk modeling approach. These values reflect an informed judgment of risk based on system knowledge or threat intelligence. These mapping highlights how SPARTA enables analysts to quantitatively assess threats associated with specific system interactions, supporting structured threat modeling aligned with frameworks such as MITRE ATT&CK.

5.5 Identified Threat Exposure

Each DFD element was systematically assessed against patterns of cyber-physical attack and aligned with the MITRE ATT&CK ICS matrix. Refer to Fig. 2, an example of DFD-to-ATT&CK mapping is provided to illustrate the DFD-to-ATT&CK mapping for a simplified Chart RADAR subsystem. The Watchkeepers' external entity demonstrates the role of human operators, and this was associated with Initial Access through Replication via Removable Media (T0847), reflecting the exposure that arises from the use of portable storage devices in shipboard operations.

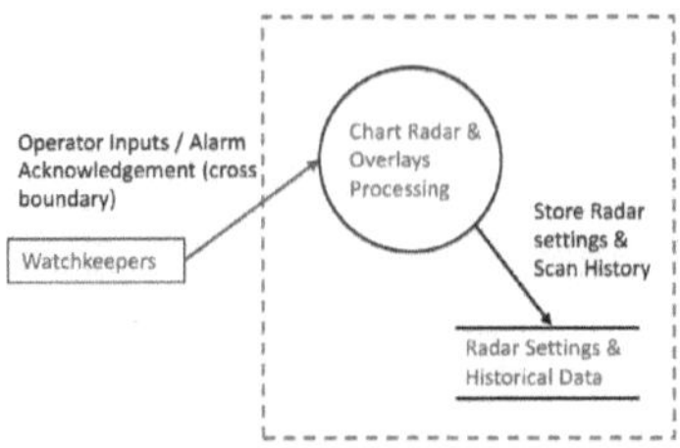

DFD Elements	ICS tactics of Mitre Att&ck	ICS technique of Mitre Att&ck
Watchkeepers (External Entity)	Initial Access	T0847: Replication Through Removable Media
Data Flow: Operator Inputs / Alarm Acknowledgement → Chart Radar & Overlays Processing	Impair Process Control	T0855: Unauthorized Command Message (adversary sends or injects command messages that the process will execute)
Chart Radar & Overlays Processing (Process)	Execution	T0807: Command-Line Interface (commands run on the host/HMI or application server controlling the radar overlay logic).
Data Flow: Chart Radar & Overlays Processing → Radar Settings & Historical Data	Impair Process Control	T0836: Modify Parameter (writes/changes configuration or set-point–like parameters via the process logic).
Radar Settings & Historical Data (Data Store)	Collection	T0893: Data from Local System (querying/reading local config and history for intelligence or staging exfiltration).

Fig. 2. Example of DFD-to-ATT&CK Mapping for Chart RADAR subsystem

The data flow from 'Operator Inputs and Alarm Acknowledgment' to the 'Chart RADAR & Overlays Processing' was linked to the tactic 'Impair Process Control', represented by Unauthorized Command Message (T0855). This captures the risk of adversaries injecting or manipulating control messages so that illegitimate commands are executed by the RADAR system. The main exposure of 'Chart RADAR & Overlays Processing' process lies under the 'Execution' tactic, specifically Command-Line Interface (T0807). This highlights the possibility of malicious actors executing system-level commands on the host or application server that runs RADAR overlay functions, thereby compromising operational integrity. The data flow from 'Chart RADAR & Overlays Processing' to 'RADAR Settings & Historical Data' is mapped to 'Impair Process Control', with Modify Parameter (T0836) illustrating the risk of unauthorized changes to stored RADAR settings or scan history, which can lead to persistent misconfigurations and wrong navigational awareness. The 'RADAR Settings & Historical Data' data store was associated with the 'Collection' tactic through Data from Local System (T0893). This reflects the threat of attackers extracting configuration files or historical records that can be used for reconnaissance or staging further attacks.

5.6 FAIR Risk Analysis in SPARTA

Considering the associated threat estimate for Adversary-in-the-Middle (T0830) on 'RADAR Echo & Target Reflection' DFD element by using values for parameters:

- Minimum = 0.2
- Probable = 0.6
- Maximum = 0.8
- Confidence = 0.5

Above values are interpreted as the PERT distribution parameters for the calculation of TC [30]. Now, calculating risk score using SPARTA and validating it numerically as follows:

(i) The vulnerability V is calculated as the probability that the TC exceeds the S, $V = P(TC > S)$. Assuming S also follows a PERT distribution (defined internally by SPARTA), we simulate:

$$TC \sim \text{PERT}(0.2, 0.6, 0.8)$$
$$S \sim \text{PERT}(0.5, 0.8, 1.0)$$

Using Monte Carlo sampling ($n = 10,000$), we compute:

$$V = \frac{\#\text{ samples where } TC > S}{n} \approx 0.9963$$

Despite the name 'vuln' as shown in Fig. 3, SPARTA uses this column as a flag with its value always equal to 1. 'vuln = 1' represents "Threat is applicable and included". The column labeled "LB" in the SPARTA's Threat Analysis table as shown in Fig. 3 does not mean "Lower Bound" in this context. Instead, in this table, LB = Likelihood of Breach, which is a semantic alias for V: $LB(intable) = P(TC > S) = Vulnerability$ and in SPARTA, that value is actually shown under the LB column, not under vuln. $LB = 0.9963$, there is a 99.63% probability that TC exceeds S, i.e. the threat will succeed almost every time it is attempted.

(ii) For $TEF = CF \times PoA$, where:
   - CF = Contact Frequency (from PERT distribution)
   - PoA = Probability of Action (from PERT distribution)

   SPARTA handles this internally by approximating the mean using their distributions.

(iii) For $LEF = V \times TEF$ with $V = 0.9963$ and $TEF \approx 1.4924$ (inferred from LEF column), SPARTA reports LEF = 1.471 (as shown in Fig. 3) and numerically calculated as $LEF = 1.4871$ as given in Table 1.

(iv) For SLE, SPARTA does not expose a defined distribution for SLE. So, from calculation, $SLE = \frac{R}{LEF} = \frac{0.77}{1.4871} \approx 0.5176 \approx 0.52$

(v) Finally, $R = LEF \times SLE = 1.4871 \times 0.52 = 0.7733 \approx 0.77$

(vi) Percentile Bounds:
   - Lower Bound (risk_LB) $\approx$ 5th % of simulated risk $\Rightarrow$ 0.01
   - Upper Bound (risk_UB) $\approx$ 95th % of simulated risk $\Rightarrow$ 4.16

(vii) Interpretation of Columns:
   - vuln is likely a Boolean indicating threat presence. It is not FAIR's V.
   - LB, UB (LEF bounds) are based on confidence intervals in the Monte Carlo simulation.
   - LEF is the direct average of simulated LEF values.

Table 1 summarizes SPARTA calculated outputs vs numerically validation results. Lastly, the ten highest-risk subsystems within the navigational bridge of a vessel are identified and ranked, as presented in Table 3 (Appendix A).

Fig. 3. SPARTA Calculated Risk Score for Adversary-in-the-Middle (T0830) Attack

Table 1. SPARTA Calculated Outputs vs Validation Results

Parameter	SPARTA Output	Validation Result	Remarks
Residual Annualized Risk	42,627.87	42,627.87	Confirmed total risk without any mitigations applied
Total Annualized Risk	42,627.87	42,627.87	No risk reduction applied yet
Maximum Single Loss Event	0.57	0.52 (from simulated SLE)	Slight deviation due to simulation sampling or tail values in Monte Carlo
Risk (R)	0.77	0.7733 (= 1.4871 × 0.52)	Matches within rounding error using $LEF \times SLE$
risk_LB (lower bound)	0.01	≈ 0.01	Based on 5th percentile in simulation output of R
risk_UB (upper bound)	4.16	≈ 4.16	Based on 95th percentile from simulation
SLE (Loss Magnitude)	0.52	0.52	Derived from reverse-calculating using $SLE = \frac{R}{LEF}$
LEF (Loss Event Frequency)	1.471	1.4871	Matches annualized $LEF = TEF \times V$
Vuln (V)	0.9963	0.9963	Computed as $P(TC > S)$ using Beta PERT distribution
LB (LEF lower bound)	0.9963	0.9963	Same as V value, shown by SPARTA as LB for LEF
UB (LEF upper bound)	1.0	∼1.0	Matches the fixed LEF percentile when both TEF and V are high and stable
LEF (mean)	1.4871	1.4871	Matches exactly

6 Discussions

This section provides a critical evaluation of SPARTA's capabilities when applied to complex cyber-physical systems, with specific focus on its use in the threat modeling of maritime shipboard systems. It outlines the practical challenges encountered during the modeling process in the maritime context, compare SPARTA with other threat modeling approaches, and discuss its potential role in supporting cybersecurity practices in the maritime domain.

6.1 Advantages of Using SPARTA for Maritime Cybersecurity

One of the most significant benefits lies in SPARTA's support for probabilistic risk modeling using FAIR. This approach enables quantitative analysis of threats, integrating uncertainty through distributions for frequency and impact. By using probabilistic inputs, SPARTA moves beyond qualitative rankings of 'high', 'medium', or 'low' to provide a more granular and defensible analysis of potential loss event frequency and magnitude. This form of risk-based prioritization offers greater fidelity than purely qualitative matrices for maritime environments where decisions often carry high operational and financial implications. Rather than producing a one-off threat list, SPARTA anchors the analysis on the architectural representation (DFD) and then allows threat elicitation and risk computation to be repeated as assumptions, mitigations, or system interfaces change [24,25,30]. This characteristic fits well with the maritime domain, where onboard configurations and integrations are adjusted during retrofits, vendor upgrades, class surveys, and operational changes.

Furthermore, SPARTA's model-driven architecture enables iterative design validation where feedback from threat simulations can be looped back into the system design process, supporting early-stage mitigation planning and architectural hardening. The framework supports custom attacker profiles, solution catalogs and threat pattern libraries, making it adaptable to the maritime context where attack vectors often combine physical proximity (e.g., port access) with remote exploitation. Importantly, SPARTA's traceability of mitigation measures across system elements and their associated threats support ongoing compliance with maritime cybersecurity guidelines, including IMO MSC.428(98) [5], IACS UR E26 [11], E27 [12], and BIMCO [4] recommendations.

6.2 Challenges Encountered

Despite SPARTA's strengths, applying it to maritime systems introduced several challenges. A key issue is related to estimating FAIR parameters in the absence of sufficient maritime-specific data. Inputs such as loss magnitude, threat event frequency, and susceptibility ranges are rarely available publicly and often need to be inferred from secondary sources, expert judgment, or analogies with other critical infrastructure domains. This reliance on indirect evidence means that the initial quantitative risk models developed in this paper should be regarded as provisional and subject to iterative refinement. The scarcity of empirical data necessitated greater reliance on expert opinion and transparent documentation of assumptions to maintain credibility. Furthermore shipboard navigation and bridge operations include human decision-making, informal workarounds, and procedural dependencies that are hard to capture using only static components and flows. Systems-theoretic approaches are useful for identifying unsafe interactions and feedback failures [19,41], but translating those insights into parametrized likelihood and impact still requires judgment [33,34].

While this reliance introduces a degree of subjectivity, it also reflects a necessary pragmatic balance. Expert judgment offers a critical qualitative foundation when reliable quantitative inputs are limited, while quantitative models help formalize these insights into a structured and repeatable framework [1,3,9,15,35]. This interplay underscores the importance of integrating qualitative expertise with quantitative rigor, where expert knowledge provides contextual depth and domain-specific understanding, and probabilistic modeling ensures consistency, comparability, and systematic prioritization of risks.

6.3 Comparison with Other Threat Modeling Tools

When compared to other tools and standards, SPARTA's capabilities stand out, particularly in its quantitative nature. The Microsoft TMT is based on the STRIDE classification model which provides ease of use through predefined STRIDE mappings and intuitive interfaces, it is primarily oriented toward software-centric applications [34]. As such, it lacks the capacity for probabilistic risk estimation, adversary behavior simulation, and iterative feedback integration into architectural decision-making. In our work, SPARTA incorporates the MITRE ATT&CK ICS matrix within its modeling process, allowing system elements defined in DFDs to be systematically evaluated against a globally recognised set of real-world adversarial tactics and techniques. Our tool combines quantitative simulation, threat pattern matching, and risk visualisation to address the multidimensional nature of maritime systems that involve embedded hardware, networked components, and human-machine interaction. As a result, SPARTA is able to elicit any threat grounded in operationally validated attack patterns, bridging the gap between abstract risk modeling and the evolving landscape of maritime cyber threats. This further allows security professionals trace between identified vulnerabilities and observed intrusion techniques, making SPARTA a robust platform for maritime threat modelling across design-time, implementation phase and operational contexts.

6.4 Future Work

Several directions remain for extending and strengthening the proposed methodology. First, further validation of the risk assessment framework on an industry-grade Maritime shipboard OT (MariOT) testbed [39] is planned. This will enable a more realistic evaluation and practical applicability of the identified risk levels under real-world conditions.

Second, while this study utilizes the MITRE ATT&CK framework (ICS domain) for adversary modeling, future work will incorporate the MariOT Matrix [27] which is a maritime-specific framework. This transition is expected to provide more domain-relevant attack patterns and improve the fidelity of threat representation within maritime operational environments.

Finally, the current implementation relies on the SPARTA framework, which is distributed as compiled binaries and does not provide access to source code, thereby limiting extensibility and customization. To address this constraint, future efforts will explore alternative open-source tools or focus on the development of a custom threat modeling and risk assessment platform. In parallel, efforts will be directed toward reducing reliance on manual inputs in the proposed model as maritime-specific cybersecurity dataset becomes available, addressing the current limitation posed by the lack of representative real-world data, and thereby enhancing overall scalability and usability.

7 Conclusion

This paper presents a structured approach to threat modeling for maritime shipboard systems, with the navigation bridge serving as a representative and safety-critical case study. The proposed methodology is centered on the SPARTA framework, which integrates DFD based system modeling, quantitative risk assessment grounded in the FAIR methodology, and adversary simulation aligned with the MITRE ATT&CK ICS matrix. By combining these elements, the approach enables comprehensive threat modeling across the entire system lifecycle, from early design stages through deployment and operational use. This facilitates a clearer understanding for system developers, operators, and stakeholders of potential vulnerabilities, relevant threat scenarios, and areas requiring enhanced security controls or design improvements. The analysis further reveals that the INS system represents the highest-risk subsystem, followed by the ECDIS and RADAR, highlighting critical components that warrant prioritized security attention.

Acknowledgment. The research is supported by the National Research Foundation, Singapore, under its National Satellite of Excellence Programme "Design Science and Technology for Secure Critical Infrastructure: Phase II" (Award No: NRF-NCR25-NSOE05-0001). Any opinions, findings and conclusions or recommendations expressed in this material are those of the author(s) and do not reflect the views of National Research Foundation, Singapore.

A Appendix

Table 2. Threat Exposure of Navigation System Subcomponents

Component	Model Name	Threat Exposure	Mapping to TTPs of ICS ATT&CK Matrix
ECDIS Processor	FMD-3300	Susceptible to USB-based malware and spoofed ENC chart files; relies on local chart updates via physical media; vulnerable to remote file copy via SMB	T1204.002: User Execution: Malicious File, T0865: Valid Accounts (Operator Console), T0882: Remote File Copy
RADAR Processor	FAR-3000	Vulnerable to MITM via Ethernet-based RADAR overlay; use of SMB/Windows services may expose legacy protocol risks. Transmits unauthenticated, unencrypted data	T0829: Man-in-the-Middle, T0804: Communication Protocols, T0883: Remote Services
AIS Transceiver	FA-170	Transmits unauthenticated, unencrypted data; spoofed targets and replayed messages can confuse operators	T0839: Spoof Reporting Message, T0813: Block Reporting Message, T0809: Wireless Communication
GPS Receiver	GP-170	Can be jammed or spoofed, impacting positional accuracy and dependent systems (ECDIS, INS, Autopilot)	T0831: Denial of Control (Jamming), T0851: Spoof Command Message
VDR System	VR-7000	May run outdated embedded OS; exposed services could be used for malware injection or RCE over LAN	T0862: Exploit Public-Facing Application, T0803: Remote System Discovery, T0842: Modify Alarm Settings [17]
INS System	Furuno VOY-AGER	Operates on a networked backbone aggregating sensor data; potential for MITM or data manipulation	T0829: Man-in-the-Middle, T0859: Replication Through Removable Media
Autopilot Control	Furuno FAP-3000	Acts on commands from ECDIS or operator; insufficient validation of route inputs could cause unsafe maneuvers	T0851: Spoof Command Message, T0806: Modify Control Logic
Display Interfaces	MU-190 / MU-231	Potential vector for social engineering or operator manipulation	T1204.001: User Execution: Malicious Link, T0820: Exploit Human-Machine Interface

Table 3. Risk Ranking Identified for Navigational Subsystems

Rank	Navigation Subsystem	Threat Types	Exposure Factors
1	INS System	Adversary-in-the-Middle, spoofing, data manipulation	Reliance on multi-sensor tracking and real-time data
2	ECDIS	Chart manipulation, malware injection, unauthorized updates	Dependence on USB and Ethernet interfaces and trust in ENC data with limited external validation
3	RADAR	Target reflection manipulation, replay, Man-in-the-Middle	Reliance on sensor-derived targets in ARPA processing
4	AIS Transceiver	Spoofing, replay, covert command-and-control	Unencrypted broadcasts
5	Autopilot Control	Lateral movement, privilege misuse	Dependence on data and access shared with other navigation systems
6	Gyro Compass	Spoofing, data integrity degradation	Reliance on a single heading source that propagates across multiple navigation subsystems
7	Speed Log	Data tampering	Limited redundancy and reliance on sensor data used by INS and autopilot
8	GPS	Jamming, spoofing	Reliance on external signals that are susceptible to RF-based interference
9	VDR	Unauthorized access, data exfiltration	Potential remote or maintenance access exposure
10	Conning Display	Deceptive display attacks	Operator reliance on visual information with limited cross-verification

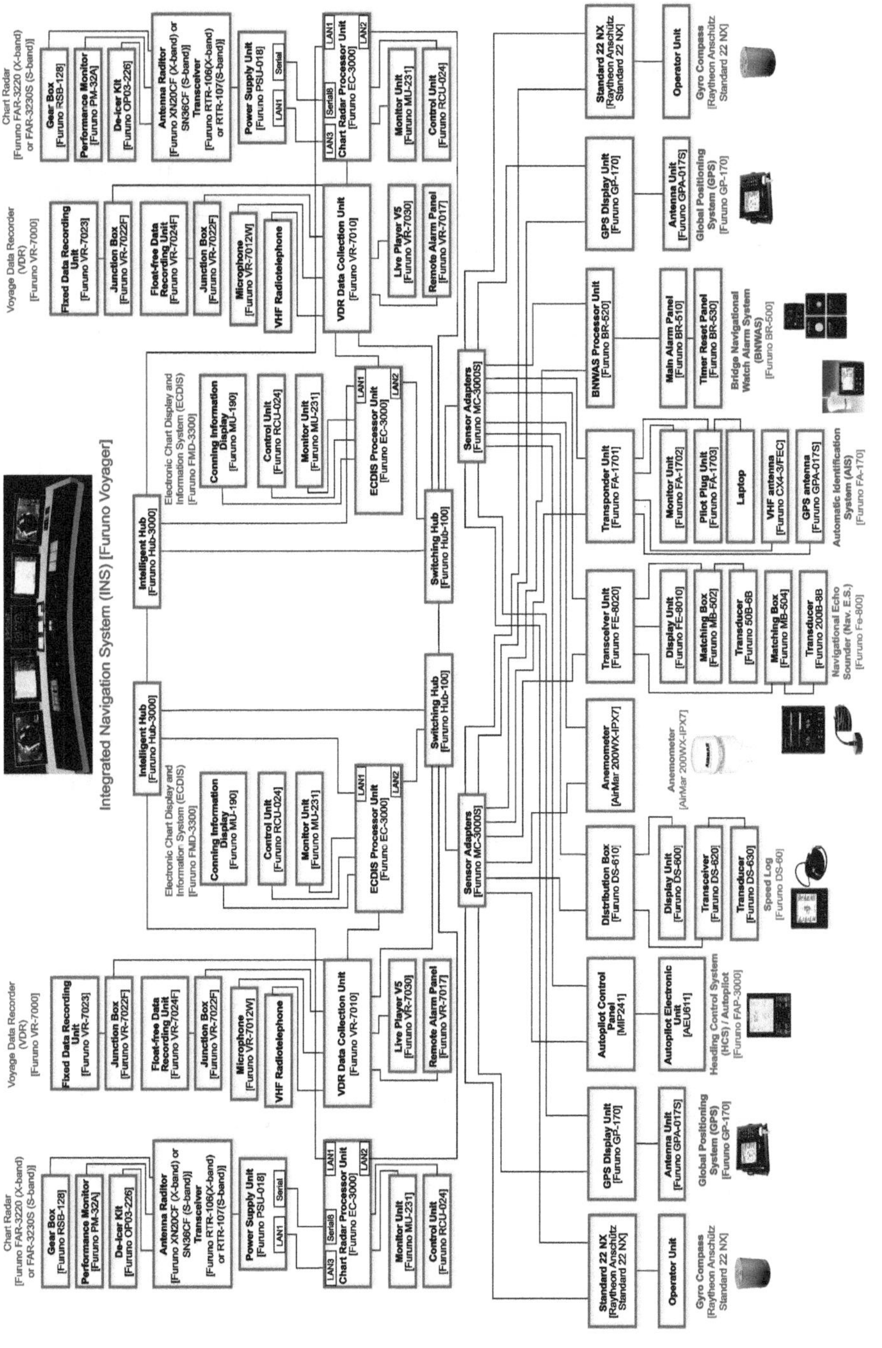

Fig. 4. System Component Diagram for Navigational Bridge

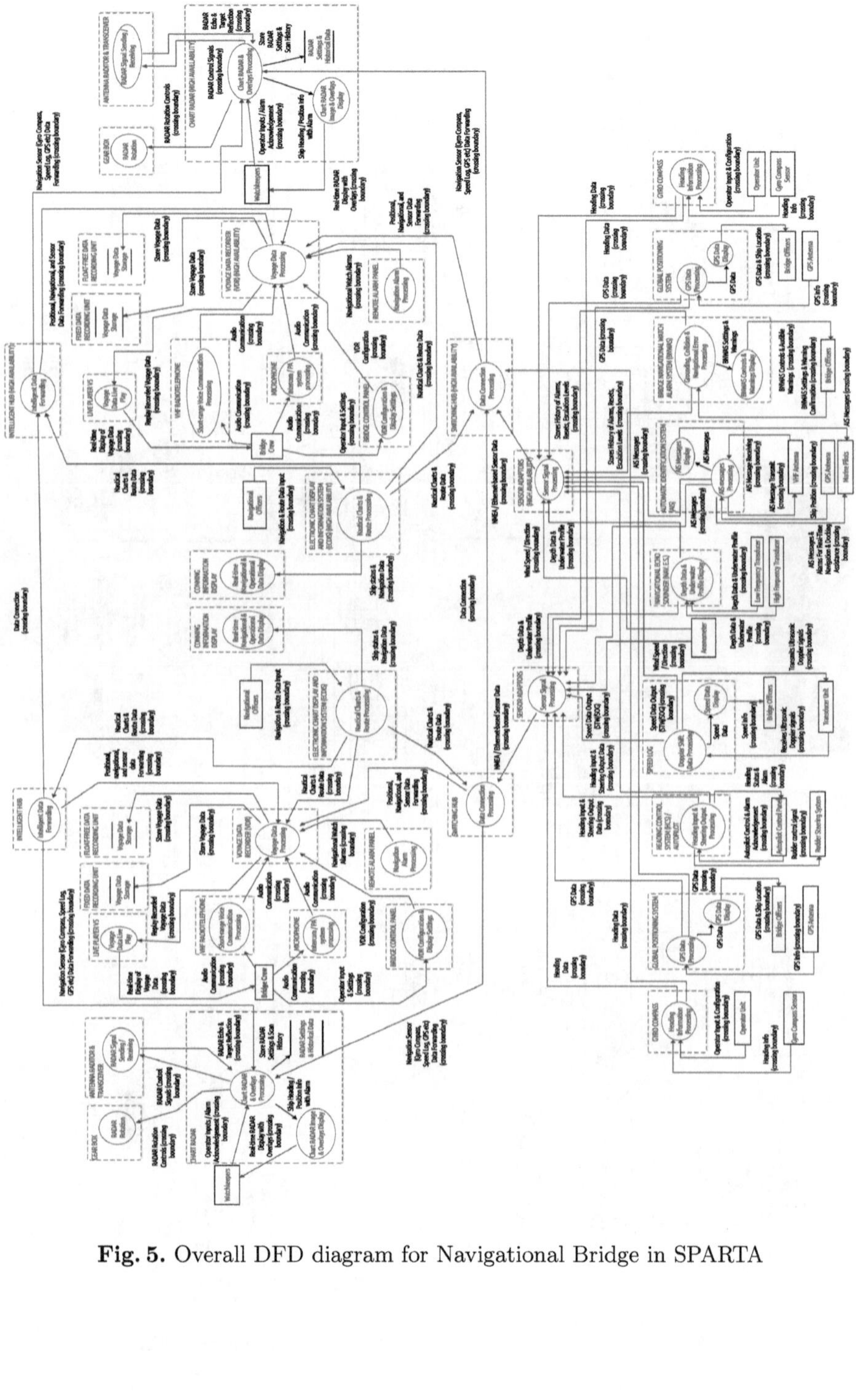

Fig. 5. Overall DFD diagram for Navigational Bridge in SPARTA

References

1. Ahmed, S., Gu, X.C.: Accident-based FMECA study of marine boiler for risk prioritization using fuzzy expert system. Results Eng. **6**, 100123 (2020). https://doi.org/10.1016/J.RINENG.2020.100123
2. Amro, A., Gkioulos, V.: From click to sink: utilizing ais for command and control in maritime cyber attacks. In: Atluri, V., Di Pietro, R., Jensen, C.D., Meng, W. (eds.) ESORICS 2022. LNCS, vol. 13556, pp. 535–553. Springer, Cham (2022). https://doi.org/10.1007/978-3-031-17143-7_26
3. Ahmed, A., Vasileios, G., Sokratis, K.: Assessing cyber risk in cyber-physical systems using the ATT&CK framework. ACM Trans. Priv. Secur. **26**, 1–33 (2023). https://dl.acm.org/doi/10.1145/3571733
4. BIMCO along with various maritime associations and cyber security consultancy companies: The guidelines on cyber security onboard ships (2024). https://www.bimco.org/news-insights/bimco-news/2024/20241114-cyber-security-guidelines
5. Badea, M., et al.: Maritime industry cybersecurity threats in 2025: advanced persistent threats (APTS), Hacktivism and vulnerabilities. Logistics **9** (2025). https://doi.org/10.3390/LOGISTICS9040178. https://www.mdpi.com/2305-6290/9/4/178
6. Basnet, S., BahooToroody, A., Chaal, M., Lahtinen, J., Bolbot, V., Banda, O.A.V.: Risk analysis methodology using STPA-based Bayesian network- applied to remote pilotage operation. Ocean Eng. **270**, 113569 (2023). https://doi.org/10.1016/J.OCEANENG.2022.113569
7. Basnet, S., Bolbot, V., Yousaf, A., Gunawan, S., Zhou, J., Banda, O.A.V.: An integrated STPA-stride-bn framework for cybersecurity risk analysis: a case study of ship remote pilotage operations. Future Gener. Comput. Syst. **183**, 108515 (2026). https://doi.org/10.1016/J.FUTURE.2026.108515. https://linkinghub.elsevier.com/retrieve/pii/S0167739X26001494
8. Battaglioni, M., Lazzari, L., Patriarca, R.: MAGIC: a method for assessing cyber incidents occurrence. Comput. Secur. **1** (2022). https://arxiv.org/abs/2206.11586
9. Bouti, A., Kadi, D.A.: A state-of-the-art review of FMEA/FMECA. Int. J. Reliab. Qual. Saf. Eng. **1**, 515–543 (1994)
10. Bygdas, E., Jaatun, L.A., Antonsen, S.B., Ringen, A., Eiring, E.: Maritime cyber attack database (MCAD) (2025). https://ieeexplore.ieee.org/document/9478215
11. International Association of Classification Societies: UR E26 cyber resilience of ships – Rev.1, Nov 2023 (2017). https://iacs.org.uk/resolutions/ur-e/ur-e26-new/ur-e26-new
12. International Association of Classification Societies: UR E27 Cyber resilience of on-board systems and equipment – Rev.1 Sep 2023 (2023). https://iacs.org.uk/resolutions/ur-e/ur-e27-rev1/ur-e27-rev1
13. The MITRE Corporation: MITRE ATT&CK for industrial control systems (2025). https://attack.mitre.org/matrices/ics/
14. Van Landuyt, D., Joosen, W.: A descriptive study of assumptions in stride security threat modeling (2022)
15. Elidolu, G., Sezer, S.I., Akyuz, E., Arslan, O., Arslanoglu, Y.: Operational risk assessment of ballasting and de-ballasting on-board tanker ship under FMECA extended evidential reasoning (ER) and rule-based Bayesian network (RBN) approach. Reliab. Eng. Syst. Saf. **231**, 108975 (2023). https://doi.org/10.1016/J.RESS.2022.108975

16. FAIR Institute: Analyst's Guide to Cyber Risk Data Sources. Fairfax, VA, USA (2025). Version May 2025
17. Furuno Electric Co., Ltd.: VR-7000/VR-7000S Voyage Data Recorder Brochure, Nishinomiya, Japan (2024)
18. Grispos, G., Mahoney, W.R.: Cyber pirates ahoy! An analysis of cybersecurity challenges in the shipping industry. J. Inf. Warfare **21**(3), 59–73 (2022). https://arxiv.org/pdf/2208.03607. preprint available at arXiv:2208.03607
19. Guzman, N.H.C., Zhang, J., Xie, J., Glomsrud, J.A.: A comparative study of STPA-extension and the UFoI-E method for safety and security co-analysis. Reliab. Eng. Syst. Saf. **211**, 107633 (2021). https://doi.org/10.1016/J.RESS.2021.107633
20. International Maritime Organization: Guidelines on maritime cyber risk management, April 2025. https://wwwcdn.imo.org/localresources/en/OurWork/Security/Documents/MSC-FAL.1-Circ.3-Rev.3.pdf
21. International Organization for Standardization: International Electrotechnical Commission: Information security, cybersecurity and privacy protection – guidance on managing information security risks. ISO/IEC 27005:2022 (2022)
22. Karahasanovic, A., Kleberger, P., Almgren, M.: Adapting threat modeling methods for the automotive industry. In: Proceedings of the 15th ESCAR Conference, pp. 1–10 (2017)
23. Keskin, O.F., Lubja, K., Bahsi, H., Tatar, U.: Systematic cyber threat modeling for maritime operations: attack trees for shipboard systems. J. Mar. Sci. Eng. **13**(4) (2025)
24. Sion, L., Verreydt, S., Van Landuyt, D., Keuppens, K.Y.T., Joosen, W.: Sparta documentation from the DistriNet research unit at KU Leuven (2025). https://docs.sparta.distrinet-research.be/main/overview/overview.html
25. Sion, L., Verreydt, S., Van Landuyt, D., Keuppens, K.Y.T., Joosen, W.: Sparta from the DistriNet research unit at KU Leuven (2025). https://sparta.distrinet-research.be/
26. Sion, L., Verreydt, S., Van Landuyt, D., Keuppens, K.Y.T., Joosen, W.: Sparta packages from the DistriNet research unit at KU Leuven (2025). https://gitlab.kuleuven.be/distrinet/research/sparta/maven/-/packages/
27. Liang, A.T.W., Yousaf, A., Feng, L., Zhou, J.: MariOT matrix for maritime cybersecurity. J. Transp. Secur. **19**(1), 33 (2026). https://doi.org/10.1007/S12198-026-00354-X. https://link.springer.com/article/10.1007/s12198-026-00354-x
28. Meixuan, L., Yousaf, A., Mark, G., Jianying, Z., Sudipta, C.: Guidelines for cyber risk management in autonomous shipping. In: Andreoni, M. (ed.) Applied Cryptography and Network Security Workshops, vol. 1, pp. 143–161. Springer Nature, Switzerland (2024). https://doi.org/10.1007/978-3-031-61489-7_9
29. National Institute of Standards and Technology: Guide for conducting risk assessments. Technical report, Special Publication (SP) 800-30 Revision 1, National Institute of Standards and Technology (2012). https://csrc.nist.gov/pubs/sp/800/30/r1/final
30. Sion, L., Landuyt, D.V., Yskout, K., Joosen, W.: SPARTA: security & privacy architecture through risk-driven threat assessment. In: 2018 IEEE International Conference on Software Architecture Companion (ICSA-C), vol. 1, pp. 89–92, April 2018. https://doi.org/10.1109/ICSA-C.2018.00032
31. Sion, L., Yskout, K., Landuyt, D.V., van den Berghe, A., Joosen, W.: Security threat modeling: are data flow diagrams enough? IEEE/ACM 42nd International Conference on Software Engineering Workshops (ICSEW 2020), vol. 1, pp. 254–257 (2020). https://doi.org/10.1145/3387940.3392221

32. Tam, K., Jones, K.: MaCRA: a model-based framework for maritime cyber-risk assessment. WMU J. Marit. Aff. **18**, 129–163 (2019). https://doi.org/10.1007/S13437-019-00162-2/METRICS. https://link.springer.com/article/10.1007/s13437-019-00162-2
33. Temple, W.G., Wu, Y., Chen, B., Kalbarczyk, Z.: Systems-theoretic likelihood and severity analysis for safety and security co-engineering. In: Fantechi, A., Lecomte, T., Romanovsky, A. (eds.) RSSRail 2017. LNCS, vol. 10598, pp. 51–67. Springer, Cham (2017). https://doi.org/10.1007/978-3-319-68499-4_4/TABLES/4
34. Young, W., Porada, R.: System-theoretic process analysis for security (STPA-SEC): cyber security and STPA. In: 2017 STAMP Conference, pp. 27–30. MIT Press, Boston, MA, USA (2017)
35. Yousaf, A., Amro, A., Kwa, P.T.H., Li, M., Zhou, J.: Cyber risk assessment of cyber-enabled autonomous cargo vessel. Int. J. Crit. Infrastruct. Prot. **46**, 100695 (2024). https://doi.org/10.1016/j.ijcip.2024.100695. https://www.sciencedirect.com/science/article/pii/S1874548224000362
36. Yousaf, A., Gunawan, S., Basnet, S., Bolbot, V., Zhou, J., Valdez Banda, O.A.: STPA-cyber: a semi-automated cyber risk assessment framework for maritime cybersecurity. Comput. Secur. **157**, 104559 (2025)
37. Yousaf, A., Loan, A., Babiceanu, R.F., Maglaras, L., Yousaf, O.: Architectural and information theoretic perspectives of physical layer intruders for direct sequence spread spectrum systems. Comput. Secur. **70**, 124–143 (2017). https://doi.org/10.1016/j.cose.2017.05.011. http://www.sciencedirect.com/science/article/pii/S0167404817301189
38. Yousaf, A., Zhou, J.: From sinking to saving: MITRE ATT&CK and D3FEND frameworks for maritime cybersecurity. Int. J. Inf. Secur. **23**, 1603–1618 (2024). https://doi.org/10.1007/s10207-024-00812-4
39. Yousaf, A., Zhou, J., Chattopadhyay, S.: Maritime operational technology shipboard testbed for cybersecurity research, training, exercises, and education. IEEE Secur. Priv. (2025). https://doi.org/10.1109/MSEC.2025.3613711
40. Zhou, J.: Maritime cybersecurity: a big change ahead. IEEE Secur. Priv. **23**, 4–7 (2025)
41. Zhou, X.Y., Liu, Z.J., Wang, F.W., Wu, Z.L.: A system-theoretic approach to safety and security co-analysis of autonomous ships. Ocean Eng. **222**, 108569 (2021). https://doi.org/10.1016/J.OCEANENG.2021.108569

From BEA to Matrix: Security Failures, an Open Standard Proposal, and Germany's Path to Federated Legal Messaging

Markus Nauroth[1,2(✉)], Nicolai Kuntze[1], and Simon Rohlmann[1]

[1] SIUE – Southern Illinois University Edwardsville, 1 Hairpin Dr, Edwardsville, IL 62026, USA
mnaurot@siue.edu,
{markus.nauroth,nicolai.kuntze,simon.rohlmann}@hs-mainz.de
[2] Hochschule Mainz – University of Applied Sciences, Lucy-Hillebrand-Street 2, 55128 Mainz, Germany

Abstract. Germany operates one of the most comprehensive mandatory electronic legal communication systems worldwide: the *besonderes elektronisches Anwaltspostfach* (BEA), which since January 2022 requires all approximately 165 000 licensed attorneys to maintain and regularly check a centrally operated mailbox for court filings and judicial notifications. This paper documents BEA for an international audience unfamiliar with the German approach and provides a security analysis showing that it suffers from four fundamental flaws: a hardware security module that breaks end-to-end encryption by design; a single point of failure responsible for a prolonged nationwide outage; a closed, proprietary codebase that prevents independent audit; and the absence of cryptographically verifiable delivery proofs. We argue that the judicial system constitutes critical infrastructure on a par with energy supply and financial services, so that mandatory digitisation of judicial communication transfers this critical function into the cyber domain—including exposure to nation-state threat actors for whom persistent access to a national legal communication system is a strategic intelligence asset. Against this backdrop, we propose an open-standard successor architecture—a gateway-based hybrid combining S/MIME encryption, TPM 2.0 hardware attestation, and decentralised delivery proofs, synthesised from published cryptographic and PKI standards—and compare it with the Matrix+MLS protocol recommended by the German Federal IT Architecture working group in 2025 [33] as a successor to the aging OSCI transport. A structured three-way comparison of BEA, the proposal, and Matrix+MLS identifies their complementary strengths and a layered migration path, yielding transferable design principles for any jurisdiction considering mandatory electronic legal communication infrastructure.

Keywords: Electronic legal communication · beA · Germany · security analysis · end-to-end encryption · open standard · Matrix protocol · MLS · critical infrastructure · nation-state threat actors · eIDAS · rule of law

S. Zonouz et al. (Eds.): DCS-CI 2026, CCIS 3109, pp. 46–68, 2027.
https://doi.org/10.1007/978-981-92-5039-4_3

1 Introduction

The digitisation of court proceedings is a global policy priority, yet few countries have progressed as far — or encountered as many difficulties — as Germany. Germany's *besonderes elektronisches Anwaltspostfach* (BEA) is one of the most extensive mandatory electronic legal communication (ELC) systems in the European Union: every licensed attorney is legally required to use it, courts increasingly depend on it for service of process, and the legal consequences of system failure — missed, non-extendable court deadlines — are severe.

For an international audience, BEA is simultaneously a valuable reference case and a cautionary tale. It demonstrates that a country can achieve near-universal adoption of a digital legal communication channel within a decade; it also demonstrates that adopting such a channel without adequate attention to security architecture, open governance, and resilience creates systemic risks that are difficult to remedy once the system is mandatory.

The Judicial System as Critical Infrastructure

Critical infrastructure protection frameworks traditionally focus on sectors such as energy supply, water distribution, transportation, and financial services [1,2]. The judicial system — courts, prosecution authorities, and the attorney profession that enables access to them — belongs in this category by the same reasoning: its disruption does not merely cause economic harm but undermines a foundational function of democratic governance, namely the adjudication of disputes under the rule of law.

The progressive digitisation of judicial services intensifies this exposure. When paper-based processes are replaced by mandatory digital channels, the attack surface of the justice system is no longer limited to physical courthouses and paper archives. It encompasses the servers, protocols, key material, and software supply chains of every system that courts and attorneys are legally required to use. A successful attack against such a system does not need to steal data to cause harm: disrupting availability — even temporarily — can extinguish legal rights that cannot be reinstated once a procedural deadline has passed.

Critically, the centralised architecture of systems like BEA creates a target profile that is qualitatively different from, and more attractive than, any individual court or law firm. A single point of compromise at the national messaging infrastructure grants access to the entire communication stream of a country's legal profession — including privileged attorney-client correspondence, evidence in ongoing criminal proceedings, and pleadings in cases involving state interests. This threat profile is not limited to criminal actors. *Nation-state and governmental threat actors* [3] have both the motivation and the technical capability to target such infrastructure, precisely because the intelligence yield of persistent access to a national legal communication system is exceptional. A compromised HSM master key, as described in Sect. 3.1, would provide retrospective access to years of decryptable archived messages — an intelligence asset of strategic value.

Germany's experience with BEA therefore illustrates a general principle: *mandatory digitisation of a critical governance function transfers that function's threat model into the cyber domain, including the threat from state-level adversaries.* The security architecture of any mandatory ELC system must be designed accordingly — not for the threat model of a commercial web application, but for the threat model of critical national infrastructure under active targeting by sophisticated adversaries.

This paper presents Germany's ELC ecosystem in sufficient depth to be useful as a guidance document for policymakers and technologists in other jurisdictions. At the same time, it situates that description within a comparative security and design analysis, proposing an open-standard successor architecture informed by an earlier unpublished requirements study [10] and comparing it with the 2025 FITKO working group recommendation to adopt the Matrix protocol for all German government communication [33].

What is New and What is Known. The individual architectural flaws discussed in Sect. 3 — the HSM re-encryption step, the 2017/2018 outage, and the closed codebase — were first publicly documented by the Chaos Computer Club in 2017–2018 [8,38]; they are not new discoveries of this paper, and we cite the original sources throughout. What is new is detailed per item in the contributions list below.

Contributions. This paper makes six contributions. First, a self-contained description of the German ELC landscape — legal framework, system architecture, and operational context — for readers outside Germany (Sect. 2). Second, a security analysis of BEA that situates four architectural flaws — three previously documented by the CCC, one (the absence of verifiable delivery proof) elaborated here — within a critical-infrastructure and nation-state threat model, including an analysis of whether the design is legally mandated (Sect. 3). Third, an open-standard gateway architecture proposal, synthesised from published cryptographic standards and structured around an earlier unpublished requirements study, including a discussion of its methodological limitations (Sect. 4). Fourth, an account of Germany's ongoing transition toward the Matrix protocol as a government messaging standard (Sect. 5). Fifth, a structured three-way comparison of BEA, the open standard proposal, and Matrix+MLS, with a discussion of their complementary roles (Sect. 6). Sixth, transferable design principles for international ELC system design (Sect. 7).

2 Germany's Electronic Legal Communication Landscape

2.1 Legal and Regulatory Framework

Electronic legal communication in Germany is governed by an interlocking set of statutes and regulations. The *Zivilprozessordnung* (ZPO, Code of Civil Procedure) [5] governs civil proceedings and specifies the requirements for legally effective electronic filing (§ 130a ZPO). The *Bundesrechtsanwaltsordnung*

(BRAO, Federal Bar Act) establishes the mandatory use of the attorney mailbox (§ 31a BRAO). The *E-Justice-Gesetz* of 2013 [6] set out a phased roadmap that culminated in: compulsory BEA use for all attorneys from 1 January 2022; mandatory electronic filing by courts from 1 January 2026.

Communications that are legally equivalent to formal service of process must be transmitted either with a *Qualified Electronic Signature* (QES) as defined by the EU eIDAS Regulation [11], or via a *sicherer Übermittlungsweg* (secure transmission channel) — a statutory category to which BEA belongs. This means that a document transmitted through BEA by an admitted attorney carries the same legal effect as a signed paper submission, without requiring a separate QES on every document.

Terminology: Passive vs. Active Obligation. German law distinguishes between (a) the *passive* duty to maintain and regularly check a BEA mailbox (§ 31a BRAO), (b) the *active* obligation to file procedural documents electronically (§ 130d ZPO [30], effective since 1 January 2022), and (c) the permissible transmission channels for electronic filings: a Qualified Electronic Signature (QES) or a *sicherer Übermittlungsweg* (§ 130a Abs. 4 ZPO). While BEA is the dominant channel in practice, it is not the *only* legally permissible one; other authorised secure channels (e.g. EGVP) also satisfy the statutory requirement. Likewise, electronic service of process (*Zustellung*) is governed by § 173 ZPO [31].

eIDAS Scope: Closed System vs. Trust Service. The eIDAS Regulation [32] does not apply to trust services used exclusively within closed systems governed by national law (Art. 2(2)). The paper's analysis relies on eIDAS primarily for the definition and legal effect of QES (Art. 25) and for the concept of electronic registered delivery services (ERDS, Art. 43–44). Whether a future ELC system falls within or outside the eIDAS scope is an *architectural* decision with significant consequences for audit obligations, trust-list integration, liability, and cross-border interoperability. This paper does not resolve that question but notes that the distinction must be explicitly addressed in any successor design (see Sect. 6).

2.2 Predecessor System: EGVP and OSCI

Before BEA, German courts communicated electronically via the *Elektronisches Gerichts- und Verwaltungspostfach* (EGVP), a federated mailbox system introduced in the early 2000s. EGVP used the OSCI (*Online Services Computer Interface*) protocol [12], an XML/SOAP-based messaging standard developed by the German federal IT standardisation body (KoSIT). OSCI provides a two-layer envelope model: an outer envelope handles transport security (TLS [22]), while an inner envelope carries the signed and encrypted payload. EGVP participation was optional for attorneys and was accessed through stand-alone client software installed on the user's workstation.

OSCI's XML/SOAP architecture is increasingly regarded as technically obsolete for reasons beyond general dissatisfaction with XML as a wire format. It signs and encrypts XML elements *within* the SOAP envelope (XML-DSig, XML-Enc) rather than treating transport and payload as an opaque whole — the same element-level design pattern underlying the well-documented *XML signature wrapping* attack family, in which a validly signed element is relocated within the document tree so that the signature still verifies while the application processes attacker-controlled content instead [40]. This attack class generalises beyond OSCI: it was used to fully bypass SAML-based single sign-on at major vendors, which share OSCI's separation of XML signing from XML parsing [41]. Beyond signature wrapping, general-purpose XML parsing carries a large attack surface (XXE entity resolution, entity-expansion denial-of-service) absent from narrow, fixed-format binary or JSON protocols; the inner envelope's XML-Enc payload encryption uses static, long-lived recipient keys with no forward secrecy, the same structural weakness as BEA's HSM re-encryption (Sect. 3.1); and interoperability depends in practice on a small number of proprietary Java implementations rather than a narrowly specified wire format, complicating independent review. A 2025 working group recommendation discussed in Sect. 5 identifies the Matrix protocol as the target successor, although a formal transition study (IT-Planungsrat Beschluss 2025/18 [35]) is still under way.

2.3 SAFE: Federated Identity Infrastructure

Secure Access to Federated eJustice/eGovernment (SAFE) is the federated identity directory that underpins both EGVP and BEA. SAFE maintains authoritative certificates and professional admission information for attorneys (managed by BRAK, the Federal Bar Association), courts (managed by the Federal Ministry of Justice, BMJ), and other participants. SAFE certificates are standard X.509 [21] certificates (RSA or ECDSA key pairs) issued under a hierarchical PKI, with BRAK and BMJ each operating a Certification Authority for their respective participant group; status is queried over TLS [22] via LDAP and OCSP, with CRLs as fallback. These are the same certificates used for QES issuance, so SAFE doubles as admission directory and trust anchor for attorneys' qualified signing keys. Since certificates are public key material, not secrets, the directory's confidentiality rests entirely on the TLS transport — the same transport-layer assumption discussed for OSCI above. A *SAFE Connector* component at each participating court or authority forwards certificate lookups and verifies admission status in real time. The SAFE infrastructure represents a significant national PKI investment and is a key asset to be preserved in any successor system.

2.4 beA: Mandatory Attorney Mailbox

BEA (*besonderes elektronisches Anwaltspostfach*, literally "special electronic attorney mailbox") was developed under contract for the *Bundesrechtsanwaltskammer* (BRAK), the Federal Bar Association — a statutory body established under public law that serves simultaneously as the regulatory authority for the

German legal profession and as the operator of the communication infrastructure that profession depends on. This governance arrangement — regulator and infrastructure operator in a single body — is a structural feature with inherent conflict-of-interest potential.

Since 1 January 2022, every licensed attorney in Germany is legally required to: (a) maintain an active BEA mailbox; (b) check it at least once per working day; (c) use it as the primary channel for receiving judicial notifications and filing procedural documents [7], although § 130d ZPO [30] frames the active filing obligation in terms of electronic transmission generally, permitting also alternative *sichere Übermittlungswege* and QES-signed submissions. In practice, no widely adopted alternative exists, making BEA the de-facto exclusive channel. Approximately 165 000 attorneys and around 2 000 courts are mandatory users.

The mandatory, exclusive status makes BEA *critical infrastructure* in the strict sense: a sustained outage does not merely inconvenience users, it disrupts the administration of justice. German procedural law defines strict, non-extendable deadlines (*Fristen*) — a 30-day period to appeal a judgment is a typical example. If an attorney cannot file because the mandatory system is unavailable, the legal right to appeal may be irreversibly lost.

BEA's principal technical components are the following. The *beA Client* is a browser-based application (originally a Java applet, subsequently an Electron-based desktop app) running on the attorney's workstation. The *Client Security Component* (CSC) is a local daemon mediating between the browser client and the attorney's smart card (QSEE — Qualified Signature Creation Environment — token) for local cryptographic operations and QES generation. *Intermediär-BRAK* is the central BEA application server operated by BRAK; *every* message between any attorney and any court passes through this single node. The *beA-HSM* is a hardware security module at BRAK's data centre, holding the *Netzwerkschlüssel* (network master key); its role — and its decisive security flaw — is described in Sect. 3.1. *SAFE-BRAK* is BRAK's instance of the SAFE directory, providing attorney certificate and admission status management.

Figure 1 illustrates the message flow through these components.

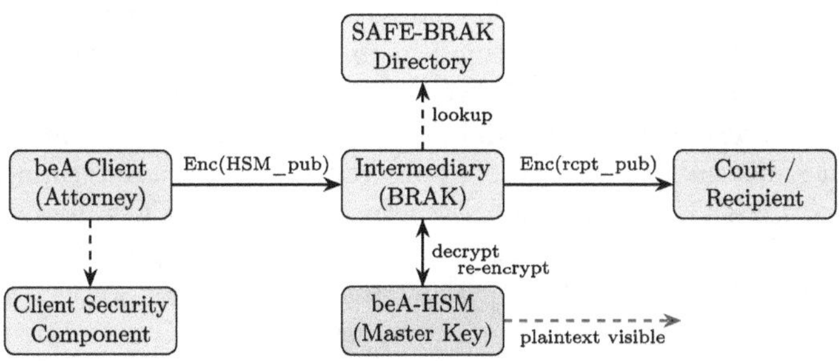

Fig. 1. BEA message flow. The HSM decrypts every message in transit and re-encrypts it for the recipient; attorney communication is transiently available as plaintext at the HSM. All traffic is funnelled through the single Intermediär node.

3 Security Analysis of BEA

3.1 Flaw 1: The HSM Re-Encryption Flaw — Broken End-to-End Encryption

End-to-end encryption (E2EE) is a cryptographic property whereby a message can be decrypted *only* by its intended recipient(s): no intermediate server — not the service operator, not any relay — ever holds the plaintext or a key capable of producing it. This property does not require one fixed mechanism. Static-key public-key encryption directly to the recipient's certificate (PGP, properly implemented S/MIME) is one valid realisation; a second establishes a shared *symmetric* session key out of band and evolves it over the conversation. Signal's Double Ratchet [42] is the most widely deployed example so that each message uses a freshly derived, rotating symmetric key, so compromising one message key exposes neither past messages (forward secrecy) nor, after a key-agreement step, future ones (post-compromise security), properties static-key S/MIME alone lacks. What both realisations share, and what matters for the analysis below, is the invariant that *no party other than sender and recipient ever possesses a key capable of recovering the plaintext.* It is this invariant that BEA's architecture violates.

BEA does *not* implement E2EE in the strict cryptographic sense, despite being marketed as a secure messaging platform. The system provides TLS transport encryption between client and server, but the central HSM re-encryption step (described below) means that message confidentiality is not preserved end-to-end. The actual message flow is as follows [8,10]. The attorney's client first encrypts the message to the **HSM's public key**, not to the recipient's key, and sends it to the Intermediär server, which forwards the ciphertext to the HSM. The HSM then **decrypts the message**, recovering attorney-client communication in plaintext, and re-encrypts it to the recipient's public key; the Intermediär finally stores and forwards the new ciphertext. The HSM decryption step is the decisive architectural flaw: the HSM holds a single network master key and uses it to decrypt *every* message in transit. Any process with access to the HSM — the legitimate BEA software, a compromised firmware update, a rogue insider, or an attacker who has exfiltrated key material via a known HSM attack surface — can read the entire communication stream of the national legal system.

The Chaos Computer Club (CCC) documented this architecture in late 2017 and characterised it as providing only the appearance of end-to-end security without substantive cryptographic protection against a determined attacker with access to BRAK's central infrastructure [8]. The CCC's analysis remains the only comprehensive public security assessment; the original report URL has been intermittently unavailable [38]. The design was adopted for legitimate operational reasons: it allows BRAK to route messages to multiple recipients without sender-side fan-out, and it enables key recovery for attorneys who lose their private key. However, this operational convenience fatally undermines the most fundamental security property the system is intended to provide. **Is the architecture legally mandated?** A natural question is whether the absence of true

E2EE follows from a legal requirement — e.g. a lawful-interception mandate obliging BRAK to retain decryption capability. We find no such mandate in the legislation governing BEA (§ 31a BRAO, § 130a/130d ZPO): the statutes mandate *use* of the mailbox and define which transmission channels satisfy formal filing requirements, but prescribe no encryption architecture. BRAK's own stated justification for the HSM is operational: law-firm representation and statutory substitution rules require that more than one authorised person be able to read a mailbox, which the HSM's re-encryption step accommodates without sender-side fan-out. When an attorney sued to compel genuine E2EE, the *Bundesgerichtshof* (BGH) held in 2021 that BRAK is under no legal obligation to provide it, finding the architecture's overall security sufficient despite conceding that message keys are re-encrypted at the HSM rather than delivered directly to the recipient [39]. The architecture is thus a discretionary, judicially tolerated operator choice, not a state-mandated surveillance requirement — so the multi-recipient and key-recovery needs that motivated it are addressable, as Sect. 4 shows, without an operator-accessible master key. The risk, in other words, is not that German law compels a backdoor, but that an operational design choice created one as a side effect. **Compounding consequences.** The HSM architecture has three cascading security implications. First, *key escrow by design*: the HSM master key is functionally a system-wide escrow key, so a single compromise exposes the confidentiality of the entire communication history. Second, the *absence of forward secrecy*: because all messages are encrypted to a single, long-lived HSM key, past traffic remains at risk indefinitely if the key is ever exposed, whereas true E2EE with ephemeral keys would limit exposure to individual sessions. Third, a *bulk interception capability*: a legal or extralegal demand targeting the HSM master key would enable retrospective decryption of all stored BEA messages — a capability that should not exist in a system designed to protect attorney-client privilege. **The nation-state threat dimension.** The HSM architecture is not only a vulnerability against criminal actors; it constitutes an exceptionally high-value target for *state-level and governmental threat actors* [3,4]. Intelligence agencies and state-sponsored actors routinely seek persistent access to communications infrastructure rather than executing disruptive one-time attacks. A single, long-lived HSM master key holding the decryption capability for the entirety of German attorney-client correspondence represents precisely the kind of target that justifies sustained, sophisticated offensive operations. This includes lawful-interception demands issued under foreign jurisdiction — a risk that is not hypothetical given documented cases of foreign intelligence services targeting national legal and judicial processes [3].

The intelligence value of such access is exceptional: legal correspondence in ongoing criminal, commercial, and administrative proceedings is among the most sensitive communication a state's citizens produce. It encompasses defence strategies in criminal trials, corporate litigation positions, whistleblower disclosures, and challenges to government action. A system architecture that concentrates decryption capability at a single operator-controlled node does not merely create a security risk; it creates a *structural invitation* for state-level targeting that is inconsistent with the rule-of-law function the system is intended to serve.

3.2 Flaw 2: Single Point of Failure and the 2017/2018 Outage

The decision to route every message through a single Intermediär node and a single HSM creates a *single point of failure* (SPOF) with nationwide scope. For most applications, a SPOF is an availability problem. For BEA, it is also a *rule-of-law* problem, because a system outage can cause attorneys to miss non-extendable procedural deadlines.

This risk is not hypothetical. In late December 2017, the CCC published a security analysis of the BEA client software revealing multiple critical vulnerabilities, including a man-in-the-middle attack that allowed an adversary to intercept and replace the client application delivered to attorneys [8]. BRAK's only available response was to take the *entire* system offline. The system remained unavailable for approximately nine months — from late December 2017 until its re-launch on 3 September 2018 [9,36].

During this extended outage, attorneys could neither receive judicial notifications electronically nor file procedural documents via the mandatory channel, even for time-critical matters, and German courts were forced to implement emergency procedures for missed electronic deadlines, creating legal uncertainty.

The outage demonstrates a fundamental mismatch between BEA's centralised architecture and the availability requirements of critical legal infrastructure. A decentralised architecture — in which each organisation operates an independent node — would confine a vulnerability to the affected node, leaving the rest of the network operational. Germany's earlier secure-email initiative De-Mail [26], France's eJustice portal, and early EU PEPPOL e-procurement implementations [27] all concentrated critical functions in centralised hubs, with analogous vulnerability profiles. The distinction in the legal domain is that the consequences of an outage are qualitatively more severe.

3.3 Flaw 3: Closed Architecture and Absence of Independent Audit

The BEA client application, Intermediär server software, and HSM firmware are all proprietary and closed-source. BRAK has not published a security architecture document, a threat model, or the results of any independent penetration test. The CCC's 2018 analysis — the only comprehensive public security assessment of BEA — was conducted entirely through reverse engineering of the client binary and analysis of network traffic [8].

This is particularly problematic for the legal profession. Attorneys are required by bar association rules to maintain attorney-client privilege. Compliance with this obligation depends entirely on the security of the communication channel — yet attorneys have no means to independently verify whether BEA provides the protection they are legally required to maintain. The closed architecture transfers this verification burden to BRAK, which is simultaneously the system's operator and the attorneys' regulatory authority.

An open, auditable architecture would allow the legal profession, academic researchers, and independent security firms to verify security properties directly, a model that has proven effective for internet security protocols.

3.4 Flaw 4: Absence of Cryptographically Verifiable Delivery Proof

German procedural law requires two forms of delivery evidence: *Zustellungsnachweis* (proof of service, analogous to certified mail return receipt) and *Empfangsbekenntnis* (explicit receipt acknowledgement by the recipient attorney). These proofs are legally significant: they establish the moment a party was notified of a judgment and start the clock on appeal deadlines.

In BEA, both forms of proof are generated and stored *centrally* by BRAK's infrastructure. This creates two problems: (1) the proofs are only as reliable as BRAK's systems, which the 2017/2018 outage demonstrated can become unavailable; (2) there is no cryptographic mechanism by which a court or opposing party can independently verify that a proof record has not been altered after the fact, since the proof is a database entry controlled by the system operator. A cryptographically signed proof generated at the recipient's endpoint would provide evidentiary strength that a centralised database record cannot.

3.5 Summary of Vulnerabilities

Table 1 summarises the four identified structural flaws and their security implications.

Table 1. Structural security flaws of BEA

Flaw	Category	Implication
HSM re-encryption	Confidentiality	No true E2EE; single key exposure compromises all messages
Single Intermediate	Availability	SPOF; nationwide outage possible
Closed codebase	Auditability	Security properties unverifiable; no community review possible
Central proof store	Integrity	Delivery proofs mutable by operator; no independent verification

4 An Open Standard Proposal for a BEA Successor

This section presents an open-standard architecture proposal for a BEA successor, synthesised by the authors from established cryptographic and PKI standards (S/MIME [14], TPM 2.0 attestation [13], MDN/AS1 [16,17], X.509 PKI [21]) and structured around requirement categories from an earlier, unpublished requirements-engineering study conducted at our institution [10].

4.1 Requirements Elicitation

The requirements underlying this proposal were elicited via semi-structured expert interviews with attorneys, court administrators, IT security specialists, and legal scholars, conducted as part of an unpublished Hochschule Mainz Master's thesis [10] that one of the present authors supervised. An open-question interview guide was used to avoid priming effects; responses were coded using qualitative content analysis. Because the underlying study is an unpublished thesis rather than a peer-reviewed work, several details are not independently verifiable, participant numbers and affiliation breakdown, sampling strategy, coding rules, and inter-coder reliability, and no sensitivity analysis of the requirement weights was performed, so marginal weight changes could alter the ranking of candidate architectures. We therefore treat the resulting categories and weights (Table 2) as a structuring device for this paper's own architecture synthesis rather than independently validated empirical evidence, reported here for transparency about provenance. Five requirement categories were identified and weighted to reflect interview participant priorities (Table 2).

Table 2. Requirement categories and weights used to structure this proposal, adapted from the requirements elicitation in [10]

Category	Description	Weight
Fachbezogen	Domain / legal compliance requirements	25%
Kommunikation	Communication reliability and protocol	25%
Integration	Integration with existing IT infrastructure	20%
Sicherheit	Security and cryptographic requirements	20%
Performance	Throughput, latency, availability	10%
Total		100%

The security category (*Sicherheit*) yielded five prioritised requirements that directly respond to the BEA flaws identified in Sect. 3: true E2EE, such that no intermediary may hold plaintext (SEC-1); non-repudiation, such that the sender cannot deny sending and the recipient cannot deny receipt (SEC-2); integrity, protecting messages against undetected modification in transit and at rest (SEC-3); availability, with no single point of failure and $\geq$99.5% monthly uptime (SEC-4); and auditability, requiring an open-source codebase or independent third-party audit with published results (SEC-5).

4.2 Architecture Selection: Three Variants

Three candidate architectures were developed and evaluated against the weighted requirements. Variant 1, *Fully Decentralised*, routes messages peer-to-peer via a distributed hash table (DHT). It has the strongest availability and E2EE properties but the lowest *Integration* score, owing to the high operational burden it places on small law firms and legacy court IT. Variant 2, *Centralised Routing and Storage*, handles routing and storage through a single central service, the structural equivalent of the current BEA, and scores highest on *Integration* (low deployment complexity) but lowest on *Sicherheit*, owing to HSM key-escrow and SPOF. Variant 3, the *Gateway Hybrid*, centralises routing while decentralising storage: a gateway layer handles protocol translation and routing centrally, messages are stored at the recipient's gateway, and delivery proofs are generated by the communicating gateways in a decentralised fashion, combining strong security with practical deployability. Table 3 shows the weighted scoring. Variant 3 achieves the highest weighted total (4.26) and is selected.

Table 3. Weighted scoring matrix for three candidate architectures, evaluated against the requirement weights of Table 2

Category	Weight	V1	V2 (beA-like)	V3 (selected)
Fachbezogen	25%	3.8	4.2	4.5
Kommunikation	25%	3.2	4.0	4.3
Integration	20%	2.5	4.5	3.8
Sicherheit	20%	4.8	2.2	4.6
Performance	10%	3.0	4.2	4.0
Weighted Total		3.47	3.74	**4.26**

4.3 Architecture Description: Gateway-Based Hybrid

The architecture has three core components, as shownin Figure 2. The *gateway*, operated by each organisation (law firm, court, or authority), handles hybrid encryption/decryption, local message storage, certificate management, and proof generation; its signing key is sealed to TPM 2.0 Platform Configuration Register (PCR) state, so the key is only released if the gateway boots an unmodified, expected software stack, and the gateway never exposes plaintext to the central router. The *central router* is a lightweight routing service that resolves attorney and court addresses to gateway endpoints and forwards encrypted payloads; it has no access to decryption keys, cannot read message content, and — unlike the BEA Intermediär — is not a cryptographic chokepoint. The *PKI / SAFE directory* is an X.509 [21] Public Key Infrastructure providing certificate issuance, revocation via Certificate Revocation Lists (CRL) and the Online Certificate Status Protocol (OCSP), and admission verification, aligned with the existing SAFE infrastructure; two parallel certificate hierarchies are proposed, with BRAK operating the attorney CA, BMJ operating the court CA, and a shared root providing cross-domain trust.

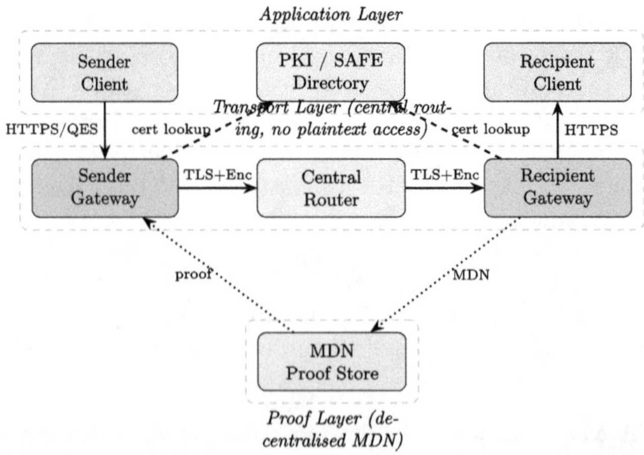

Fig. 2. Variant 3: gateway-based hybrid architecture proposed in this paper. The central router forwards encrypted payloads without decryption; messages are stored at the recipient's gateway; delivery proofs are generated and held as signed MDN records.

4.4 Key Design Decisions

Encryption: S/MIME over OpenPGP. Messages use hybrid encryption: a fresh AES-256-GCM [23] session key per message, encrypted asymmetrically to each recipient's X.509 [21] certificate (RSA-4096 [24] or ECDH P-384 [25]) per the S/MIME standard [14]. S/MIME is selected over OpenPGP [15] because its hierarchical PKI model provides built-in admission verification through CA issuance policy and aligns with the existing SAFE/LDAP infrastructure. It is important to note that eIDAS trust service lists (Art. 22 [32]) pertain specifically to *qualified* trust service providers and their qualified certificates. The bar-association admission verification performed through SAFE is a *professional identity check*, not an eIDAS qualification. A future design must therefore explicitly decide whether to operate an internal PKI within a closed system (eIDAS Art. 2(2)) or to seek eIDAS-qualified trust service provider status, with corresponding implications for audit, liability, and cross-border interoperability. OpenPGP's Web-of-Trust model cannot offer the required hierarchical certificate management. **Delivery Proof: MDN/AS1 over Blockchain.** A signed Message Disposition Notification [16] (MDN), following the AS1 profile [17] used in B2B legal messaging, is selected over blockchain-based proof of existence. The MDN is generated by the recipient's gateway upon successful decryption and contains the original message identifier, recipient identity, timestamp, content hash, and the gateway's S/MIME signature. MDN/AS1 requires no external infrastructure, has zero confirmation latency, publishes no message metadata publicly, and is an already-auditable open standard. When the gateway signature is bound to the TPM attestation state, the proof is anchored to both the message content and the attested hardware state of the receiving gateway at the time of delivery — a stronger evidential chain than a centralised database record. **Map-**

ping MDN/AS1 to eIDAS Art. 44. Were the delivery mechanism operated as an eIDAS-qualified electronic registered delivery service, Art. 44 [32] would require high-confidence sender/recipient identification, an advanced signature or seal securing the data, a qualified timestamp, and modification detection. The gateway's S/MIME signature and content hash address the latter two; SAFE admission verification partially addresses identification but would need formal eIDAS-level assurance, and a qualified time-stamping authority is not yet integrated. **Hardware Trust: TPM 2.0 Attestation.** Each gateway's signing key is sealed to the TPM's Platform Configuration Register (PCR) state via the TPM 2.0 Measured Boot chain [13]. The remote attestation protocol allows all participants to verify that a gateway is running an unmodified software stack before admitting it to the network. This replaces the BEA centralised HSM with a *distributed* trust model in which every gateway provides hardware-attested integrity guarantees.

5 Germany's Move to Matrix

In 2025, a working group of federal and state IT architects, coordinated through FITKO (*Föderale IT-Kooperation*) under the ZaPuK target architecture process, published a recommendation to adopt the **Matrix protocol** [19] paired with **Messaging Layer Security** (MLS, RFC 9420 [20]) as the cryptographic group-encryption layer [18,33,34]. The IT-Planungsrat formalised this direction through Beschluss 2025/28 (ZaPuK concept) [33], which identifies Matrix+MLS as the target communication layer for federated government mailbox services. Separately, Beschluss 2025/18 [35] commissions a study on the future of the OSCI/XTA transport standards; accordingly, the relationship between OSCI and Matrix is best characterised as under active, binding review rather than a completed replacement decision.

The scope is broad: all German government communication channels that are currently fragmented across platforms (BundID, OZG-PLUS, MeinJustizpostfach, Elster, and De-Mail) are targeted for consolidation. Courts, tax offices, health departments, and immigration authorities are all within scope. A binding decision by the *IT-Planungsrat* (IT Planning Council) is expected in 2026.

5.1 Why Matrix?

Matrix is an open, federated, real-time messaging protocol with properties directly relevant to the ELC problem. Under *federation*, each organisation operates a homeserver, and messages are routed between homeservers without a central intermediary, eliminating the SPOF that afflicts BEA. *MLS group encryption* [20] provides scalable, asynchronous E2EE with forward secrecy and post-compromise security — cryptographic properties that static S/MIME alone does not offer. Both Matrix and MLS are *open standards*, openly specified under permissive terms and enabling independent audit and multi-vendor implementations, addressing the closed-architecture flaw of BEA. Finally,

Matrix's *decentralised identity* model — homeserver-scoped identifiers of the form `@user:homeserver.tld` — can be bound to PKI certificates via external verification, providing a path for SAFE integration.

5.2 What Matrix Does Not Yet Address for ELC

The FITKO recommendation targets *general* government messaging and does not define a domain profile for the legal-specific requirements that distinguish ELC from inter-agency communication: Qualified Electronic Signatures per eIDAS for individual documents, bar-association admission verification (SAFE integration), legally binding, tamper-evident delivery proofs, and attorney-client privilege protections that exceed ordinary confidentiality requirements.

These gaps do not make Matrix unsuitable for ELC but indicate that an *ELC domain profile*, specifying how Matrix primitives are to be used/extended for legal-grade communication, is needed before Matrix homeservers can serve as a compliant replacement for BEA. The legal-domain layer of the architecture proposed in Sect. 4 (PKI policy, admission connector, MDN proof module, TPM attestation) is precisely the specification work that would constitute such a profile.

6 Three-Way Comparison

The comparison was constructed property-by-property against primary sources: for BEA, the CCC report [8], BRAK's public statements, and the 2021 BGH ruling [39] (Sect. 3.1); for the proposed architecture, the design in Sect. 4; for Matrix+MLS, the Matrix specification [19], RFC 9420 [20], and the ZaPuK concept paper [33,34]. A property is marked fulfilled (✓) only where a system's specification defines a mechanism for it, not fulfilled (×) where it explicitly does not, and partial (∼) where it depends on a domain profile that does not yet exist — which applies to several rows for Matrix+MLS, since the FITKO recommendation defines a general government messaging layer, not an ELC-specific profile. The result is a conservative comparison: Matrix+MLS is credited only with what its published specification already provides, not with what an ELC profile could in principle add on top of it. Table 4 compares the three systems across the security and domain properties most relevant to ELC.

The table compresses each property to a single symbol; we expand on the rows where the comparison is least obvious. *Forward secrecy* is partial for the proposed standard because S/MIME, as specified, encrypts to a long-lived recipient certificate, so a single private-key compromise still exposes every message ever sent to that key even though no operator-side key exists; MLS resolves this by ratcheting a fresh tree secret per epoch, the strongest argument for layering the proposed standard's legal-domain semantics on an MLS transport rather than treating S/MIME as sufficient alone (Sect. 6.1). *Decentralised delivery proof* is fulfilled by the proposed standard's MDN/AS1 mechanism (Sect. 4.4) but only partial for Matrix, whose `m.receipt` event is unsigned and not designed for

Table 4. Comparison: BEA vs. proposed open standard (Sect. 4) vs. Matrix+MLS

Property	beA	Proposed Std.	Matrix+MLS
True end-to-end encryption	×	✓	✓
No key escrow at operator	×	✓	✓
Forward secrecy	×	∼[a]	✓
No single point of failure	×	✓	✓
Independent auditability	×	✓	✓
Decentralised delivery proof	×	✓	∼[b]
Hardware-attested gateway	×	✓	×[c]
QES for individual documents	✓	✓	∼[d]
SAFE / admission integration	✓	✓	∼[e]
§ 130a Abs. 4 ZPO *sicherer Weg*	✓	✓	×[f]
eIDAS ERDS (Art. 43–44)	×[h]	∼[i]	×[j]
Migration path from OSCI/beA	n/a	Medium	High[g]

✓ = fulfilled, × = not fulfilled, ∼ = partial or conditional

[a] Static-key S/MIME has no forward secrecy; combining with MLS transport resolves this.

[b] Matrix `m.receipt` is unsigned; not yet a defined MDN equivalent.

[c] Not part of the Matrix spec; addable as a custom state event.

[d] Possible via `m.file`; a profile must specify the requirement.

[e] Join-policy hooks can enforce SAFE verification; requires explicit profile definition.

[f] Matrix homeservers are not yet a *sicherer Übermittlungsweg*; legislative amendment required.

[g] FITKO recommendation creates a political mandate for Matrix in courts.

[h] beA is not an eIDAS ERDS; it is a closed system under national law (eIDAS Art. 2(2)).

[i] MDN/AS1 addresses several Art. 44 requirements; a formal compliance mapping is not yet complete.

[j] Matrix homeservers are not registered as eIDAS ERDS; would require a dedicated legal-domain profile.

legally admissible disposition evidence, and *hardware-attested gateway* is unique to the proposed standard, since Matrix has no analogous concept of attested homeserver hardware. *SAFE/admission integration* and *QES for individual documents* are achievable for Matrix in principle (via join-policy hooks and file-event attachments) but partial because the specification treats both as implementation choices rather than normative requirements, leaving room for inconsistent vendor behaviour unacceptable in a mandated channel. Finally, *migration path from OSCI/beA* is the one property where Matrix+MLS outranks the proposed standard — not technically, but because the 2025 FITKO recommendation gives Matrix a standing political and budgetary mandate that a freestanding ELC standard would have to acquire independently; this asymmetry, technical completeness against institutional momentum, is the basis for the layered migration strategy in Sect. 6.1.

6.1 Complementary Strengths

Document-Layer Security. Table 4's "QES for individual documents" row depends on the signature and on research on PDF, OOXML, and ODF signatures demonstrating shadow-content injection and signature-spoofing attacks that produce documents appearing validly signed while containing attacker-controlled content. Any ELC system relying on QES must therefore specify admissible document profiles (e.g. PDF/A-3, PAdES baseline), authoritative validators, long-term validation and evidence preservation over archival periods spanning decades, and mitigations for shadow-content attacks.

The comparison reveals that the proposed open standard and Matrix+MLS are not competing alternatives but *complementary layers* addressing different aspects of the ELC problem. **Matrix+MLS** provides the transport and federation layer: an open standard, a multi-vendor homeserver ecosystem, MLS-grade forward secrecy, federated routing without SPOF, and the political momentum of the FITKO recommendation. **The proposed standard's legal domain layer** provides the ELC-specific semantics: S/MIME-based admission-bound certificates, TPM-attested gateway integrity, legally admissible MDN delivery proofs, and the SAFE connector integration (see § 130a ZPO / § 31a BRAO).

A practical migration path follows naturally: courts and law firms that deploy Matrix homeservers for general government communication under the FITKO recommendation automatically have the transport infrastructure for ELC, requiring only the addition of the legal-domain profile. This avoids the need for a separate ELC-specific deployment and aligns the proposed standard with an already-recommended technology, improving its prospects for adoption.

6.2 beA as the Baseline: What Must Not Be Repeated

Both successor approaches share the insight that BEA's central architectural decisions — a master-key HSM, a single routing node, and a closed proprietary codebase — are individually correctable but collectively constitute a pattern that must not be reproduced in any successor system. Any future ELC standard, whether gateway-based, Matrix-based, or hybrid, must satisfy five minimum constraints: no operator-accessible plaintext at any point in the message path; no system-wide key whose compromise exposes the entire communication history; a federated or distributed architecture that eliminates nationwide SPOFs; cryptographically verifiable delivery proofs independent of the operator's database; and an open, auditable specification and implementation.

7 Lessons for International Audiences

Germany's BEA experience and the open-standard proposal in Sect. 4 yield several transferable observations:

Mandatory Use Raises the Stakes for Security and Resilience. A voluntary system that fails inconveniences its users. A mandatory system that fails disrupts the administration of justice. The decision to mandate a particular channel should be accompanied by legally defined fallback procedures, measurable availability SLAs enforced by an independent body, and clear liability allocation in the event of operator failure. Germany's experience suggests that the absence of these mechanisms amplified the harm of the 2017/2018 outage significantly. **Design for the CI threat model, not the web-application threat model.** When a judicial communication system moves from optional to mandatory, its threat model changes categorically. Optional systems compete in the market; their primary adversaries are opportunistic criminals seeking financial gain. Mandatory systems that concentrate access to an entire profession's privileged communications are catalogued by national intelligence services and state-sponsored threat actors as strategic targets [3,4]. The appropriate design reference is therefore not a commercial SaaS platform but the security engineering practices applied to other critical national infrastructure: power grid control systems, financial clearing networks, or telecommunications interception-resistance requirements under national security law.

Concretely, this means that any mandatory ELC system should be evaluated against the question: *what does full compromise of this system's central infrastructure yield to a competent nation-state adversary?* If the answer includes mass decryption of attorney-client privilege, access to evidence in pending criminal proceedings, or the ability to disrupt judicial deadlines nationwide, the architecture requires the same resilience and confidentiality controls that protect genuinely critical infrastructure — including the elimination of any operator-accessible master key. **Governance structure matters as much as technical architecture.** BEA's operator (BRAK) is simultaneously the regulator of its mandatory users (attorneys). This arrangement creates audit capture: the party best positioned to criticise the system's security is also the party whose regulatory standing depends on the system's legitimacy. Separation of the operator role from the regulatory role — or independent technical oversight — is a governance principle that should be specified before, not after, a mandatory system goes live. **E2EE and operational convenience must be reconciled in the design, not traded off.** BEA's HSM was introduced to solve a legitimate operational need: message routing to multiple recipients and key recovery for attorneys. These are real requirements that any successor must address. The architecture proposed in Sect. 4 demonstrates that they can be satisfied without breaking E2EE — through gateway-side key management sealed to hardware attestation, with a notarised key backup held by a neutral trustee for recovery. The lesson is that operational convenience requirements must be listed explicitly in the design phase so that solutions that do not require an operator-accessible master key can be evaluated. **Open standard adoption is both a technical and a political decision.** The FITKO recommendation for Matrix demonstrates that an open protocol can achieve political momentum within a large, complex governmental structure when a compelling consolidation narrative is available.

For ELC specifically, a domain profile of an already-recommended open protocol is far more likely to be adopted than a bespoke standard requiring separate governance and deployment. Countries designing ELC systems should identify which existing open protocols are already in use or under consideration in their government IT architecture and design ELC as a domain profile rather than a standalone system. **Federation is compatible with data sovereignty if jurisdiction is bound to the homeserver, not to the protocol.** Federation is sometimes seen as in tension with data sovereignty, since messages cross organisational and, in cross-border deployments, potentially national boundaries. For a domestic ELC deployment this tension is smaller than it first appears: sovereignty is a property of *where data is stored and who controls the keys*, not of the federation protocol itself. Each law firm, court, and authority would operate its own homeserver under German jurisdiction, on infrastructure procured and audited under German public-sector rules — the same trust boundary that already exists for BEA's SAFE-affiliated participants, just distributed rather than centralised; federation changes *who* can reach a homeserver, not where its data resides or which law governs it. Germany already operates this model at national scale in *TI-Messenger*, the federated Matrix-based layer for the statutory healthcare telematics infrastructure, which requires every homeserver to be run by an approved provider under German data-protection law while still federating across the entire sector [45]. An ELC domain profile would adopt the same approach — homeserver operation restricted to vetted or BRAK/court-affiliated providers, with federation scoped to the closed set of ELC participants — making sovereignty a matter of deployment and accreditation policy layered on the protocol, exactly as it already is for the existing SAFE PKI (Sect. 2). **Federation widens the software supply chain that must be secured.** The same property that removes BEA's single point of failure — many independently operated homeservers and gateways rather than one Intermediär — also multiplies the software stacks, client libraries, and update channels an attacker could target, and a compromise of a widely used implementation could propagate to many participants at once. This is the generic supply-chain risk profile of federated systems, not something Matrix-specific, and an ELC domain profile should mandate the corresponding generic mitigations: published software bills of materials and a recognised integrity framework such as SLSA [43] or NIST's SSDF [44]; binding release artifacts to the TPM 2.0 attestation mechanism of Sect. 4.4, so tampered software fails attestation and is excluded rather than silently admitted; an allow-list of accredited operators (as in TI-Messenger) rather than open federation with arbitrary homeservers, bounding the blast radius of any one compromise; and independent security audits before accreditation, consistent with the auditability requirement already identified for the proposed standard (SEC-5, Sect. 4.1). **Legal validity requirements must drive technical specifications.** eIDAS QES requirements, the admissibility of electronic delivery proofs, and the definition of a *sicherer Übermittlungsweg* (a national statutory concept under § 130a Abs. 4 ZPO, distinct from eIDAS trust service categories) are all legal concepts with direct technical implications. The weighted-requirements methodology underlying Sect. 4 — beginning with requirements derived from legal and technical experts and made explicit and falsifiable rather than implicit in the architecture — provides a replicable approach for any jurisdiction seeking to align technical design with legal validity requirements, provided the elicitation itself is conducted as a documented, peer-reviewable study rather than repeating the methodological limitations noted in Sect. 4.1.

8 Conclusion

This paper has presented Germany's mandatory electronic legal communication system BEA as a reference case and cautionary tale for an international audience. BEA demonstrates that near-universal adoption of a digital legal channel is achievable within a decade, and simultaneously that a centralised, closed architecture with a master-key HSM creates systemic security risks whose consequences — in the legal domain — extend beyond data loss to the disruption of judicial proceedings. Against this background, we have proposed an open-standard successor architecture: a gateway-based hybrid that achieves true E2EE through S/MIME hybrid encryption at locally operated gateways, replaces the centralised HSM with TPM 2.0 hardware attestation, and provides legally admissible delivery proofs through a decentralised MDN/AS1 mechanism. The proposal was selected from three candidate architectures through a weighted requirements scoring process structured around an earlier, unpublished, non-peer-reviewed expert-interview study whose methodological limitations we have reported explicitly; it achieves the highest score in that scoring process while remaining deployable within the existing German legal IT ecosystem, but the scoring itself should be treated as a structuring device for the proposal rather than as independently validated empirical evidence. The 2025 FITKO recommendation to adopt the Matrix protocol for all German government communication creates a political and infrastructural opportunity. Matrix+MLS addresses the transport and federation layer; the proposed standard's legal domain layer addresses the ELC-specific semantics that Matrix alone does not define. Layering the proposed open standard on a Matrix transport offers a migration path that avoids separate deployment effort and aligns with a technology already recommended by the federal IT architecture process. For international policymakers and technologists, the five minimum constraints identified in Sect. 6 — no operator plaintext access, no system-wide escrow key, federated architecture, cryptographic delivery proofs, and an open specification — provide a technology-independent checklist for evaluating any proposed ELC system. Germany's experience, both its successes and its failures, is a resource that other jurisdictions can learn from without having to repeat. Future work includes a formal Matrix ELC domain profile, a prototype implementation with a SAFE connector, penetration testing against the full proposed stack, an interoperability roadmap with analogous systems such as Austria's ERV [28] and France's RPVA [29], a formal compliance matrix mapping MDN/AS1 to eIDAS Art. 44, an analysis of how the forthcoming EUDI Wallet and eIDAS 2.0 attribute attestation mechanisms [32] may reshape the admission-verification and PKI layer, a document-layer security model covering admissible document profiles and long-term evidence preservation, and an explicit eIDAS scope decision — closed system (Art. 2(2)) vs. qualified trust service — with a documented impact assessment on audit, liability, and cross-border interoperability.

Disclosure of Interests. The authors have no competing interests to declare that are relevant to the content of this article.

References

1. European Parliament and Council: Directive (EU) 2022/2555 on measures for a high common level of cybersecurity across the Union (NIS2 Directive). Official J. Eur. Union, L 333 (2022). https://eur-lex.europa.eu/legal-content/EN/TXT/?uri=CELEX:32022L2555
2. Bundesamt für Bevölkerungsschutz und Katastrophenhilfe (BBK): Kritische Infrastrukturen — Sektoren und Branchen. https://www.bbk.bund.de/DE/Themen/Kritische-Infrastrukturen/kritische-infrastrukturen_node.html
3. European Union Agency for Cybersecurity (ENISA): ENISA Threat Landscape 2024. ENISA, Heraklion (2024). https://www.enisa.europa.eu/publications/enisa-threat-landscape-2024
4. Bundesamt für Sicherheit in der Informationstechnik (BSI): Die Lage der IT-Sicherheit in Deutschland 2024. BSI, Bonn (2024). https://www.bsi.bund.de/DE/Service-Navi/Publikationen/Lagebericht/lagebericht_node.html
5. Zivilprozessordnung (ZPO) § 130a: Elektronisches Dokument; Verordnungsermächtigung. Bundesministerium der Justiz (2021). https://www.gesetze-im-internet.de/zpo/__130a.html
6. Gesetz zur Förderung des elektronischen Rechtsverkehrs mit den Gerichten (E-Justice-Gesetz). Bundesgesetzblatt I, Nr. 47 (2013)
7. Bundesrechtsanwaltsordnung (BRAO) § 31a. Bundesministerium der Justiz (2021). https://www.gesetze-im-internet.de/brao/__31a.html
8. Chaos Computer Club (CCC): Analyse des besonderen elektronischen Anwaltspostfachs (beA). CCC Security Analysis Report (2017). https://www.ccc.de/de/updates/2018/bea
9. Bundesrechtsanwaltskammer (BRAK): Pressemitteilung zur Abschaltung des beA, December 2017. Re-launch (2018)
10. Kriechel, D., Kummer, A.: Konzeption eines offenen Standards für den elektronischen Rechtsverkehr in Deutschland. Master's thesis, Hochschule Mainz - University of Applied Sciences (2020)
11. Parliament, E.: Council: Regulation (EU) No 910/2014 on electronic identification and trust services for electronic transactions in the internal market (eIDAS). Off. J. Eur. Union (2014)
12. KoSIT (Koordinierungsstelle für IT-Standards): OSCI Transport 1.2. Specification (2016)
13. Trusted Computing Group: TPM Library Specification, Family 2.0. ISO/IEC 11889 (2015)
14. Schaad, J., Ramsdell, B., Turner, S.: Secure/Multipurpose Internet Mail Extensions (S/MIME) Version 4.0 Message Specification. RFC 8551, IETF (2019)

15. Callas, J., Donnerhacke, L., Finney, H., Shaw, D., Thayer, R.: OpenPGP Message Format. RFC 4880, IETF (2007)
16. Hansen, M., Vaudreuil, G.: Message Disposition Notification. IETF 3798, IETF (2004)
17. Group, E.-I.W.: MIME-based secure peer-to-peer business data interchange over the internet (AS1). RFC 3335, IETF (2002)
18. netzpolitik.org: Verwaltungsdigitalisierung — Arbeitsgruppe empfiehlt Matrix-Protokoll für Behördenkommunikation (press report on the ZaPuK recommendation) https://netzpolitik.org/2025/verwaltungsdigitalisierung-arbeitsgruppe-empfiehlt-matrix-protokoll-fuer-behoerdenkommunikation/
19. Matrix.org Foundation: the Matrix Specification, Version 1.x. https://spec.matrix.org/
20. Barnes, R., Beurdouche, B., Robert, R., Millican, J., Omara, E., Cohn-Gordon, K.: The messaging layer security (MLS) protocol. RFC 9420, IETF (2023)
21. Cooper, D., Santesson, S., Farrell, S., Boeyen, S., Housley, R., Polk, W.: Internet X.509 public key infrastructure certificate and certificate revocation list (CRL) profile. RFC 5280, IETF (2008)
22. Rescorla, E.: The transport layer security (TLS) protocol version 1.3. RFC 8446, IETF (2018)
23. National institute of standards and technology (NIST): advanced encryption standard (AES). FIPS PUB 197 (2001). https://doi.org/10.6028/NIST.FIPS.197
24. Rivest, R.L., Shamir, A., Adleman, L.: A method for obtaining digital signatures and public-key cryptosystems. Commun. ACM **21**(2), 120–126 (1978)
25. National institute of standards and technology (NIST): recommendation for pair-wise key-establishment schemes using discrete logarithm cryptography. SP 800-56A Rev. 3 (2018). https://doi.org/10.6028/NIST.SP.800-56Ar3
26. Bundesministerium des Innern und für Heimat (BMI): De-Mail-Gesetz. Bundesgesetzblatt I, Nr. 26, 3. Mai 2011; see also: https://www.bsi.bund.de/EN/Themen/Verbraucherinnen-und-Verbraucher/Informationen-und-Empfehlungen/Digitale-Verwaltung/De-Mail/de-mail_node.html
27. OpenPeppol AISBL: Peppol Technical Documentation — eDelivery and Transport Infrastructure Specifications. https://peppol.org/documentation/technical-documentation/
28. Bundesministerium für Justiz (Austria): Elektronischer Rechtsverkehr (ERV). https://www.justiz.gv.at/service/digitale-justiz/elektronischer-rechtsverkehr-(erv).967.de.html
29. Conseil National des Barreaux (CNB): Réseau Privé Virtuel des Avocats (RPVA) — description and governance. https://fr.wikipedia.org/wiki/R%C3%A9seau_priv%C3%A9_virtuel_des_avocats
30. Zivilprozessordnung (ZPO) § 130d: Nutzungspflicht für den elektronischen Rechtsverkehr. Bundesministerium der Justiz. https://www.gesetze-im-internet.de/zpo/__130d.html
31. Zivilprozessordnung (ZPO) § 173: Zustellung elektronischer Dokumente. Bundesministerium der Justiz. https://www.gesetze-im-internet.de/zpo/__173.html
32. European Parliament and Council: Regulation (EU) No 910/2014 on electronic identification and trust services for electronic transactions in the internal market (eIDAS) — consolidated text incorporating amendments through Regulation (EU) 2024/1183 (eIDAS 2.0). https://eur-lex.europa.eu/legal-content/EN/TXT/?uri=CELEX:02014R0910-20241018

33. IT-Planungsrat: Beschluss 2025/28 — ZaPuK-Konzept (Zielarchitektur Postfach- und Kommunikationsdienste; Matrix+MLS). https://www.it-planungsrat.de/beschluss/beschluss-2025-28
34. FITKO/ZaPuK: Konzeptpapier — Matrix+MLS als Kommunikationsschicht. https://www.it-planungsrat.de/fileadmin/beschluesse/2025/Beschluss_2025_28_ZaPuK_Konzept.pdf
35. IT-Planungsrat: Beschluss 2025/18 — Eskalationsentscheidung OSCI und XTA (commissioning a study on the future of OSCI/XTA transport standards). https://www.it-planungsrat.de/beschluss/beschluss-2025-18
36. Bundesrechtsanwaltskammer (BRAK): Sicherheit im beA — Sicherheitsgutachten und Wiederinbetriebnahme (2018). https://www.brak.de/anwaltschaft/bea-erv/sicherheitsgutachten/
37. Bundesrechtsanwaltskammer (BRAK): Aktive Nutzungspflicht des elektronischen Rechtsverkehrs ab 1.1.2022. https://www.brak.de/newsroom/news/aktive-nutzungspflicht-des-elektronischen-rechtsverkehrs-ab-112022/
38. Golem.de: beA — noch mehr Sicherheitslücken im Anwaltspostfach. Press report corroborating the CCC findings (2018). https://www.golem.de/news/bea-noch-mehr-sicherheitsluecken-im-anwaltspostfach-1801-131942.html
39. Bundesgerichtshof (BGH), Anwaltssenat: Beschluss vom 22. März 2021, Az. AnwZ (Brfg) 2/20 — kein Anspruch auf Ende-zu-Ende-Verschlüsselung beim besonderen elektronischen Anwaltspostfach (beA). https://www.lto.de/recht/juristen/b/besonderes-elektronisches-anwaltspostfach-bea-bgh-anwzbrfg2-20-ende-zu-ende-verschluesselung-brak-rechtsanwaelte-sicherheit
40. McIntosh, M., Austel, P.: XML signature element wrapping attacks and countermeasures. In: Proceedings of the 2005 Workshop on Secure Web Services (SWS 2005), pp. 20–27. ACM (2005). https://doi.org/10.1145/1103022.1103026
41. Somorovsky, J., Mayer, A., Schwenk, J., Kampmann, M., Jensen, M.: On breaking SAML: l. In: Proceedings of the 21st USENIX Security Symposium, pp. 397–412. USENIX (2012). https://www.usenix.org/conference/usenixsecurity12/technical-sessions/presentation/somorovsky
42. Perrin, T. (ed.), M. Marlinspike: The Double Ratchet Algorithm. Signal Specification, Revision 1, 2016 (revised continuously). https://signal.org/docs/specifications/doubleratchet/
43. Open source security foundation (OpenSSF)/Linux foundation: supply-chain levels for software artifacts (SLSA), Version 1.0. https://slsa.dev/
44. National institute of standards and technology (NIST): secure software development framework (SSDF), Version 1.1. NIST SP 800-218 (2022). https://csrc.nist.gov/pubs/sp/800/218/final
45. gematik GmbH: TI-Messenger — federated matrix-based messaging for the German healthcare telematics infrastructure. https://www.gematik.de/anwendungen/ti-messenger

UDE-Based Cyber Twin for Residual-Based Anomaly Detection on Sensor and Driver Channels in Data-Center Infrastructure

Karthik Pappu(✉) and Chad Fenner

Dakota State University, Madison, SD, USA
karthik.pappu@trojans.dsu.edu, chad.fenner@dsu.edu

Abstract. Data centers are critical to national infrastructure; outages cascade to finance, healthcare, government, and artificial intelligence services. The telemetry that mediates their tightly coupled thermal–power dynamics constitutes an operational observability surface on which anomalies, whether benign disturbance, controller fault, or adversarial tampering, must be flagged early. We ask whether a Universal Differential Equation (UDE), trained as a cyber twin of rack-level thermal behavior, produces residuals that reveal such anomalies. Using six months of telemetry from a real tropical data-center testbed, we train a UDE that combines a two-parameter physics Ordinary Differential Equation (ODE) with a neural residual term, and evaluate its residual stream on six sensor-channel tampering scenarios and three driver-channel telemetry perturbations using static, rolling-window, and Cumulative Sum (CUSUM) detectors at a fixed false-positive rate.

The UDE achieves test root-mean-squared error of 0.934 °C against the unperturbed observation, compared with 1.061 °C for the calibrated two-parameter physics ablation. Against sensor-channel tampering, the UDE matches or modestly outperforms the ablation across all six scenarios, with the largest position-aggregate AUC gains on negative-offset tampering (+0.058) and sensor freeze (+0.030). Under driver-channel perturbations of airflow and supply-air temperature, the UDE residual separates anomaly windows more strongly than the ODE ablation, with rolling-window area under the Receiver Operating Characteristic curve (ROC-AUC) of 0.860 vs. 0.750 for cooling degradation and 0.888 vs. 0.813 for supply-air manipulation, flagging the anomalies 3.7 h and 3.2 h sooner respectively at the same false-positive rate. Across four random seeds the recovered physics parameters and test RMSE are reproducible (test RMSE 0.98 ± 0.07 °C). The neural residual inside the UDE is most beneficial for out-of-distribution driver-channel telemetry perturbations and for sensor-channel scenarios where the residual envelope shifts with the perturbation.

Keywords: Cyber Twin · Universal Differential Equations · Anomaly Detection · Data Centers · Cyber-Physical Security

S. Zonouz et al. (Eds.): DCS-CI 2026, CCIS 3109, pp. 69–92, 2027.
https://doi.org/10.1007/978-981-92-5039-4_4

1 Introduction

Data centers are critical national infrastructure. They underpin financial transactions, healthcare records, government services, and the artificial intelligence platforms that increasingly mediate public information [4]. Their thermal–power behavior is tightly coupled [10,33]: temperature, airflow, and electrical consumption interact through computational load, and any disruption of cooling capacity or workload distribution can drive equipment outside safe operating envelopes within minutes. The telemetry that mediates this coupling [38] (rack-level power, supply and return-air temperature, airflow) constitutes an operational observability surface for anomaly monitoring. Early detection of benign disturbances, controller faults, or adversarial tampering enables timely response. National guidance from the National Institute of Standards and Technology (NIST) [35] frames data-center operational technology within the broader scope of cyber-physical risk management for industrial systems.

The cyber-physical security literature has developed system-theoretic detectability bounds for attacks on linear time-invariant systems [12,28], replay-attack detectors [25], and adaptive secure estimators for false-data injection [7]; surveys converge on residual-based detection as the dominant physics-grounded approach [15,24,36], benchmarked on the Secure Water Treatment (SWaT) and Water Distribution (WADI) testbeds [1,16,20,23]. Open challenges remain for sensor-signal tampering on operational systems; we survey this lineage in Sect. 2.

Within this lineage, digital twins have been adopted as normal-behavior reference models for anomaly monitoring across industrial control, manufacturing, and electrical-power infrastructure [3,14,32,40,41]. In this paper we use the term *cyber twin* for a digital twin deployed primarily as an anomaly-reference model rather than as a control optimizer, in contrast with operations-focused twins [5,37,39] built for data-center cooling optimization.

For a cyber twin to be useful in the data-center context, it must capture the physics of rack-level heat exchange while remaining computationally light enough for online residual scoring. We instantiate the twin as a Universal Differential Equation (UDE) [29], a hybrid scientific-machine-learning architecture that embeds a small neural network as an additive vector-field term inside a known ordinary differential equation, supported by the Julia `DifferentialEquations.jl` [30] ecosystem. The additive-residual placement, formalized in Sect. 3.2, allows the linear physics to absorb in-distribution variation and the neural term to behave as a training-distribution-bound residual; this asymmetry sits behind our class-conditional anomaly-detection finding. The UDE framework has not previously been used as an anomaly-monitoring layer for real data-center infrastructure.

The UDE is trained on six months of telemetry from the NTU Tropical Data Center 2.0 testbed [21], an air-cooled facility instrumented with rack-level power meters, supply and return-air thermal sensors, airflow probes, and humidity sensors. Prior physics-informed machine learning work on data-center thermal and airflow prediction [8,9,17,22,38] has focused on forecasting and cooling control rather than anomaly monitoring on the telemetry stream.

This paper makes the following contributions.

1. We construct a UDE-based cyber twin for rack-level data-center telemetry from six months of NTU Tropical Data Center 2.0 measurements, combining a two-parameter convective heat-transfer ordinary differential equation with a neural residual term.
2. We present a residual-based anomaly-detection pipeline for sensor- and driver-channel telemetry, using static z-score, rolling-window z-score, and CUSUM detectors, with thresholds calibrated once on clean test residuals at a fixed false-positive rate and applied unchanged to every anomaly scenario.
3. We construct a two-part anomaly benchmark on the testbed, comprising six sensor-channel tampering scenarios that perturb only the observed back-rack temperature, and three driver-channel perturbation scenarios that affect cooling airflow, supply-air temperature, and rack power.
4. We carry out a physics-only ablation study, contrasting the UDE residual against the residual produced by removing the neural component, isolating when the neural correction contributes detection power and when it does not.
5. We report a class-conditional anomaly-detection finding. Against sensor-channel tampering the UDE residual matches or modestly outperforms the calibrated physics ablation across all six scenarios, with the largest position-aggregate gains on negative-offset tampering and sensor freeze; against driver-channel perturbations of cooling airflow and supply-air temperature it produces a 0.075 to 0.110 absolute AUC improvement and a 3.2 to 3.7 h reduction in detection delay at the same false-positive rate. We also report multi-seed reproducibility of the cyber-twin fit across four random seeds.

The remainder of the paper is organized as follows. Section 2 surveys related work in data-center thermal modeling, scientific machine learning, cyber-physical anomaly and attack detection, and digital twins. Section 3 describes the dataset, the ordinary differential equation and UDE formulations, the synthetic anomaly injection protocol, and the three residual-based detectors. Section 4 reports the cyber-twin fit on the test split, the sensor-channel and driver-channel anomaly-detection results, and the physics-only ablation. Section 5 interprets the conditional finding through the lens of out-of-distribution driver telemetry. Section 6 discusses limitations, and Sect. 7 concludes.

2 Related Work

2.1 Data-Center Thermal Modeling and Physics-Informed Machine Learning

Thermal modeling for data-center racks has historically relied on computational fluid dynamics, reduced-order resistor-capacitor networks, or coupled cyber-physical models. Sharma et al. introduced dynamic thermal-management strategies that exploit workload migration to balance heat distribution [33]; Moore et al. proposed temperature-aware workload scheduling to reduce cooling cost [26].

Parolini et al. formalized the data center as two coupled networks, a computational network for the cyber side and a thermal network for the physical side, and showed that coordinated control of both yields measurable energy gains [27]. Dayarathna et al. [10] surveyed the corresponding energy-consumption modeling literature, and Barroso et al. [4] provide the canonical systems-level reference for warehouse-scale computing.

A recent strand of work applies physics-informed machine learning to data-center thermal and airflow prediction [8,9,17,22], with the closest peer being the multiphysics-informed framework of Wang et al. [38]; that line of work uses energy-balance residual learning rather than the embedded-neural-network architecture we adopt and addresses neither cyber-physical anomaly detection nor residual-based security monitoring. Operations-focused digital twins for data centers [5,37,39] are designed for thermal forecasting, capacity planning, and cooling-control optimization rather than anomaly monitoring. The cyber twin in this research produces a residual stream as its primary output rather than a forecast or a control signal.

2.2 Scientific Machine Learning and Universal Differential Equations

The Universal Differential Equation framework of Rackauckas et al. [29], implemented in the Julia `DifferentialEquations.jl` ecosystem [30], embeds a neural network as one term inside a known differential equation, in contrast to Physics-Informed Neural Networks [19,31], which use the network as the solution surrogate. The present work adapts the framework to anomaly detection on data-center telemetry; the formal contrast between UDE and PINN architectures, and its consequence for the residual stream, is deferred to Sect. 3.2.

2.3 Cyber-Physical Anomaly and Attack Detection

Attack detection for cyber-physical systems has developed a substantial system-theoretic toolbox. Major challenges remain beyond the linear time-invariant setting. The system-theoretic foundations were laid by Pasqualetti et al., who characterized the limits of detectability and identifiability of attacks against linear time-invariant cyber-physical systems through a graph- and geometry-theoretic framework [28], and by Fawzi et al., who quantified the maximum number of corrupted sensors a secure estimator can correctly account for [12]. Mo and Sinopoli established the conditions under which replay attacks evade chi-squared detectors layered over Kalman filters [25]. Chattopadhyay and Mitra developed adaptive secure estimators against false-data injection [7]. Zhou et al. explicitly distinguish sensor attacks from actuator attacks and propose a learning-based framework that handles both [42], while Hosseinzadeh and Sinopoli design active detectors for prevented-actuation attacks [18]. Standardized threat-modeling frameworks include the National Institute of Standards and Technology Special

Publication 800-82 [35] and the MITRE ATT&CK knowledge base for industrial control systems [2], both of which catalogue sensor manipulation, command injection, and impair-response techniques.

Surveys by Mitchell and Chen [24], Giraldo et al. [15], and Tan et al. [36] converge on residual-based detection as the dominant physics-grounded approach; Giraldo et al. taxonomize residual methods into stateless tests over instantaneous residuals and stateful tests such as CUSUM, placing this work in the stateful branch [6,15]. The Secure Water Treatment (SWaT) and Water Distribution (WADI) testbeds at the Singapore University of Technology and Design [1,23] have driven much of the empirical industrial-control-system anomaly-detection literature, on which deep learning detectors [16,20] and rule-mined invariants [13] have been benchmarked. The neighboring building-automation, HVAC-security, and sensor-FDI literature for industrial control systems [11,34] is also relevant. Etalle has noted, in discussion following his ESORICS 2017 keynote, that intrusion detection is not always well-suited to manipulated sensor signals because the sensors continue to operate as intended. This caveat informs the positioning of the present work as anomaly detection on the telemetry stream rather than intrusion detection.

2.4 Digital Twins for Cyber-Physical Security

Digital twins have been adopted in the cyber-physical-security literature as normal-behavior reference models that support anomaly detection by comparing predicted and observed plant behavior. Xu et al. proposed digital-twin-based anomaly detection with timed-automaton plus generative-adversarial backends, evaluated on SWaT, WADI, and BATADAL [40], and extended the framework with curriculum learning [41]. Gao et al. developed a digital-twin anomaly-detection framework combining clustering and martingale tests on the Tennessee Eastman benchmark [14]. Balta et al. adapted the digital-twin paradigm to cyber-physical manufacturing systems with model-based controllers [3]. Salvi et al. extended the same idea to electrical-power critical infrastructure with a focus on cyber-resilience and situational awareness [32]. None of these prior works targets data-center rack-level thermal dynamics, and none uses a Universal Differential Equation as the twin's mathematical backbone.

3 Methodology

3.1 Dataset

The research uses the air-cooled NTU Tropical Data Center 2.0 testbed dataset [21], a public release of telemetry from a tropical Singapore facility comprising 521 servers under direct-expansion cooling, served by computer room air conditioning (CRAC) units. The testbed is instrumented with rack-level power meters, rack-mounted front and back-of-rack temperature and humidity sensors at three heights, supply, return, and outside-air thermal sensors, supply and

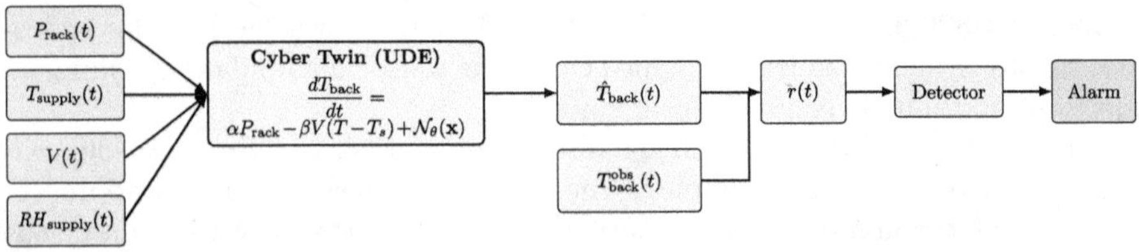

Fig. 1. Cyber-twin architecture. Driver telemetry feeds the trained UDE, which integrates forward to produce a predicted back-of-rack temperature $\hat{T}_{back}(t)$. The residual $r(t) = T_{back}^{obs}(t) - \hat{T}_{back}(t)$ is consumed by static, rolling-window, and CUSUM detectors that raise an alarm when the score exceeds a threshold calibrated once on clean test residuals.

return-air humidity sensors, and supply-airflow probes. The present study targets rack R0503-PB across the period 2022-09-01 to 2023-02-28, comprising six months of operation. The processed dataset contains $N = 26063$ samples and ten variables (used in the present study): rack power P_{rack} (kW), back-of-rack temperature T_{back} (°C) and humidity, supply-air temperature T_{supply} (°C) and humidity, return-air temperature T_{return} (°C) and humidity, outside-air temperature $T_{outside}$ (°C) and humidity, and supply-airflow velocity V (m/s).

The research uses a strict chronological 70/30 split, training on the first 18244 samples and reserving the final 7819 samples for evaluation. No shuffling is applied; all temporal continuity is preserved within each split.

Data Preprocessing Pipeline. Raw 30-second telemetry from each sensor stream is converted into a continuous-time representation suitable for differential-equation integration via three steps:

1. *Sensor alignment.* Timestamps from the rack power meter, thermal/humidity sensors, and airflow probes are aligned to a common clock; records missing any primary field (P_{rack}, T_{back}, T_{supply}, V) are dropped and sub-30-second gaps linearly interpolated.
2. *Temporal binning.* Aligned 30-second records are aggregated into uniform 10-minute bins by mean reduction, matching the dominant thermal time constant of rack-level dynamics while suppressing high-frequency sensor noise. This yields the $N = 26063 \times 10$ matrix described above with no remaining missing values.
3. *Continuous-time interpolation.* Piecewise-linear interpolation of each driver signal ($P_{rack}(t)$, $T_{supply}(t)$, $V(t)$, $RH_{supply}(t)$) provides continuous-time access for the Tsit5 solver at internal sub-step times without being constrained to the 10-minute grid.

3.2 Cyber-Twin Formulation

The cyber twin of the rack is defined as a dynamical model that consumes driver telemetry and produces an estimate $\hat{T}_{back}(t)$ of the back-of-rack temperature. Figure 1 shows the end-to-end pipeline, with driver inputs feeding the trained model, model output compared against observation to form a residual, and the residual consumed by static, rolling-window, and CUSUM detectors. The forecasting target is the same rack temperature observation $T^{\text{obs}}_{\text{back}}(t)$, and the residual stream

$$r(t) \;=\; T^{\text{obs}}_{\text{back}}(t) - \hat{T}_{\text{back}}(t) \tag{1}$$

serves as the cyber twin's anomaly-detection residual.

Physics-only ablation (ODE). The physics ablation captures convective heat transfer between the rack and the supply airflow:

$$\frac{dT_{\text{back}}}{dt} \;=\; \alpha\, P_{\text{rack}}(t) \;-\; \beta\, V(t)\,\big(T_{\text{back}}(t) - T_{\text{supply}}(t)\big). \tag{2}$$

The parameter α acts as a heating coefficient that scales the contribution of rack power to the temperature derivative; β is a cooling coefficient that combines the rack thermal capacitance with the heat-transfer coefficient against the supply airstream. Both parameters are reported alongside their fitted numerical values in Sect. 4.

Universal Differential Equation (UDE). The cyber twin we evaluate adds a learned residual term $\mathcal{N}_\theta(\mathbf{x}(t))$ to Eq. 2:

$$\frac{dT_{\text{back}}}{dt} \;=\; \alpha\, P_{\text{rack}}(t) \;-\; \beta\, V(t)\,\big(T_{\text{back}}(t) - T_{\text{supply}}(t)\big) \;+\; \mathcal{N}_\theta\big(\mathbf{x}(t)\big). \tag{3}$$

The state-and-input vector

$$\mathbf{x}(t) \;=\; \big(P_{\text{rack}}, T_{\text{back}}, T_{\text{supply}}, T_{\text{back}} - T_{\text{supply}}, V, RH_{\text{supply}}\big) \in \mathbb{R}^6$$

includes the two physics drivers, the predicted state, the temperature gap that drives the convective cooling term, and the supply-air humidity. The learned function $\mathcal{N}_\theta : \mathbb{R}^6 \to \mathbb{R}$ is a feed-forward neural network with two hidden layers of 32 and 16 units, tanh activations, and a linear scalar output. Including the two physics parameters, the UDE has 771 trainable parameters.

The design choice of placing $\mathcal{N}_\theta$ as an additive vector-field component, rather than replacing the right-hand side outright, follows the Universal Differential Equation framework [29]; this distinguishes the approach from a Physics-Informed Neural Network [19,31], in which the network is itself the proposed solution rather than a residual augmentation of the dynamics. The additive-residual placement is what allows the linear physics to absorb in-distribution variation, leaving the neural term to behave as a training-distribution-bound residual that is sensitive to off-manifold driver perturbations. The asymmetry produces the class-conditional anomaly-detection finding (Sect. 4). Algorithms 1 and 2 summarize the two integration step functions used by the solver.

3.3 Training

Both the ODE ablation and the UDE are trained by minimizing mean-squared error between the integrated trajectory and the observed back-of-rack temperature on the training split:

Algorithm 1. ODE ablation: vector field at time t given state T and parameters (α, β).

```
input: state T, time t, parameters (α, β), driver interpolators P_rack(·), V(·), T_supply(·)
P ← P_rack(t) {rack-power driver}
V_t ← V(t) {airflow driver}
T_s ← T_supply(t) {supply-air driver}
Ṫ ← αP − βV_t(T − T_s) {Eq. 2}
return Ṫ
```

Algorithm 2. UDE: vector field at time t given state T, physics parameters (α, β), and neural-network parameters θ.

```
input: state T, time t, parameters (α, β, θ), driver interpolators P_rack(·), V(·), T_supply(·), RH_supply(·)
P ← P_rack(t)
V_t ← V(t)
T_s ← T_supply(t)
RH ← RH_supply(t)
ΔT ← T − T_s {temperature gap drives convective cooling}
x ← (P, T, T_s, ΔT, V_t, RH)
c ← N_θ(x) {neural residual; 6→32→16→1, tanh activations}
Ṫ ← αP − βV_t(T − T_s) + c {Eq. 3}
return Ṫ
```

$$\mathcal{L}(\alpha, \beta, \theta) = \frac{1}{N_{\text{train}}} \sum_{i=1}^{N_{\text{train}}} \left(T_{\text{back}}^{\text{obs}}(t_i) - \hat{T}_{\text{back}}(t_i)\right)^2 + \lambda \, \mathcal{R}(\alpha, \beta), \tag{4}$$

where $\mathcal{R}$ is a soft penalty enforcing physical bounds on α and β, with $\lambda = 100$. Optimization proceeds in two stages: Adam warm-up followed by Broyden–Fletcher–Goldfarb–Shanno (BFGS) quasi-Newton refinement with initial step norm 0.01. The ODE ablation uses 100 Adam iterations at learning rate 10^{-2} followed by 50 quasi-Newton iterations; the UDE uses 20 Adam iterations at learning rate 10^{-3} followed by 10 BFGS iterations. The forward integrator is the explicit Tsitouras 5(4) Runge–Kutta solver (Tsit5) with absolute and relative tolerance 10^{-6} and a maximum-iteration cap of 10^7. Reverse-mode automatic differentiation through the ODE solver is provided by SciMLSensitivity within the Julia `DifferentialEquations.jl` ecosystem [30].

3.4 Anomaly Model: Scope, Capabilities, and Assumptions

The research characterizes the perturbation source against the cyber twin in terms of capabilities, knowledge, and the two perturbation classes evaluated. The framing is adapted from the cyber-physical attack-detection literature [7,12,28]; the research uses it here to bound what is in and out of scope for the anomaly-detection experiments, not to claim a complete attacker model.

Capabilities: The research models perturbations as write events on one of two cyber-side telemetry channels, namely the back-of-rack temperature observation $T_{\text{back}}^{\text{obs}}(t)$ or one of the driver inputs (P_{rack}, T_{supply}, V). Alteration is restricted to the cyber side; the underlying physical state of the rack is not perturbed. The trained UDE weights, detector thresholds, and physics-coefficient values are not exposed, since the cyber twin is hosted on a separate, hardened analytics endpoint. Channel compromise does not require physical site access: paths in the cyber-physical security literature include lateral movement from an adjacent building-management or DCIM sub-network, supply-chain compromise of a sensor-firmware update, and a misconfigured monitoring API.

Knowledge: For the present anomaly-detection scope, the source is assumed to know that residual-based monitoring is in operation and has access to the sensor schedule and telemetry rates, but not to the calibrated detector thresholds or the per-driver weighting inside the neural residual. Stronger white-box-adversary settings, which would require adversarial-robustness analysis of the trained weights, are out of scope and are addressed in the follow-up intrusion-detection direction outlined in Sect. 7.

Perturbation Classes: The research evaluates sensor-channel tampering, in which the back-of-rack temperature observation $T_{\text{back}}^{\text{obs}}$ is altered while the underlying physics is held fixed, and driver-channel perturbation, in which a driver signal supplied to the cyber twin is altered while the observation is held fixed. The class definitions are adapted from constructions in the cyber-physical attack-detection literature [7,12,25,28]; these synthetic, deterministic alterations to the telemetry stream serve as our anomaly scenarios. The question of whether a given anomaly arises from benign disturbance, controller fault, or adversarial action is out of scope and is left to a follow-up intrusion-detection study.

Assumptions: We assume the data-stitching, feature-alignment, and training pipelines are unaltered, consistent with the standard secure-base-station assumption in the cyber-physical security literature. Physical perturbation of cooling infrastructure (actuator manipulation, refrigerant alteration) is not modeled. Replay-style anomalies against full driver vectors [25] would constitute a third class to be evaluated in future work.

3.5 Synthetic Anomaly Injection

Anomalies are injected by deterministic offline modification of the test-split data stream prior to integration and detection scoring. No live actuator or controller is perturbed. We evaluate the cyber twin's residual stream under nine synthetic anomalies grouped into two classes. Sensor-channel tampering anomalies modify only the observation $T_{\text{back}}^{\text{obs}}$ that the operator sees; driver-channel telemetry perturbations modify the driver signals that the cyber twin integrates over. In both cases the underlying physical reality is held fixed; the anomalies simulate alteration of the cyber-side telemetry channel rather than physical actuator manipulation. The sensor-vs-actuator dichotomy is adopted from the control-engineering literature [12,42] and is used in cyber-physical security.

Table 1. Sensor-channel tampering scenarios. All scenarios modify only the observed back-rack temperature $T_{\text{back}}^{\text{obs}}$. Window column gives the perturbation duration.

Scenario	Form on $[t_s, t_e]$	Window
`step_spoof_+3C`	$T_{\text{back}}^{\text{obs}}(t) \leftarrow T_{\text{back}}^{\text{obs}}(t) + 3\,^\circ\text{C}$	12 h
`step_spoof_+5C`	$T_{\text{back}}^{\text{obs}}(t) \leftarrow T_{\text{back}}^{\text{obs}}(t) + 5\,^\circ\text{C}$	12 h
`slow_drift_+5C_12h`	$T_{\text{back}}^{\text{obs}}(t) \leftarrow T_{\text{back}}^{\text{obs}}(t) + 5\,(t - t_s)/(t_e - t_s)$	12 h
`fast_ramp_+5C_2h`	same linear-ramp form as slow drift	2 h
`sensor_freeze_6h`	$T_{\text{back}}^{\text{obs}}(t) \leftarrow T_{\text{back}}^{\text{obs}}(t_s)$	6 h
`negative_offset_-3C`	$T_{\text{back}}^{\text{obs}}(t) \leftarrow T_{\text{back}}^{\text{obs}}(t) - 3\,^\circ\text{C}$	12 h

Note: scenario identifiers are inherited from the analysis pipeline and retained for reproducibility; `spoof` denotes the data-stream alteration form, not impersonation in the security-literature sense.

Table 2. Driver-channel perturbation scenarios. Each scenario perturbs one driver input the cyber twin integrates over; the underlying physical reality and the observation $T_{\text{back}}^{\text{obs}}$ are unchanged.

Scenario	Form on $[t_s, t_e]$	Window
`cooling_degradation`	$V(t) \leftarrow 0.5\,V(t)$	12 h
`supply_air_manipulation`	$T_{\text{supply}}(t) \leftarrow T_{\text{supply}}(t) + 4\,^\circ\text{C}$	12 h
`power_surge`	$P_{\text{rack}}(t) \leftarrow 1.4\,P_{\text{rack}}(t)$	12 h

Let $[t_s, t_e]$ denote the perturbation window. Across all scenarios we set t_s to 30 % of the test split and t_e to provide either a 12-h, 6-h, or 2-h duration depending on the scenario.

Tables 1 and 2 list the two anomaly groups. Sensor-channel anomalies leave the model's prediction stream unchanged because the driver inputs are unmodified; the residual under tampering equals the clean residual plus the additive perturbation signal. Driver-channel perturbations require re-integrating the UDE and the ODE ablation forward in time under the altered driver interpolators; the observation $T_{\text{back}}^{\text{obs}}$ is held at its clean value, since the anomaly alters only the cyber-side driver signal that the model sees.

3.6 Detectors and Threshold Calibration

We instantiate three residual-based detectors, all consuming the residual stream $r(t)$ defined in Eq. 1. Let μ_r and σ_r denote the mean and standard deviation of r on the training trajectory.

Static z-score: The instantaneous score

$$s_t^{\text{stat}} = \frac{|r_t - \mu_r|}{\sigma_r} \tag{5}$$

is a stateless test that flags individual residuals exceeding a fixed standardized magnitude.

Algorithm 3. Residual-based anomaly-detection pipeline.

input: trained model $\mathcal{M} \in \{\text{ODE}, \text{UDE}\}$, train residual stream r_{train}, clean test residual stream r_{clean}, anomaly scenarios $\mathcal{A}$, target false-positive rate $\rho = 0.01$
$\mu_r \leftarrow \text{mean}(r_{\text{train}})$, $\sigma_r \leftarrow \text{std}(r_{\text{train}})$ {train residual statistics}
for detector $D \in \{\text{static}, \text{rolling}, \text{cusum}\}$ **do**
  $s^D_{\text{clean}} \leftarrow D(r_{\text{clean}}; \mu_r, \sigma_r)$ {Eqs. 5–7}
  $\tau^D \leftarrow \text{quantile}(s^D_{\text{clean}}, 1 - \rho)$ {calibrate threshold once}
end for
for anomaly $a \in \mathcal{A}$ **do**
  inject a into the test stream and obtain residual r^a
  for detector $D \in \{\text{static}, \text{rolling}, \text{cusum}\}$ **do**
    $s^D_a \leftarrow D(r^a; \mu_r, \sigma_r)$
    alarm $\leftarrow s^D_a > \tau^D$ {apply unchanged threshold}
    record ROC-AUC, TPR at FPR ρ, and detection delay
  end for
end for

Rolling-window z-score: For window length $W = 18$ samples (three hours at 10-minute resolution), the rolling mean is $\bar{r}_t = (1/W)\sum_{i=t-W+1}^{t} r_i$ and the score

$$s_t^{\text{roll}} = \frac{|\bar{r}_t - \mu_r|}{\sigma_r/\sqrt{W}} \tag{6}$$

tests whether the windowed mean has drifted from the training distribution beyond what clean variability would predict.

Two-sided Cumulative Sum: The two-sided CUSUM [15] accumulates departures from μ_r in both directions:

$$\begin{aligned} S_t^+ &= \max\big(0,\ S_{t-1}^+ + (r_t - \mu_r) - k\big), \\ S_t^- &= \max\big(0,\ S_{t-1}^- - (r_t - \mu_r) - k\big), \\ s_t^{\text{cusum}} &= \max\big(S_t^+,\ S_t^-\big). \end{aligned} \tag{7}$$

with the slack parameter $k = 0.5\,\sigma_r$.

Threshold Calibration: To avoid leaking anomaly-window information into the threshold, we calibrate each detector's threshold once on the *clean* test residual stream, with no anomaly injected, by selecting the value at which the false-positive rate on clean data equals one percent. The resulting threshold is then applied unchanged across every anomaly scenario. Algorithm 3 formalizes the end-to-end detection pipeline, including the threshold-calibration step and the per-anomaly scoring loop.

3.7 Reported Metrics

For each (anomaly, detector) pair the research reports the ROC-AUC swept across all decision thresholds and the true-positive rate at the calibrated one-percent false-positive rate. The detection delay is measured as the time from

Table 3. Forecasting fit on the 7 819-sample test split. Both models are trained on identical 18 244-sample training trajectories with Adam warm-up followed by quasi-Newton refinement, the same Tsit5 solver, and tolerance 10^{-6}.

Model	Train RMSE (°C)	Test RMSE (°C)	Test MAE (°C)	Parameters
ODE ablation (UDE without $\mathcal{N}_\theta$)	1.038	1.061	0.903	2
UDE (physics + neural residual)	**0.884**	**0.934**	**0.791**	771

anomaly onset to the first detector alarm during the perturbation window. Detection delay is reported in minutes. A no-detection (ND) is recorded if no alarm is raised within the perturbation window. Driver-channel perturbations additionally report within-window root-mean-squared error of each model's prediction against the unperturbed observation. The perturbation displacement quantifies how far each model's prediction diverges from the actual rack trajectory.

4 Results

The research organizes the results around the three research questions: cyber-twin fit on the held-out test split (RQ1), sensor-channel anomaly detection under synthetic tampering (RQ2), and driver-channel anomaly detection under synthetic telemetry perturbations (RQ3). All numbers reported in this section trace to logged outputs of the training and detection pipelines described in Sect. 3.

4.1 Cyber-Twin Fit on the Test Split (RQ1)

Table 3 summarizes the trained UDE alongside its physics-only ODE ablation on the 7 819-sample test split. The UDE achieves a test root-mean-squared error (RMSE) of 0.934 °C and a test mean absolute error (MAE) of 0.791 °C, with a small train–test gap consistent with no overfitting. The ODE ablation, calibrated under an analogous Adam→quasi-Newton protocol described in Sect. 3.3, achieves a test RMSE of 1.061 °C and an R^2 of 0.615. The UDE improves test RMSE by 0.13 °C over the calibrated physics ablation; on smooth, low-volatility 10-minute thermal telemetry the two-parameter physics already captures most of the forecasting signal, and the neural residual contributes a measurable additional reduction.

The learned physics coefficients differ between the two models because the neural component in the UDE absorbs part of what the bare physics would otherwise have to fit. The ODE ablation converges to $\alpha_{\text{ODE}} = 0.464\,°\text{C/kW}$ and $\beta_{\text{ODE}} = 0.0343$, while the UDE converges to $\alpha_{\text{UDE}} = 0.505\,°\text{C/kW}$ and $\beta_{\text{UDE}} = 0.073$. Both heating coefficients are within an operationally plausible range for a single-rack thermal model. Figure 2 consolidates the cyber-twin fit and residual diagnostics: panels (a) and (b) show the UDE forecast tracking the diurnal envelope and high-amplitude transients on train and test; panel (c) confirms the train and test residual envelopes overlap (no overfitting); panel

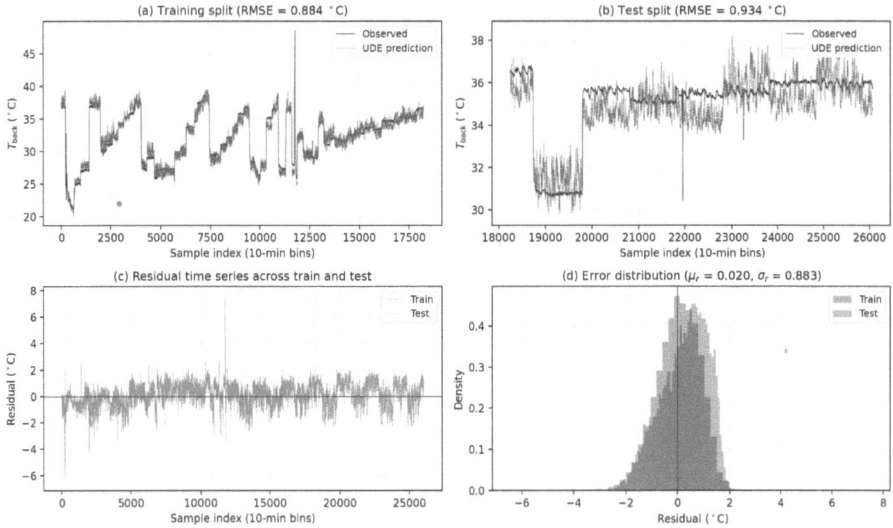

Fig. 2. Cyber-twin fit and residual diagnostics on the NTU TDC 2.0 testbed. (a) Training fit on the 18 244-sample chronological train split (RMSE 0.884 °C). (b) Test fit on the 7 819-sample held-out split (RMSE 0.934 °C). (c) Residual time series across train (red) and test (blue) on the absolute sample timeline; the train/test boundary is marked by the dotted line. (d) Density histogram of train and test residuals ($\mu_r = 0.020$, $\sigma_r = 0.883$), supporting the standardization assumption used by the residual-based detectors of Sect. 3.6 (Color figure online).

(d) shows the residual distribution is approximately zero-mean with finite variance ($\mu_r = 0.020$, $\sigma_r = 0.883$ on the training trajectory), consistent with the assumption underlying the residual-based detectors of Sect. 3.6.

To assess sensitivity to neural-network initialization, the UDE is retrained under three additional random seeds while holding the data, network architecture, and optimizer settings fixed. Table 4 reports per-seed parameters and forecasting performance, with the four-seed mean and standard deviation summarized in the bottom row.

The four seeds converge to consistent physics parameters (α coefficient of variation 4.2%, β coefficient of variation 26.6%) and forecasting performance (test RMSE coefficient of variation 7.2%), with all four test RMSE values below 1.06 °C, without any per-seed re-tuning of optimizer hyperparameters.

4.2 Sensor-Channel Anomaly Detection (RQ2)

Sensor-channel anomalies modify only the observed back-of-rack temperature. The UDE and the ODE ablation produce unchanged predictions under tampering, so the residual under tampering equals the clean residual plus the additive perturbation signal. Detection therefore depends on how cleanly the residual distribution separates the perturbation window from the surrounding clean opera-

Table 4. Multi-seed reproducibility of the UDE cyber twin under four random seeds. Identical training procedure (Adam warm-up of 20 iterations at lr= 10^{-3}, followed by 10 BFGS iterations with initial step norm 0.01). The seed-42 row corresponds to the canonical results in Table 3.

Seed	α	β	Train RMSE (°C)	Test RMSE (°C)
7	0.517	0.146	1.005	1.027
11	0.525	0.116	0.986	0.906
42	0.505	0.073	0.884	0.934
123	0.557	0.127	1.001	1.053
mean ± std	0.526 ± 0.022	0.116 ± 0.031	0.969 ± 0.057	0.980 ± 0.071

tion. The research evaluates the rolling-window z-score detector at the threshold calibrated on clean test residuals at a 1 % false-positive rate, and reports ROC-AUC, true-positive rate at that operating point, and the time-to-first-alarm (detection delay) within the perturbation window.

Table 5 reports the detection metrics at the canonical perturbation-start position of 30 % into the test split, alongside the mean and standard deviation across five perturbation-start positions ($\{0.10, 0.20, 0.30, 0.50, 0.70\}$) to defend against the concern that a single position might be lucky. At the canonical position the UDE residual matches or modestly exceeds the ODE residual on all six scenarios. Position-aggregated, the UDE-minus-ODE rolling-window AUC is positive on the negative-offset scenario (0.928 ± 0.034 vs 0.873 ± 0.037) and on the sensor-freeze scenario (0.630 ± 0.211 vs 0.600 ± 0.311); on the four remaining scenarios the ODE ablation matches or slightly exceeds the UDE residual. The sensor-freeze case is the hardest for both detectors because freezing the observation suppresses the variability that the rolling-window standardization relies on.

At the canonical perturbation-start position (0.30 into the test split) the UDE matches or slightly outperforms the calibrated ODE ablation on all six sensor-channel scenarios, with the largest absolute gain on the negative-offset scenario (+0.027 AUC). The position-aggregate UDE-minus-ODE rolling-window AUC is positive on two of six scenarios (negative-offset at +0.058 and sensor-freeze at +0.030) and negative on the four others, with magnitude between −0.016 and −0.088. The conclusion is therefore nuanced: the UDE neural residual produces measurable AUC improvements on a subset of sensor-channel scenarios rather than a uniform improvement. Figure 4 visualizes the canonical-position ROC curves; Fig. 3 shows the per-scenario residual time series, with the additive perturbation shift visible during the shaded perturbation window.

Table 5. Sensor-channel anomaly detection at threshold calibrated to a 1 % false-positive rate on clean test residuals. "0.30" columns report the canonical-position metrics; "mean ± std across positions" aggregates over $\{0.10, 0.20, 0.30, 0.50, 0.70\}$. Higher AUC is better; lower delay is better. ND indicates no detection within the perturbation window.

Scenario	AUC at pos. 0.30		AUC mean±std (5 pos.)		Delay at 0.30 (min)	
	UDE	ODE	UDE	ODE	UDE	ODE
step_spoof_+3C	**0.997**	0.995	0.945 ± 0.063	**0.983 ± 0.020**	**30**	40
step_spoof_+5C	**0.999**	0.998	0.972 ± 0.032	**0.991 ± 0.010**	**10**	20
slow_drift_+5C_12h	**0.947**	0.945	0.879 ± 0.112	**0.895 ± 0.114**	**220**	230
fast_ramp_+5C_2h	**0.925**	0.903	0.703 ± 0.290	**0.791 ± 0.220**	**70**	80
sensor_freeze_6h	**0.682**	0.663	**0.630 ± 0.211**	0.600 ± 0.311	ND	ND
negative_offset_-3C	**0.900**	0.873	**0.931 ± 0.048**	0.873 ± 0.037	**140**	170

Table 6. Driver-channel anomaly detection at the 30 % perturbation-start position with rolling-window z-score detector. *In-window RMSE* measures how far each model's prediction diverges from the unchanged observation during the 12 h perturbation window; larger divergence yields a stronger residual signal. ROC-AUC threshold is calibrated to a 1 % false-positive rate on clean test residuals.

Scenario	In-window RMSE		Rolling-z AUC		Delay (min)		ΔAUC
	UDE	ODE	UDE	ODE	UDE	ODE	
cooling_degradation	2.977	1.950	**0.860**	0.750	**210**	430	+0.110
supply_air_manipulation	3.101	2.084	**0.888**	0.813	**160**	350	+0.075
power_surge	0.540	0.549	0.264	0.368	ND	ND	−0.104

4.3 Driver-Channel Anomaly Detection (RQ3)

Driver-channel perturbations modify the input telemetry that the cyber twin integrates over while leaving the observation $T_{\text{back}}^{\text{obs}}$ unchanged. Both the UDE and the ODE are re-integrated forward in time under the altered driver interpolators, and the residual against the unchanged observation is consumed by the same three detectors. Table 6 reports the detection metrics at the canonical 30 % perturbation-start position; the within-window root-mean-squared error of each model's prediction against the clean observation is reported alongside, quantifying how far the perturbation displaces each model's prediction from the rack trajectory.

The two scenarios where the perturbation produces a sizeable prediction divergence (cooling degradation and supply-air manipulation) extend the sensor-channel observation: the UDE residual achieves rolling-z AUC 0.860 versus the ODE's 0.750 on cooling degradation, and 0.888 versus 0.813 on supply-air manipulation. The detection delay at the same operating point drops from 430 min to 210 min on cooling degradation (the UDE flags the anomaly 3.7 h sooner) and from 350 min to 160 min on supply-air manipulation (the UDE flags the anomaly

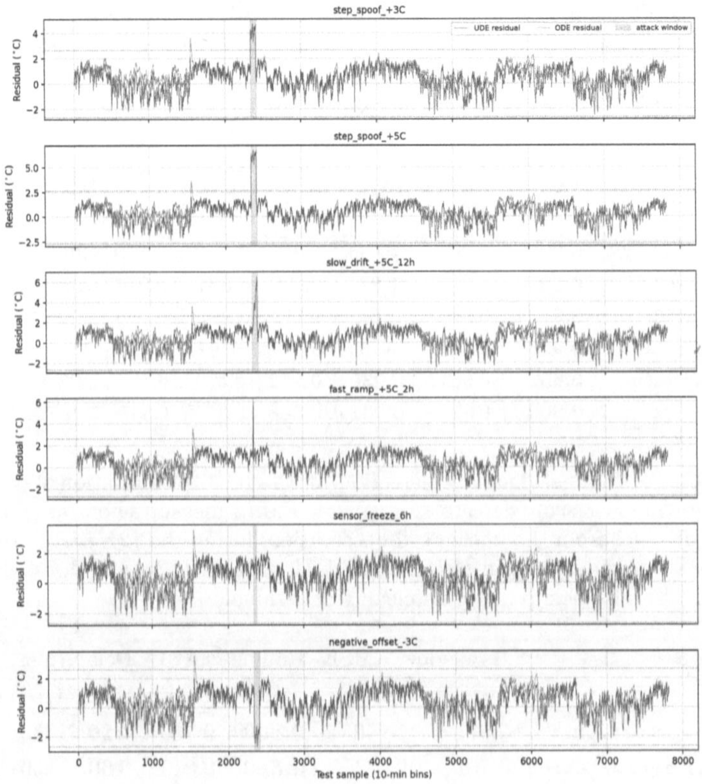

Fig. 3. Residual time series under each of the six sensor-channel tampering scenarios at perturbation-start position 0.30. Shaded band marks the perturbation window. Dashed lines mark the $\pm 3\sigma_r$ envelope of the train residual distribution.

3.2 h sooner). The within-window prediction RMSE is 48.8 % to 52.7 % larger for the UDE than for the ODE on these two scenarios; this larger divergence under altered drivers is what produces the cleaner detection signal. The third driver-channel scenario, the power surge, is too small in absolute thermal effect to produce a residual that exceeds the clean-residual envelope. Both models score below a useful AUC. Figure 5 consolidates the driver-channel residual time series (top row) and the corresponding rolling-window ROC curves (bottom row) for all three scenarios.

4.4 Synthesis Across Anomaly Classes

Reading the two anomaly groups together: the neural residual matches or modestly outperforms the calibrated physics ablation across the six sensor-channel tampering scenarios. Measurable gains were identified on negative-offset and sensor-freeze, and the neural residual produces a 0.075 to 0.110 absolute AUC improvement on the two large-magnitude driver-channel perturbations. The ben-

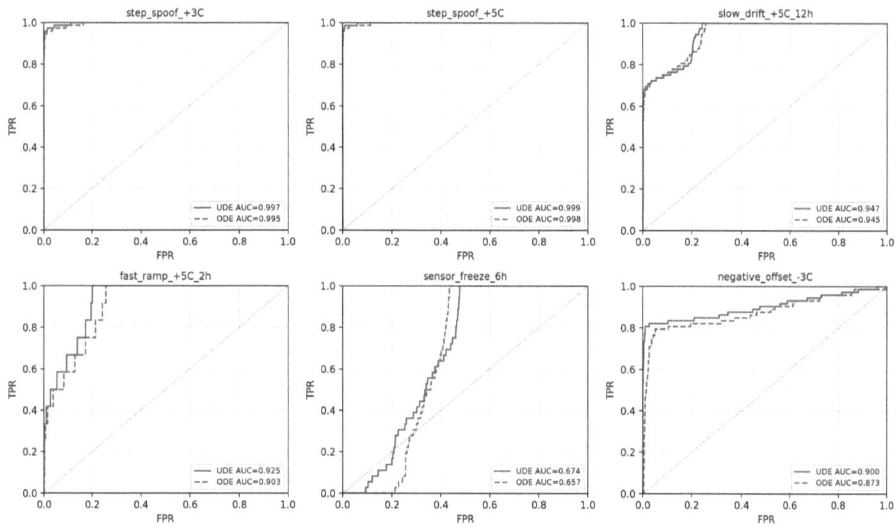

Fig. 4. Receiver Operating Characteristic curves for the rolling-window z-score detector under each sensor-channel tampering scenario at the canonical perturbation-start position. UDE residual (red) versus ODE-ablation residual (blue). The UDE curve matches or modestly dominates the ODE curve across all six scenarios. (Color figure online)

efit is not uniform. Out-of-distribution driver-channel perturbations and sensor-channel scenarios where the residual envelope shifts were specifically informative.

5 Discussion

The benefit of the neural residual inside the UDE depends on anomaly class. It is specifically large for out-of-distribution driver-channel telemetry perturbations, with smaller incremental gains on a subset of sensor-channel scenarios. Three mechanistic observations support this finding.

The research does not claim the residual alarms are, by themselves, evidence of intrusion or adversarial intent. The contribution is narrower: an anomaly-sensitive residual stream for rack-level data-center telemetry. It covers sensor-channel and driver-channel perturbations consistent with tampering-style patterns in the cyber-physical security literature. Determining whether a flagged anomaly reflects a benign sensor fault, a controller misconfiguration, or an adversarial manipulation of the telemetry channel requires a downstream intrusion-detection and attribution stage. The downstream stage is outlined as future work (Sect. 7).

First, sensor-channel tampering does not engage the neural residual through the dynamics: both models predict from the same clean driver inputs, so the residual under tampering is the clean residual plus the additive bias, and detection is governed by the clean-residual variance relative to that bias. The UDE's

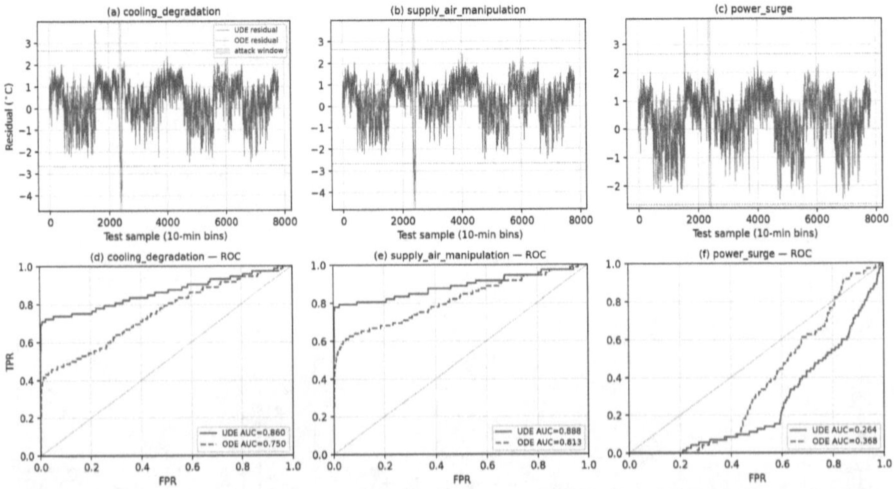

Fig. 5. Driver-channel perturbation diagnostics. Top row (a)–(c): residual time series under cooling degradation, supply-air manipulation, and power surge respectively; UDE residual (red), ODE-ablation residual (blue), shaded band marks the 12 h perturbation window, gray dashed lines mark the $\pm 3\sigma_r$ envelope of the train residual distribution. Bottom row (d)–(f): rolling-window z-score Receiver Operating Characteristic curves for the same three scenarios. The UDE curve dominates the ODE curve for cooling degradation and supply-air manipulation; both curves are below the diagonal for the power surge.

tighter clean residuals ($\sigma_r = 0.883\,°\mathrm{C}$ vs $1.034\,°\mathrm{C}$) yield modest AUC gains on the most variability-sensitive scenarios (negative-offset and sensor-freeze) while leaving the step-tampering and ramp scenarios within the same regime.

Second, driver-channel perturbations push the model's input vector $\mathbf{x}(t)$ off the training manifold. The neural network does not extrapolate gracefully when $V(t)$ is halved or $T_{\text{supply}}(t)$ shifted by $4\,°\mathrm{C}$ beyond its training envelope; the UDE prediction diverges further from the unchanged observation than the linear ODE prediction does, by between 48.8% and 52.7% in within-window RMSE. The additive vector-field placement ensures the linear physics absorbs what it can, leaving the neural term as a training-distribution-bound residual.

Third, the position-sweep aggregate confirms that the qualitative ordering of UDE versus ODE residuals holds across perturbation-window placement, with detection performance varying more for harder anomaly classes. This suggests an operational deployment would benefit from regime-aware threshold adaptation rather than a single global threshold, as production security operations centers already do for legacy intrusion-detection systems [35].

The detection-delay reduction on driver-channel perturbations has direct operational consequence: an earlier alarm at the same false-positive rate gives operators a meaningful margin to prevent thermal-envelope excursions. This gain follows from the larger residual divergence under perturbation, not from detector tuning, so it is model-driven rather than threshold-dependent.

Beyond detection, the cyber twin lends itself to both shift-left (prevention) and shift-right (recovery) workflows that are sketched here and left to future work. Shift-left: proposed cooling-setpoint or workload-distribution changes can be integrated through the trained twin during change management to check whether the projected trajectory remains inside the learned residual envelope, before any deployment. Shift-right: the live residual stream can feed an anomaly-monitoring pipeline that gates downstream recovery actions such as workload migration, supply-air setpoint relaxation, or controller rollback. Closed-loop automation, alert-fatigue management, and regime-aware thresholding are beyond the present scope.

The research deliberately did not foreground a forecasting horse race against deep learning baselines (multi-layer perceptron, long short-term memory, gated recurrent unit). Lower forecast root-mean-squared error does not imply better anomaly detection: a deep regressor with no physics constraint adapts to whatever pattern is present, including stealthy tampering, leaving its residual small under perturbation. The cyber-twin posture instead requires a physics-constrained model whose residual term refuses to extrapolate off-distribution.

6 Limitations

The findings of this paper are bounded in five ways that we state explicitly so that future work and operational deployments can plan accordingly.

First, the cyber twin is trained on a single rack (R0503-PB) at the NTU Tropical Data Center 2.0 testbed, and has not been transferred to other racks or facilities with different cooling architectures, climates, or load profiles. Cross-rack and cross-site validation are natural next steps; the two physics coefficients α and β should provide a stable transfer prior, but the neural residual is rack-specific and would need re-training on each new asset.

Second, all six sensor-channel tampering scenarios and three driver-channel perturbation scenarios are synthetic injections constructed from analytical attack models in the cyber-physical security literature; no labeled real-world anomaly traces are available for the testbed, and detection performance against operational anomalies in the wild may differ. Future work should evaluate against red-team exercises or live-deployment traces if such data become available. The injection set covers parametric tampering (offset, scale, freeze, drift, surge) but omits replay-style scenarios in which past sensor data is recycled to deceive a closed-loop controller in the manner of Stuxnet; evaluating those requires a controller-in-the-loop simulator and is left to future work.

Third, the 40% rack-power surge over 12 h sits below the noise floor: it produces an in-window prediction RMSE of only 0.540 °C for the UDE and 0.549 °C for the ODE ablation, both close to the clean-residual standard deviation of 0.883 °C, leaving the rolling-window detector unable to discriminate the perturbation from natural thermal variability at a 1% false-positive rate. Detection of small-amplitude driver perturbations would require either a higher-resolution model with tighter clean residuals or a regime-aware detector that lowers the threshold during quiescent periods.

Fourth, the pipeline reports anomalies, not attacks; attack attribution and intent inference are out of scope (see Sect. 5 and Sect. 7).

Fifth, the cyber twin runs as an open-loop residual computer over historical predictions; it has not been integrated into an anomaly-monitoring pipeline or a real-time control loop. Such integration is the natural next phase and would require attention to streaming-residual latency, alert-fatigue management, and threshold-adaptation policies under regime drift.

7 Conclusion

The research presented a Universal Differential Equation cyber twin for rack-level data-center thermal dynamics, trained on six months of telemetry from the NTU Tropical Data Center 2.0 testbed, and evaluated its residual stream as a detector for two classes of telemetry-channel anomaly: six sensor-channel tampering scenarios and three driver-channel perturbations.

The paper's central finding is that the value of the neural-residual component depends on anomaly class. Against sensor-channel tampering the UDE residual matches or modestly outperforms the calibrated physics ablation, with the largest position-aggregate gains on negative-offset (+0.058 AUC) and sensor freeze (+0.030). On the two large-magnitude driver-channel perturbations it achieves rolling-window ROC-AUC of 0.860 and 0.888 against 0.750 and 0.813 for the ablation, flagging the anomalies up to 3.7 h sooner at the same false-positive rate, and the fit is reproducible across four random seeds (test RMSE $0.98 \pm 0.07\,^{\circ}\mathrm{C}$).

Future work spans four directions: cross-rack and cross-facility transfer with the physics coefficients as transfer prior; integration into a security-operations-center streaming pipeline with regime-aware thresholding; replication with red-team or live-anomaly traces; and extension from anomaly detection toward intrusion detection through downstream attack-pattern classification, telemetry provenance reasoning, and adversarial robustness evaluation, building on the residual stream this paper produces.

Ethical Considerations

This work uses the publicly released NTU Tropical Data Center 2.0 telemetry dataset under its associated open-data terms; no proprietary or confidential operational data is involved. All evaluated anomaly scenarios are synthetic injections constructed from analytical models in the cyber-physical security literature and are evaluated offline against logged telemetry; no real data-center systems were perturbed, no live actuators were manipulated, and no operational availability was affected. The work does not involve human subjects, user data, or personally identifiable information, and therefore did not require institutional review board approval. The cyber twin and detector pipeline are positioned as defensive tooling: the analyses identify residual signatures of well-known cyber-physical anomaly classes cataloged in the public threat-modeling literature and do not introduce new tampering capabilities or zero-day vulnerabilities. As a result, no responsible-disclosure process applies. We will release our analysis pipeline publicly at https://github.com/kp7425/ude-power-dcs after presentation at DCS-CI 2026 to support reproduction and to encourage adoption of residual-based monitoring in operational data-center settings.

Generative AI Usage

Multiple large language model assistants (Anthropic Claude, OpenAI ChatGPT, and Overleaf's built-in AI assistant) supported code development and manuscript preparation. The research idea, experimental design, results, and conclusions are the authors'; the assistants supported code generation and review, prose drafting, LaTeX formatting, and consistency auditing. The assistants did not propose citations, design experiments, generate numerical results, or interpret findings. Citations were chosen and verified by the authors directly against the original reference paper PDFs. The results, equations, algorithms, and tables were verified against source result files and training scripts. The authors reviewed, edited, and assume full responsibility for the content, claims, and conclusions of this manuscript. The models are not listed as authors, in line with the ACM Policy on Authorship.

Disclosure of Interests. The authors have no competing interests to declare that are relevant to the content of this article.

References

1. Ahmed, C.M., Palleti, V.R., Mathur, A.P.: Wadi: a water distribution testbed for research in the design of secure cyber physical systems. In: Proceedings of the 3rd International Workshop on Cyber-Physical Systems for Smart Water Networks, CySWATER 2017, pp. 25–28 (2017). https://doi.org/10.1145/3055366.3055375
2. Alexander, O., Belisle, M., Steele, J.: MITRE ATT&CK for industrial control systems: design and philosophy. The MITRE Corporation **29**, 21–85 (2020)
3. Balta, E.C., Pease, M., Moyne, J., Barton, K., Tilbury, D.M.: Digital twin-based cyber-attack detection framework for cyber-physical manufacturing systems. IEEE Trans. Autom. Sci. Eng. **21**(2), 1695–1712 (2024). https://doi.org/10.1109/TASE.2023.3243147
4. Barroso, L., Hölzle, U., Ranganathan, P.: The Data Center as a Computer: Designing Warehouse-Scale Machines. Springer Cham (2026).https://doi.org/10.1007/978-3-031-99489-0
5. Cao, Z., Wang, R., Zhou, X., Wen, Y.: Reducio: model reduction for data center predictive digital twins via physics-guided machine learning. In: Proceedings of the 9th ACM International Conference on Systems for Energy-Efficient Buildings, Cities, and Transportation, pp. 1–10. Association for Computing Machinery (2022). https://doi.org/10.1145/3563357.3564050

6. Chandola, V., Banerjee, A., Kumar, V.: Anomaly detection: a survey. ACM Comput. Surv. **41**(3) (2009). https://doi.org/10.1145/1541880.1541882
7. Chattopadhyay, A., Mitra, U.: Security against false data-injection attack in cyber-physical systems. IEEE Trans. Control Netw. Syst. **7**(2), 1015–1027 (2020). https://doi.org/10.1109/TCNS.2019.2927594
8. Chen, D., Chui, C.K., Lee, P.S.: Adaptive physically consistent neural networks for data center thermal dynamics modeling. Appl. Energy **377**, 124637 (2025). https://doi.org/10.1016/j.apenergy.2024.124637, https://www.sciencedirect.com/science/article/pii/S0306261924020208
9. Chen, D., Chui, C.K., Lee, P.S.: Physics informed machine learning based predictive control for intelligent operation of edge datacenters. Appl. Energy **402**, 126975 (2026). https://doi.org/10.1016/j.apenergy.2025.126975, https://www.sciencedirect.com/science/article/pii/S0306261925017052
10. Dayarathna, M., Wen, Y., Fan, R.: Data center energy consumption modeling: a survey. IEEE Commun. Surv. Tutorials **18**(1), 732–794 (2016). https://doi.org/10.1109/COMST.2015.2481183
11. Elnour, M., Noorizadeh, M., Shakerpour, M., Meskin, N., Khan, K., Jain, R.: A machine learning based framework for real-time detection and mitigation of sensor false data injection cyber-physical attacks in industrial control systems. IEEE Access **11**, 86977–86998 (2023). https://doi.org/10.1109/ACCESS.2023.3303015
12. Fawzi, H., Tabuada, P., Diggavi, S.: Secure estimation and control for cyber-physical systems under adversarial attacks. IEEE Trans. Autom. Control **59**(6), 1454–1467 (2014). https://doi.org/10.1109/TAC.2014.2303233
13. Feng, C., Palleti, V.R., Mathur, A., Chana, D.: A systematic framework to generate invariants for anomaly detection in industrial control systems. In: NDSS, pp. 1–15 (2019).nhttps://doi.org/10.14722/ndss.2019.23265
14. Gao, C., Park, H., Easwaran, A.: An anomaly detection framework for digital twin driven cyber-physical systems. In: Proceedings of the ACM/IEEE 12th International Conference on Cyber-Physical Systems, ICCPS 2021, pp. 44–54. (2021). https://doi.org/10.1145/3450267.3450533
15. Giraldo, J., et al.: A survey of physics-based attack detection in cyber-physical systems. ACM Comput. Surv. **51**(4) (2018). https://doi.org/10.1145/3203245
16. Goh, J., Adepu, S., Tan, M., Lee, Z.S.: Anomaly detection in cyber physical systems using recurrent neural networks. In: 2017 IEEE 18th International Symposium on High Assurance Systems Engineering (HASE), pp. 140–145 (2017). https://doi.org/10.1109/HASE.2017.36
17. Hashemi, Z., Gholampour, M., Su, H.H., Chiang, C.Y., Chang, H.H., Wang, C.C.: Hybrid physics-informed neural networks for airflow and thermal modeling in data center environments. Int. Commun. Heat Mass Transfer **168**, 109511 (2025). https://doi.org/10.1016/j.icheatmasstransfer.2025.109511, https://www.sciencedirect.com/science/article/pii/S0735193325009376
18. Hosseinzadeh, M., Sinopoli, B.: Active attack detection and control in constrained cyber-physical systems under prevented actuation attack. In: 2021 American Control Conference (ACC), pp. 3242–3247 (2021). https://doi.org/10.23919/ACC50511.2021.9483322
19. Karniadakis, G.E., Kevrekidis, I.G., Lu, L., Perdikaris, P., Wang, S., Yang, L.: Physics-informed machine learning. Nat. Rev. Phys. **3**(6), 422–440 (2021). https://doi.org/10.1038/S42254-021-00314-5
20. Kravchik, M., Shabtai, A.: Detecting cyber attacks in industrial control systems using convolutional neural networks. In: Proceedings of the 2018 Workshop on

Cyber-Physical Systems Security and PrivaCy, CPS-SPC 2018, pp. 72–83 (2018). https://doi.org/10.1145/3264888.3264896
21. Le, V.D.: An Air-Cooled Tropical Data Center (TDC2.0) Dataset (2023). https://doi.org/10.21979/N9/BLBQ2T
22. Loi, K., Song, C., Chao, H., Chen, L., Guo, A.: High temporal-resolution hvac control in gpu-centric data centers via reinforcement learning and a knowledge-embedded physics-informed neural network. J. Build. Eng. **123**, 115499 (2026). https://doi.org/10.1016/j.jobe.2026.115499, https://www.sciencedirect.com/science/article/pii/S2352710226003207
23. Mathur, A.P., Tippenhauer, N.O.: Swat: a water treatment testbed for research and training on ics security. In: 2016 International Workshop on Cyber-physical Systems for Smart Water Networks (CySWater), pp. 31–36 (2016). https://doi.org/10.1109/CySWater.2016.7469060
24. Mitchell, R., Chen, I.R.: A survey of intrusion detection techniques for cyber-physical systems. ACM Comput. Surv. **46**(4) (2014). https://doi.org/10.1145/2542049
25. Mo, Y., Sinopoli, B.: Secure control against replay attacks. In: 47th Annual Allerton Conference on Communication, Control, and Computing (Allerton), pp. 911–918 (2009). https://doi.org/10.1109/ALLERTON.2009.5394956
26. Moore, J., Chase, J., Ranganathan, P., Sharma, R.: Making scheduling "cool": temperature-Aware workload placement in data centers. In: 2005 USENIX Annual Technical Conference (USENIX ATC 2005). USENIX Association, Anaheim, CA (Apr 2005). https://www.usenix.org/conference/2005-usenix-annual-technical-conference/making-scheduling-cool-temperature-aware-workload
27. Parolini, L., Sinopoli, B., Krogh, B.H., Wang, Z.: A cyber-physical systems approach to data center modeling and control for energy efficiency. Proc. IEEE **100**(1), 254–268 (2012). https://doi.org/10.1109/JPROC.2011.2161244
28. Pasqualetti, F., Dörfler, F., Bullo, F.: Attack detection and identification in cyber-physical systems. IEEE Trans. Autom. Control **58**(11), 2715–2729 (2013). https://doi.org/10.1109/TAC.2013.2266831
29. Rackauckas, C., et al.: Universal differential equations for scientific machine learning. arXiv preprint (2020). https://doi.org/10.48550/arXiv.2001.04385
30. Rackauckas, C., Nie, Q.: Differentialequations.jl-a performant and feature-rich ecosystem for solving differential equations in julia. J. Open Res. Softw. **5**(1), 15 (2017). https://doi.org/10.5334/jors.151
31. Raissi, M., Perdikaris, P., Karniadakis, G.: Physics-informed neural networks: a deep learning framework for solving forward and inverse problems involving nonlinear partial differential equations. J. Comput. Phys. **378**, 686–707 (2019). https://doi.org/10.1016/j.jcp.2018.10.045
32. Salvi, A., Spagnoletti, P., Noori, N.S.: Cyber-resilience of critical cyber infrastructures: Integrating digital twins in the electric power ecosystem. Comput. Secur. **112**(C) (2022). https://doi.org/10.1016/j.cose.2021.102507
33. Sharma, R., Bash, C., Patel, C., Friedrich, R., Chase, J.: Balance of power: dynamic thermal management for internet data centers. IEEE Internet Comput. **9**(1), 42–49 (2005). https://doi.org/10.1109/MIC.2005.10
34. Sheikh, A., Kamuni, V., Patil, A., Wagh, S., Singh, N.: Cyber attack and fault identification of hvac system in building management systems. In: 2019 9th International Conference on Power and Energy Systems (ICPES), pp. 1–6 (2019). https://doi.org/10.1109/ICPES47639.2019.9105438

35. Stouffer, K., Lightman, S., Pillitteri, V., Abrams, M., Hahn, A.: Nist sp 800-82 rev. 3: guide to operational technology security. Gaithersburg: NIST (2023). https://doi.org/10.6028/NIST.SP.800-82r3
36. Tan, S., Guerrero, J.M., Xie, P., Han, R., Vasquez, J.C.: Brief survey on attack detection methods for cyber-physical systems. IEEE Syst. J. **14**(4), 5329–5339 (2020). https://doi.org/10.1109/JSYST.2020.2991258
37. Wang, R., Xia, D., Cao, Z., Wen, Y., Tan, R., Zhou, X.: Toward data center digital twins via knowledge-based model calibration and reduction. ACM Trans. Model. Comput. Simul. **33**(4) (2023).https://doi.org/10.1145/3604283
38. Wang, R., Zhang, Q., Jia, J., Wen, Y., Kennedy, S.: Toward multiphysics-informed machine learning for sustainable data center operations: intelligence evolution with deployable solutions for computing infrastructure. IEEE Energy Sustainability Mag. **2**(1), 69–80 (2026). https://doi.org/10.1109/ESM.2025.3626704
39. Wang, R., et al.: Kalibre: knowledge-based neural surrogate model calibration for data center digital twins. In: Proceedings of the 7th ACM International Conference on Systems for Energy-Efficient Buildings, Cities, and Transportation, pp. 200–209 (2020). https://doi.org/10.1145/3408308.3427982
40. Xu, Q., Ali, S., Yue, T.: Digital twin-based anomaly detection in cyber-physical systems. In: 2021 14th IEEE Conference on Software Testing, Verification and Validation (ICST), pp. 205–216 (2021). https://doi.org/10.1109/ICST49551.2021.00031
41. Xu, Q., Ali, S., Yue, T.: Digital twin-based anomaly detection with curriculum learning in cyber-physical systems. ACM Trans. Softw. Eng. Methodol. **32**(5) (2023). https://doi.org/10.1145/3582571
42. Zhou, Y., Vamvoudakis, K.G., Haddad, W.M., Jiang, Z.P.: A secure control learning framework for cyber-physical systems under sensor and actuator attacks. IEEE Trans. Cybern. **51**(9), 4648–4660 (2021). https://doi.org/10.1109/TCYB.2020.3006871

Physical Processes, Digital Risks: Component-Level STRIDE Risk Assessment for an Industrial CPS

Suleiman Muhammad Sabo(✉), Chuadhry Mujeeb Ahmed, and Rajiv Ranjan

School of Computing, Newcastle University, Newcastle upon Tyne, UK
{s.m.sabo2,Mujeeb.Ahmed,Raj.Ranjan}@newcastle.ac.uk

Abstract. Component-level risk assessment enables the identification of specific vulnerabilities in individual components of Industrial Cyber-Physical Systems (CPS). This paper presents a component-level STRIDE-based risk assessment of an industrial-scale water treatment process. Using Data Flow Diagrams (DFD), we analyse Programmable Logic Controllers (PLCs), sensors, actuators, Supervisory Control and Data Acquisition (SCADA), Human-Machine Interface (HMI), communication flows, and historian components to identify 20 threat consequences. Six underlying vulnerability classes (labelled as VC-1–VC-6) are identified and mapped to MITRE Common Weakness Enumeration (CWE) identifiers, enabling cross-system comparison with broader CPS vulnerability databases. Risks are prioritised using the National Institute of Standards and Technology (NIST) SP 800-30r1 qualitative risk matrix, producing differentiated ratings from Moderate to High across nine consolidated threat events. We structure mitigation strategies by vulnerability class and align them with NIST SP 800-53 Rev. 5 controls. The findings show that component-level STRIDE analysis can support actionable, standards-aligned security recommendations for a realistic water-treatment CPS.

Keywords: Cyber-Physical Systems · ICS Security · STRIDE · Threat Modelling · NIST Controls

1 Introduction

Industrial control systems (ICS) are critical in sectors such as water treatment, power generation, and heavy industries. These systems, including Distributed Control Systems (DCS), SCADA systems, and PLCs, have traditionally relied on isolated, mechanical, and electrotechnical equipment [6]. However, the rapid digitisation brought about by Industry 4.0 has transformed ICS into interconnected Cyber-Physical Systems, integrating operational technologies (OT) with information technologies (IT) [5,19]. This growing interconnection enhances system functionality but introduces new cybersecurity challenges, particularly as these systems are increasingly targeted by cyberattacks [27].

S. Zonouz et al. (Eds.): DCS-CI 2026, CCIS 3109, pp. 93–114, 2027.
https://doi.org/10.1007/978-981-92-5039-4_5

CPS components are highly interconnected and dependent on one another [29]. For example, power is essential for operating water treatment and distribution systems, which require a connection to a reliable power distribution source. This interdependence implies that a malfunction or cyberattack in one part of the system can cascade, potentially disrupting other parts or even the entire network [28]. CPS attacks are among the most pressing emerging threats in the control community [34,37].

We conduct this assessment on the Secure Water Treatment (SWaT) testbed, a scaled-down replica of operational water treatment plants developed by the iTrust Centre at Singapore University of Technology and Design [13]. The existing research on the SWaT testbed has primarily focused on security aspects such as attack detection and anomaly detection, leveraging the SWaT and Water Distribution (WaDi) datasets to evaluate various defence mechanisms [5,31]. These studies have advanced the understanding of cybersecurity within water systems by improving detection capabilities and introducing novel defensive strategies. However, they often lack a comprehensive risk assessment framework that evaluates both individual component vulnerabilities and their specific security requirements [3,9,18,35].

Our work addresses this gap by providing a detailed component-level risk assessment of an industrial CPS. Component-level assessments focus on individual components, such as sensors, PLCs, controllers, and actuators, identifying specific vulnerabilities such as weak authentication mechanisms, insufficient data integrity checks, and insecure firmware configurations. Unlike system-level assessments that evaluate broader systemic risks and cascading failures, component-level analysis enables targeted, high-impact security improvements. This paper makes the following contributions:

1. We conduct a component-level risk assessment of a water treatment CPS through systematic STRIDE threat modelling of individual components.
2. We identify 20 component-level threat consequences through a hybrid process that combines STRIDE analysis of Data Flow Diagram (DFD) elements with evidence from the SWaT attack dataset called A6_2019.
3. We distinguish between threat consequences that are log-confirmed and STRIDE/DFD-derived, thereby separating observed attack evidence from analytical threat modelling.
4. We produce a risk register by mapping vulnerabilities to MITRE CWE codes and applying NIST SP 800-30r1 risk matrices, enabling practitioners to prioritise security improvements based on likelihood and impact.
5. We propose actionable mitigation strategies aligned with specific NIST SP 800-53 Rev. 5 control identifiers.

2 Related Work

2.1 Cyber-Attacks on Critical Infrastructure

Cyberattacks on critical infrastructure have evolved dramatically over the past four decades. While early attacks targeted isolated mechanical systems [38], the

Industry 4.0 digitisation of critical infrastructure, particularly water treatment facilities, has significantly expanded the attack surface. Cyber-attacks on CPS are among the most pressing emerging threats facing the control community [34].

High-profile attacks on power systems, such as the 2015 Ukraine blackout and 2023 European grid outages, have received considerable attention [7]. However, water infrastructure has historically received less attention despite its critical importance. Recent events underscore the emerging threat to water systems. In 2024, a ransomware attack on a US water treatment facility caused operational disruptions, putting thousands of citizens' access to clean water at risk [20]. Despite existing security measures, attackers exploited network weaknesses, resulting in severe financial losses and operational paralysis [20]. This incident demonstrates the shortcomings of current cybersecurity protocols and the urgent need for more proactive, robust defences.

This highlights a critical gap: while extensive research has addressed attack detection in water systems [9,18], comprehensive component-level risk assessment frameworks that identify specific vulnerabilities within individual system components remain limited. This work addresses that gap by providing a detailed component-level risk assessment of water CPS.

2.2 Threat Modelling Frameworks

Threat modelling determines which system vulnerabilities exist and how they might affect operational procedures [16]. It is the initial stage in risk assessment, with a methodical approach to recognising and categorising possible risks to both existing systems and those under development [17]. The approach generally entails assessing attack vectors, thoroughly characterising the underlying architecture, and proposing security measures to defend against identified risks. Threat modelling thus identifies what will be targeted, by whom, and through what means [21].

Multiple threat modelling frameworks exist, including Damage, Reproducibility, Exploitability, Affected users, and Discoverability (DREAD) [32], Attack Trees (hierarchical attack path modelling) [32], and the Cyber Kill Chain (seven-stage attack model) [12]. However, Microsoft's STRIDE method provides structured threat categorisation across six threat categories: Spoofing (S), Tampering (T), Repudiation (R), Information Disclosure (I), Denial of Service (D), and Elevation of Privilege (E) [33].

Other frameworks approach this differently. DREAD collapses everything into a single risk score per threat, which makes it difficult to trace findings back to specific system components. Attack Trees work top-down from an attacker's goal, which is useful for modelling attack paths but risks overlooking vulnerabilities in components that do not sit on a known attack path. STRIDE-per-element works bottom-up: each DFD element is directly interrogated against all applicable threat categories. For component-level CPS assessment, where the objective is to surface specific weaknesses in individual PLCs, sensors, and data flows, this systematic per-element structure provides more comprehensive and traceable coverage.

2.3 Risk Assessment Methodologies

Risk assessment is a fundamental component of risk management per ISO 31000 [30], encompassing risk identification, analysis, and evaluation. Approaches range from probabilistic methods using fault trees, Bayesian networks, and Markov chains [8] to qualitative frameworks suited to environments where full probabilistic data is unavailable. Methodologies are broadly categorised as top-down, leveraging historical data from analogous systems, or bottom-up, building estimates from detailed component-level data [26].

This study employs the NIST SP 800-30r1 qualitative risk matrix [24], which is particularly suited to component-level CPS assessment because qualitative ratings (Low, Moderate, High) are understood by diverse stakeholders, it is scalable from individual components to system-wide assessment, and it requires less data than full probabilistic methods while still providing actionable prioritisation.

3 System Description and Methodology

This section outlines the systematic approach taken to identify, assess, and mitigate risks, providing the frameworks and processes used in the component-level risk assessment.

3.1 SWaT Testbed Overview

The Singapore University of Technology and Design's iTrust Centre for Research in Cyber Security developed the Secure Water Treatment testbed (SWaT) in 2015 to conduct cybersecurity research and provide educational resources. It is a scaled-down, fully operational, six-stage plant that simulates the behaviour of actual water treatment facilities. It enables researchers to create and test sophisticated algorithms to minimise risks, ensure the security and dependability of CPS, and advance understanding of CPS security properties. Raw water is fed into SWaT and proceeds through six stages, each designated P1–P6 [22].

Raw water enters at P1 (raw water intake) and progresses through P2 (chemical dosing), P3 (ultrafiltration), P4 (UV dechlorination), and P5 (reverse osmosis), with P6 (membrane backwash) managing the cleaning cycle to maintain filtration efficiency. Each stage is controlled by dual PLCs that monitor sensors and adjust actuators, communicating locally over a fault-tolerant Ethernet ring using the Device Level Ring (DLR) protocol. Inter-stage communication connects all PLCs to the HMI, SCADA system, and historian via a Level 1 Ethernet star topology for centralised monitoring and control [22].

3.2 Data Collection and Analysis

This study uses the publicly available SWaT security dataset, obtained through a formal data access request submitted to the iTrust Centre for Research in Cyber Security at the Singapore University of Technology and Design [13]. Specifically, we analyse the `SWaT_Dataset_Attack` logs, which contain time-series sensor and actuator readings captured during controlled attack scenarios executed by the iTrust research team on the physical SWaT testbed [22]. The attack log used corresponds to the A6_2019 scenario, which comprises two sequential attack phases documented in the dataset:

- **Phase 1 (10:30–11:20 AM):** Malware introduced to the SCADA workstation via USB, exfiltrating historian data in cycles of 5 min active attack followed by 10 min inactivity.
- **Phase 2 (12:30–1:25 PM):** A command-and-control (C2) server used to disrupt sensor readings and actuator behaviour in cycles of 3 min active attack followed by 10 min inactivity. Post-attack data was recorded from 1:30 to 1:45 PM.

3.3 STRIDE Methodology

We employ the STRIDE threat modelling framework to conduct the component-level risk assessment and address the specific vulnerabilities of individual components within the SWaT testbed. We adopt and instantiate the seven-step process established by [16], adapting it specifically to the CPS context of the SWaT testbed. The seven steps are: (i) identify assets and security objectives; (ii) decompose the system into logical and structural components; (iii) visualise each system component's functionalities by creating a Data Flow Diagram (DFD); (iv) locate STRIDE threats for every system component in the DFD; (v) investigate vulnerabilities creating threats; (vi) develop appropriate mitigation solutions; (vii) regularly review and update the threat model (Fig. 1).

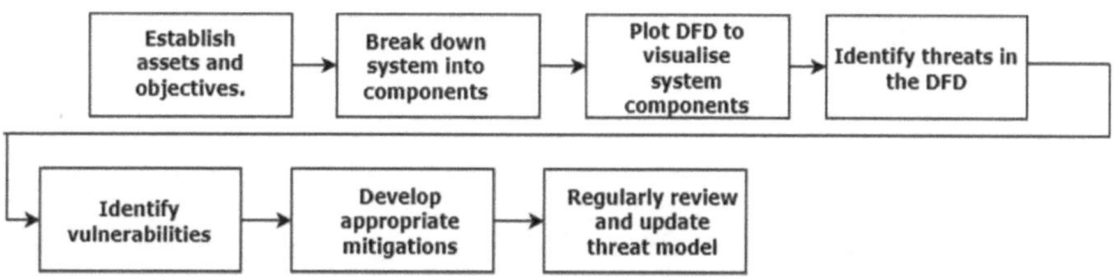

Fig. 1. STRIDE-based threat modelling methodology.

To apply the STRIDE threat modelling approach to the SWaT testbed, we begin by identifying our assets and security objectives. The next step is decomposing the system into logical and structural components. Internal processes and elements interacting within the system, or external factors interacting with the system, are both considered components, with the internal components being the PLCs, sensors, actuators, and the SCADA System. In contrast, the external entities are the network interfaces.

The next stage is to visualise each system component's functionalities, both internal and external to the system, by creating a DFD. In this stage, four elements are considered: (i) External Entity, an object that receives or emits information; (ii) Process, an action that processes data received as input, producing an output; (iii) Data Flow, the connections between DFD elements; (iv) Data Store, elements like databases that store data.

Each DFD element is only vulnerable to a few or all the STRIDE threats, as shown in Table 1. The next step is to locate STRIDE threats for every system component in the DFD. Depending on each component's functionality, some STRIDE threats may not apply. Investigating the vulnerabilities that create threats is necessary once threats have been identified. Based on the vulnerabilities found, appropriate mitigation solutions are developed. The final step is to regularly review and update the threat model to adapt to new threats and changes in the system.

As proposed by Microsoft, STRIDE can be applied per element or per interaction [14]. Since STRIDE-per-element evaluates the behaviours and functions of every system component, it is more straightforward to implement. Threats might arise in specific situations where system components interact. Still, STRIDE-per-element's defence mechanisms are sufficient to keep the system safe because cyberattacks normally entail malicious interactions between system elements [21].

Table 1. DFD Element's Likelihood to STRIDE Threats.

DFD Element	S	T	R	I	D	E
Entity	X		X			
Data Store		X		X	X	
Data Flow		X	X	X	X	
Process	X	X	X	X	X	X

Establish Assets and Objectives. To decompose the SWaT testbed into its components, we identify several key components, each playing a vital role in the water treatment process. The SWaT testbed comprises six main stages (P1–P6), each controlled by dual PLCs for redundancy. We identify the critical assets for component-level assessment:

Control Assets: 12 PLCs (dual per stage P1–P6).

Sensing Assets: Water level sensors, pressure sensors, flow rate sensors per stage.

Actuation Assets: Pumps, control valves, chemical dosing systems.

Data Assets: Real-time sensor data, control commands, historian database.

System Assets: SCADA server, HMI workstation, network infrastructure (Ethernet ring topology with DLR protocol, industrial switch, Level 1 Ethernet star topology).

The SWaT testbed operates through a distributed control approach in which each stage is managed by its local PLCs. The primary function of the PLCs is to receive data from the sensors and control actuators such as pumps and valves, ensuring seamless operation. The various stages of PLCs communicate via a Level 1 network using an Ethernet star topology controlled by an industrial switch. The HMI, SCADA, and historian provide centralised monitoring and control. The threat analysis focuses on cyber-physical components, specifically PLCs, sensors, actuators, network communication systems, SCADA, HMI, and historian, which constitute the exploitable attack surface of the SWaT testbed.

Data Flow Diagram (DFD) for System Components. DFDs are crucial for visualising data flow and control within the system, helping to identify vulnerabilities. Unlike flowcharts, DFDs focus on data transformation without including control elements [14]. The Level 0 DFD provides a high-level overview, showing the SWaT testbed as a single process with interactions with SCADA, HMI, actuators, sensors, and the historian. The Level 1 and 2 DFDs offer detailed breakdowns that identify the six stages of water treatment (P1 to P6), managed by dual PLCs, and map data flows among them. They also visualise interactions between local PLCs, sensors, actuators, SCADA, and the historian, highlighting critical data and control exchanges.

Level 0 DFD. The central process is labelled as SWaT System Operations, as shown in Fig. 2. This central process represents the core operational activities of the SWaT testbed. It is connected to four key entities and two data stores. The entities are:

- **SENSORS:** Continuously monitor various parameters of the water treatment process, such as water levels, pressure, and chemical composition. The sensor data flows into the SWaT testbed, providing the necessary real-time monitoring and control inputs.
- **ACTUATORS:** Receive control commands from the SWaT testbed to perform actions such as opening or closing valves and turning pumps on or off. The actuators execute decisions made by the system based on the control logic and sensor data.
- **HMI:** Allows an operator to control the SWaT testbed manually. It provides a user interface through which operators can send manual control commands and receive feedback on the system's status. This interaction enables an operator to intervene and adjust the system's operations when necessary.
- **SCADA SYSTEM:** Monitors and controls the entire SWaT testbed. It processes real-time sensor data and provides processed data for monitoring and control purposes. Ensuring efficient and safe water treatment is crucial to the SCADA system.

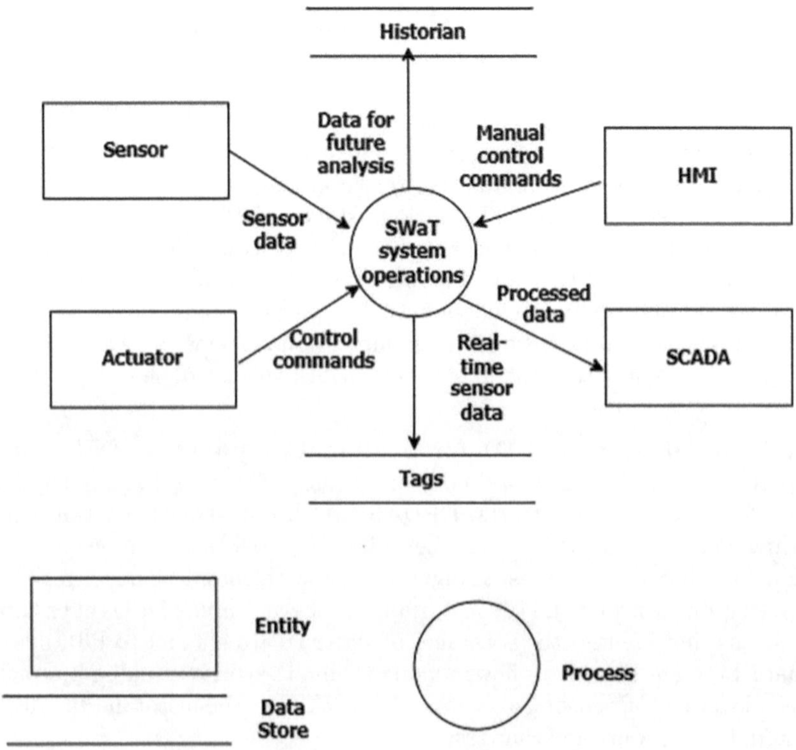

Fig. 2. Level 0 DFD.

Level 1 DFD. The DFD shown in Fig. 3, along with the interactions among processes, entities, and data stores in Table 2, provides a clear and structured view of the water treatment process. The system is divided into six main stages (P1–P6), representing the key water treatment operations, from raw water intake to chemical dosing, filtration, dechlorination, reverse osmosis and backwashing. These stages rely on sensors (E-6) to monitor critical parameters like water levels, flow rates, and pressure. The sensor data flows into the stages and is processed and sent to PLCs for decision-making. The PLCs play a central role by analysing the data and sending control commands to actuators (E-2), which perform physical actions such as opening valves, turning pumps on or off, and regulating flow. The processed data is also sent to the SCADA system (E-3), which is the central hub for monitoring and managing the entire process. SCADA provides real-time feedback to operators via the HMI (E-4), allowing manual intervention when needed, such as sending override commands or adjustments. The treated water, the final output, flows to the Treated Water Output (E-5) as part of the system's primary objective. Additionally, all sensor data and control decisions are logged into the Tags data store (DS-1), ensuring the information is available for future analysis and performance evaluation.

Table 2. Interaction for Level 1 and Level 2 DFD.

Interaction	Description
DF-1	Sensor data to Processes.
DF-2	Control commands from PLC to Actuators.
DF-3	Processed sensor data from PLC to SCADA.
DF-4	Commands and updates between SCADA and HMI.
DF-5	Processed sensor data from Processes to PLC.
DF-6	Raw water flow from Raw Water Source to Stage P1.
DF-7	Sensor data stored from PLCs to Tags.
DF-8	Log sensor data from SCADA to Historian.
DF-9	Historical data access from Historian to SCADA.
DF-10	Control signal (Pump) from PLC to Raw Water Pump.
DF-11	Control signal (Valve) from PLC to Inlet Valve.
DF-12	Pumped water flow from Raw Water Pump to Stage P2.
P-1 to P-6	Processes.
E-1 to E-10	Entities.
DS-1	Tags.
DS-2	Historian.

Level 2 DFD. To illustrate the Level 2 decomposition methodology, Stage P1 (Raw Water Intake) is selected as a representative stage. P1 was chosen because it serves as the entry point of the water treatment process and contains the full complement of DFD element types, sensors, actuators, PLCs, data flows, and data stores, making it the most complete representative of the component interactions across all six stages. The methodology applied here is directly extensible to P2–P6, which share the same structural pattern of dual PLCs, local sensors, actuators, and SCADA connectivity. Figure 4 provides a detailed breakdown of Stage P1: Raw Water Intake, highlighting the internal processes, data flows, and components involved in managing the inflow of raw water into the SWaT system. The process begins with the Raw Water Source (E-7) supplying water to Stage P1. A Water Level Sensor (E-8) monitors the water level and sends Water Level Data (DF-1) to the PLC for processing. The PLC (E-1) is the central control unit, analysing the sensor data and generating control signals to manage the actuators. Specifically, the Control Signal for the Inlet Valve (E-9) (DF-11) and Control Signal for the Raw Water Pump (E-10) (DF-10) ensure that water flow is regulated and sent to the next stage (Stage P2) through the Raw Water Pump. The PLC also stores processed sensor data in the Tags Data Store (DS-1) (DF-7) for future analysis and retrieval. Simultaneously, the processed sensor data flows to the SCADA system (E-3) (DF-3), the centralised monitoring and control platform.

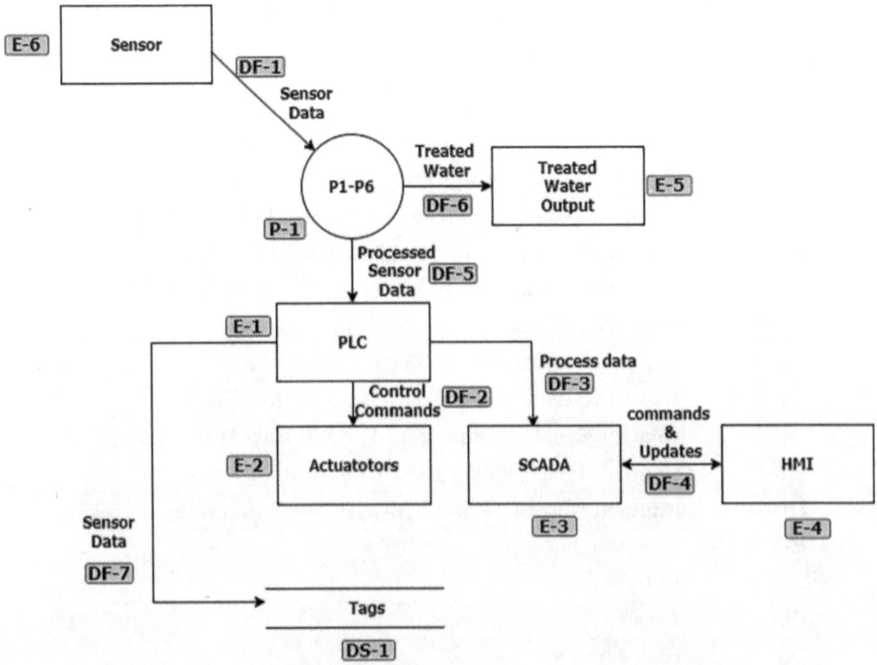

Fig. 3. Level 1 Data Flow Diagram of the SWaT Testbed.

SCADA sends commands and updates (DF-4) to the HMI (E-4), allowing operators to view system status and intervene manually if necessary. Additionally, SCADA logs sensor data (DF-8) into the Historian (DS-2) for archival purposes, enabling historical data access (DF-9) for performance analysis and future decision-making.

Possible Threat Consequences (TC) and Vulnerabilities. After plotting the DFD as specified in the STRIDE methodology, the next step is to identify threats and associated vulnerabilities. Vulnerabilities are defined as "the intrinsic properties of something resulting in susceptibility to a risk source that can lead to an event with a consequence" [36]. All identified vulnerabilities are linked to their respective MITRE CWE codes [1].

Table 3 presents 20 identified threat consequences (TC-1 through TC-20) derived through a hybrid process combining STRIDE-per-element analysis of the DFD model with evidence from the A6_2019 attack timeline. The Source column distinguishes consequences directly evidenced by the attack log (log-confirmed) from those derived analytically through STRIDE/DFD analysis. The threat consequences are later mapped to STRIDE categories and component vulnerabilities in Sect. 4.1.

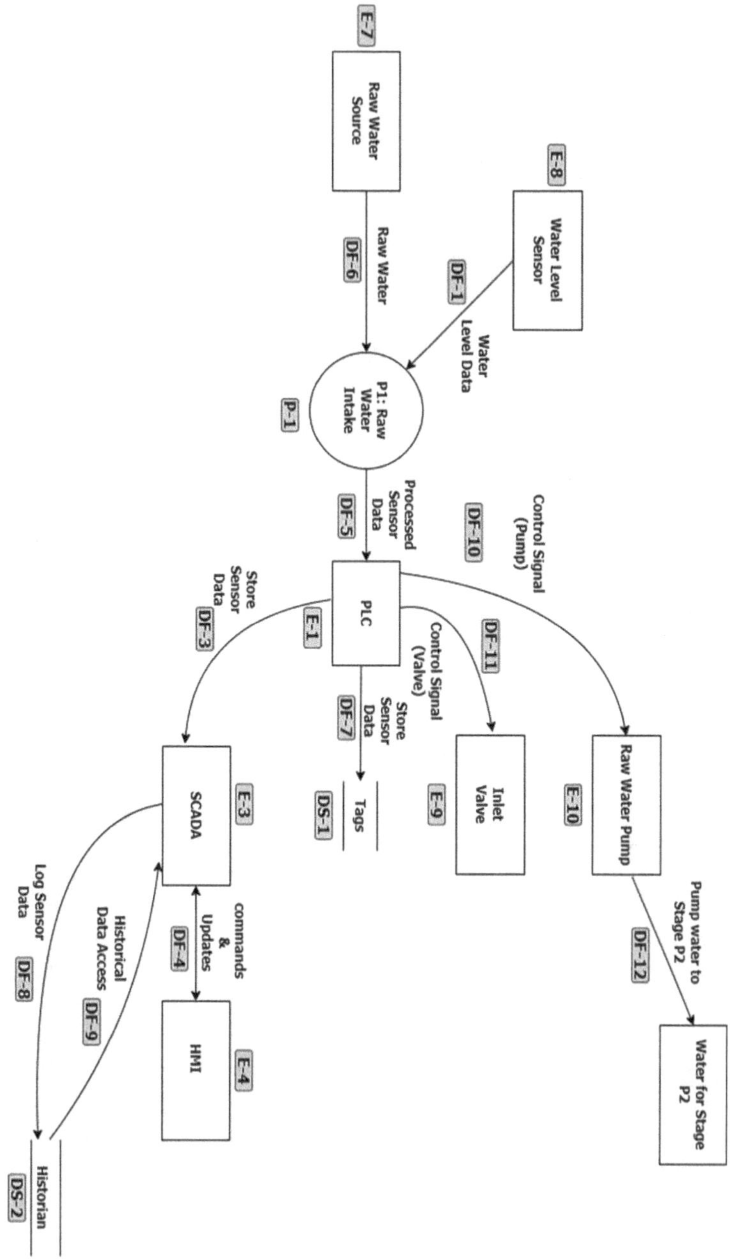

Fig. 4. Level 2 Data Flow Diagram of the SWaT Testbed (Stage P1: Raw Water Intake).

Table 3. Component-Level Possible Threat Consequences (TCs).

Code	Threat Consequence	Source
TC-1	Unauthorised access to SCADA workstation.	Log-confirmed
TC-2	Exfiltration of historian data.	Log-confirmed
TC-3	Disruption of sensor readings and process measurements.	Log-confirmed
TC-4	Manipulation of actuator commands.	Log-confirmed
TC-5	Injection of malicious PLC firmware or logic.	STRIDE/DFD
TC-6	Address Resolution Protocol (ARP) spoofing and man-in-the-middle traffic control.	STRIDE/DFD
TC-7	Infiltration of SCADA through weak security practices.	STRIDE/DFD
TC-8	Disruption of SCADA–PLC communication.	STRIDE/DFD
TC-9	Unauthorised connection to system components.	STRIDE/DFD
TC-10	Compromise of HMI-based manual control.	STRIDE/DFD
TC-11	Raw water tank overflow due to manipulated valves.	STRIDE/DFD
TC-12	Alteration of water chemical properties.	STRIDE/DFD
TC-13	Reconnaissance of network structure and configurations.	STRIDE/DFD
TC-14	Disruption of UF membrane backwash cycle.	STRIDE/DFD
TC-15	Continuous manipulation of sensor readings.	Log-confirmed
TC-16	Reduction in treated-water output.	STRIDE/DFD
TC-17	Undetected intermittent attacks.	STRIDE/DFD
TC-18	Sustained manipulation of LIT101.Pv and FIT101.Pv.	Log-confirmed
TC-19	Delayed post-attack sensor recovery.	Log-confirmed
TC-20	Coordinated disruption of sensors and actuators.	STRIDE/DFD

Identified Risks and Associated Matrices: While effective in categorising threats, the STRIDE methodology lacks a standardised approach to quantify the risk associated with each identified threat. By incorporating the risk matrix from NIST SP 800-30r1 [24], we assign qualitative values to the impact and probability of each threat, thereby creating a detailed and actionable risk profile for the SWaT testbed. The risk register includes the following headers: Threat Event, Threat Source, Threat Source Characteristics, Relevance, Likelihood of Attack Initiation, Vulnerabilities and Predisposing Conditions, Severity and Pervasiveness, Likelihood Initiated Attack Succeeds, Overall Likelihood, Level of Impact, and Risk.

Framework for Mitigation: To address Step 6 of the STRIDE methodology, we employ NIST SP 800-53 Revision 5 [25], which provides a comprehensive set of security and privacy controls designed to safeguard information systems and organisations from a wide range of security threats. This framework was utilised to address the specific needs of the SWaT testbed and guided the development and implementation of robust mitigation strategies. The framework is structured to provide a systematic approach to selecting security controls that is scalable and applicable across diverse sectors, including industrial control systems such as SWaT. The NIST framework's detailed catalogue of controls covers aspects from access control to incident response and system and communications protection. Given the critical nature of water treatment infrastructure and the significant potential consequences of security breaches, we adopted NIST SP 800-53 Revision 5 to provide depth and breadth in security controls.

4 Evaluation and Results

This section comprehensively analyses the findings from the component-level risk assessment conducted on the SWaT testbed, synthesising data gathered through DFD analysis and threat modelling using the STRIDE framework. The results highlight critical vulnerabilities within the SWaT testbed, indicate potential threat consequences, and propose mitigation approaches.

4.1 Threat Analysis in the DFD

We identify threat consequences through systematic analysis of DFD elements, attack logs, and the six STRIDE threat categories. Table 4 systematically maps each of these threat categories to the specific DFD elements within the SWaT testbed and their corresponding threat consequences.

Spoofing: Spoofing involves making communication from an unidentified source appear legitimate by exploiting protocols such as ARP, Internet Protocol (IP), and Domain Name System (DNS) [4]. In the SWaT testbed, spoofing processes (P1–P6) can lead to unauthorised commands and controls. Spoofing SCADA (E-3) and HMI (E-4) elements can result in unauthorised access to critical control systems.

Tampering: Tampering, which is unauthorised alteration of data [15], can disrupt SWaT operations by manipulating data flows and control commands. For example, tampering with data in transit or within processes can result in severe disruptions, including incorrect operations, process failures, and unauthorised actions.

Repudiation: In the SWaT testbed, repudiation involves denying actions or events, which can lead to accountability issues and operational ambiguity [4]. Repudiation threats arise when actions or commands executed by SCADA (E-3) and HMI (E-4) elements are not correctly logged.

Information Disclosure: Information disclosure in the SWaT testbed involves unintended or unauthorised exposure of system-related data [10]. Information disclosure threats are linked to unauthorised access to sensitive data, arising from data flows (DF-1–DF-8) and data stores (DS-1, DS-2).

Denial of Service: DoS attacks in the SWaT testbed disrupt essential data flows and control systems, leading to operational failures [11]. Attacks on data flows (DF-1–DF-10), PLCs (E-1), and processes (P1–P6) can disrupt critical communications and operations, causing widespread system failures.

Elevation of Privilege: Elevation of privilege in the SWaT testbed allows attackers to gain unauthorised access or higher-level permissions, controlling critical system components. Exploiting vulnerabilities, these entities (P1–P6), (E-3), (E-4), and data stores (DS-1, DS-2) enable attackers to send unauthorised commands, disrupting the system.

Table 4. Threat Modelling Using STRIDE-Per-Element Approach.

Type	DFD Element	Threat Consequences (TC)
S	P1–P6	TC-1, TC-4, TC-5, TC-7, TC-9, TC-10
	E-3, E-4	TC-1, TC-7, TC-9
T	DF-1–DF-10	TC-1, TC-3, TC-4, TC-5, TC-8, TC-10, TC-11, TC-12, TC-14, TC-16, TC-18, TC-20
	P1–P6	TC-1, TC-3, TC-4, TC-5, TC-11, TC-12, TC-14, TC-16, TC-18, TC-20
	E-3, E-4	TC-7
R	E-3, E-4	TC-7
I	DF-1, DF-2, DF-3, DF-4, DF-6, DF-7, DF-8	TC-2, TC-6, TC-13, TC-15
	DS-1, DS-2	TC-2, TC-6, TC-13, TC-15
D	DF-1, DF-3, DF-5, DF-6, DF-7, DF-9, DF-10	TC-3, TC-8, TC-11, TC-12, TC-14, TC-16, TC-20
	E-1	TC-8, TC-11
	P1–P6	TC-3, TC-8, TC-11, TC-12, TC-14, TC-16, TC-20
E	P1–P6	TC-1, TC-5, TC-7, TC-9, TC-10
	E-3, E-4	TC-1, TC-7, TC-9, TC-10
	DS-1, DS-2	TC-1, TC-5, TC-7, TC-9, TC-10

4.2 Component Vulnerability Classes and CWE Mapping

The STRIDE analysis revealed six underlying vulnerability classes (VC-1 through VC-6) spanning all SWaT components. Each is exploitable across multiple DFD elements. Each class is linked to one or more MITRE CWE identifiers [1], enabling cross-system comparison and structured mitigation strategies.

VC-1 – Weak Authentication (CWE-287, CWE-294): PLC-to-PLC, PLC-to-SCADA, and operator-to-HMI communications provide no mutual authentication, enabling impersonation and replay attacks. Enables TC-1, TC-7, TC-9, TC-10 (spoofing and elevation of privilege).

VC-2 – Insufficient Data Integrity (CWE-345, CWE-20): Sensor-to-PLC data flows carry no cryptographic integrity protection. Directly confirmed in A6_2019 Phase 2, where C2-driven sensor value manipulation produced erroneous control actions without triggering any alarm. Enables TC-3, TC-4, TC-12, TC-15, TC-18 (tampering).

VC-3 – Firmware Vulnerabilities (CWE-494, CWE-693): PLC firmware lacks integrity verification for updates, allowing malicious logic injection via the update pathway. Enables TC-5, TC-17 (firmware tampering and evasion).

VC-4 – Inadequate Logging (CWE-778, CWE-223): Historian (DS-2) and SCADA (E-3) do not enforce tamper-evident audit logging. Confirmed in A6_2019 Phase 1, where historian exfiltration proceeded undetected over 50 min. Enables TC-2, TC-7, TC-13, TC-19 (repudiation and information disclosure).

VC-5 – Insufficient Access Controls (CWE-284, CWE-269): No Role-Based Access Control (RBAC) is enforced across SCADA, HMI, or historian; any authenticated user can issue PLC commands or modify historian records without additional authorisation. Enables TC-1, TC-7, TC-9, TC-10 (elevation of privilege and spoofing).

VC-6 – Weak Network Security (CWE-319, CWE-923): The Level 1 Ethernet ring and inter-stage network operate without Virtual Local Area Network (VLAN) segmentation or encrypted communications, permitting lateral movement and man-in-the-middle interception. Enables TC-6, TC-8, TC-11, TC-14, TC-16, TC-20 (DoS and information disclosure).

4.3 Identified Component-Level Risks

The NIST SP 800-30r1 risk register evaluation identified nine consolidated threat events, each characterised by threat source, attacker capability, intent level, and targeting propensity, as detailed in Table 5. The analysis reveals that all identified threat events exhibit High Capability and Intent ratings from both external attackers and insider threats, indicating significant risk from well-resourced adversaries. This uniform characterisation reflects the critical nature of water treatment infrastructure as a target for both opportunistic and nation-state actors.

Table 6 presents the complete risk assessment, mapping the likelihood of attack initiation against identified vulnerabilities and predisposing conditions to determine overall risk levels. The risks are rated as either High or Moderate, indicating that current defensive measures across the SWaT testbed provide insufficient protection against the threat landscape. The distribution of Moderate versus High likelihood ratings, combined with differentiated impact assessments, resulted in five High risk events and four Moderate risk events. Threat events directly enabling loss of operational capability are rated High, while those causing security degradation without direct operational disruption are rated Moderate. Likelihood of attack initiation is derived from threat source characteristics in Table 5 via NIST Table G-2 [24]. Level of impact is differentiated per NIST Table H-3, and risk is determined per NIST Table I-2 as a combination of overall likelihood and impact [24].

Table 5. Component-Level Threat Events and Characteristics.

Threat Event	Threat Sources	Capability	Intent	Targeting
Unauthorised access due to spoofing of inter-PLC communications.	External attackers, insider threats	High	High	High
Erroneous control actions due to tampering with sensor data to processes.	External attackers, insider threats	High	High	High
System compromises due to vulnerable firmware and software.	External attackers, insider threats	High	High	High
Inability to track and deny malicious activities due to inadequate logging and monitoring.	External attackers, insider threats	Moderate	High	Moderate
Targeted attacks due to the exposure of sensitive data.	External attackers, insider threats	High	High	High
Unauthorised access due to weak access controls.	External attackers, insider threats	High	High	High
Operational halts due to network and communication vulnerabilities.	External attackers, insider threats	High	High	High
Unauthorised access due to privilege escalation vulnerabilities.	External attackers, insider threats	High	High	High
Unanticipated system operations due to insecure firmware and software configurations.	External attackers, insider threats	Moderate	High	Moderate

4.4 Component-Level Proposed Mitigation Approaches

To address vulnerabilities identified in Step 6 of the proposed STRIDE methodology, several mitigation measures consistent with NIST SP 800-53 Revision 5 can be implemented [25], each mapped to specific control identifiers with implementation constraints noted where relevant to operational water treatment environments.

VC-1 (Weak Authentication, CWE-287, CWE-294): Deploy mutual authentication on all inter-PLC and PLC-to-SCADA communications and enforce Multi-Factor Authentication (MFA) on HMI and SCADA workstations. When legacy PLCs preclude MFA, timing-based authentication [2] protects against replay and masquerade. Controls: IA-2, IA-3. Constraint: legacy PLCs may require network segmentation as a compensating control.

Table 6. Component-Level Likelihood, Vulnerabilities, and Risks.

Likelihood of Attack Initiation	Vulnerabilities and Predisposing Conditions	Level of Impact	Risk
High	Weak Authentication Mechanisms, Unsecured Network Communications.	High	High
High	Insufficient Data Integrity Checks.	High	High
High	Outdated/Insecure Firmware and Software.	High	High
Moderate	Lack of Comprehensive Logging and Monitoring Mechanisms.	Moderate	Moderate
High	Unsecured/Improperly Secured Data Flows.	Moderate	Moderate
High	Insufficient Access Controls.	High	High
High	Weak Network Infrastructure.	Moderate	Moderate
High	Weak Access Controls and Misconfigured Permissions.	High	High
Moderate	Weak Access Controls and Insecure Firmware and Software.	Moderate	Moderate

VC-2 (Insufficient Data Integrity, CWE-345, CWE-20): Apply Message Authentication Codes (MACs) or digital signatures to all sensor-to-PLC data flows and implement physics-based invariant checking at the PLC level to reject physically implausible readings—directly addressing the A6_2019 Phase 2 attack vector. Controls: SI-7, SC-13. Constraint: integrity checking latency must be validated against process safety margins.

VC-3 (Firmware Vulnerabilities, CWE-494, CWE-693): Require cryptographic signing of all PLC firmware updates, maintain a baseline firmware inventory with automated deviation alerting, and apply patches via staged rollout within maintenance windows. Controls: SI-2, CM-3, CM-7. Constraint: firmware updates require process shutdown; coordinate with operational schedules.

VC-4 (Inadequate Logging, CWE-778, CWE-223): Deploy tamper-evident audit logging with cryptographic log chaining across historian (DS-2), SCADA (E-3), and HMI (E-4), with real-time forwarding to an isolated Security Information and Event Management (SIEM), directly addressing the undetected 50-min exfiltration in A6_2019 Phase 1. Controls: AU-2, AU-6, AU-9. Constraint: define tiered log retention policies to manage historian storage capacity.

VC-5 (Insufficient Access Controls, CWE-284, CWE-269): Enforce RBAC across SCADA, HMI, and historian with least-privilege principles, separating read and write permissions on historian data stores and conducting quarterly access reviews. Controls: AC-2, AC-3, AC-6. Constraint: role definitions must be developed with operations staff to avoid restricting emergency response procedures.

VC-6 (Weak Network Security, CWE-319, CWE-923): Segment the flat network using VLANs to isolate process stages, deploy OT-aware industrial firewalls and an Intrusion Detection System (IDS) capable of parsing Modbus, EtherNet/IP, and DLR traffic, and encrypt historian data flows DF-8 and DF-9. Controls: SC-7, SC-8, SI-4. Constraint: VLAN configuration on the DLR ring must be validated for fault-tolerance impact before deployment.

5 Discussion

Component-level STRIDE assessment proved effective in identifying specific, actionable vulnerabilities within individual SWaT components. Through DFD analysis and empirical validation against the A6_2019 dataset, we identified six CWE-grounded vulnerability classes (VC-1–VC-6) yielding 20 threat consequences and nine consolidated threat events rated Moderate to High per NIST SP 800-30r1 [24].

Unauthorised access due to spoofing of inter-PLC communications (VC-1), erroneous control actions due to tampering with sensor data (VC-2), and unauthorised access due to weak access controls (VC-5) are rated High due to high exploitation likelihood and direct physical process impact. Erroneous control actions due to tampering with sensor data to processes were directly confirmed in A6_2019 Phase 2, where C2-driven sensor manipulation produced erroneous control actions without triggering any integrity alarm, demonstrating that VC-2 is an actively exploited vector, not a theoretical risk. Mutual authentication and MFA (VC-1), physics-based invariant checking (VC-2), and RBAC enforcement (VC-5) represent the highest-priority remediations. CWE grounding enables practitioners to cross-reference identified weaknesses against ICS vulnerability databases and Common Vulnerabilities and Exposures (CVE) records for their specific deployed components, improving transferability beyond SWaT.

However, component-level assessment has inherent limitations. Evaluating components in isolation may miss compounded risks; for example, when VC-2 and VC-4 co-occur, it creates a scenario in which sensor manipulation goes both undetected and unrecorded, a systemic vulnerability neither class exposes alone [23]. Resource intensity also limits scalability across large systems. Mitigations should be sequenced by risk rating, High before Moderate, and component-level hardening must be complemented by system-level analysis such as cyber-extended Hazard and Operability study (HAZOP) or attack tree modelling to capture cascading failures across the full treatment process.

6 Conclusion

Component-level risk assessment using STRIDE threat modelling provides a systematic framework for identifying and mitigating specific vulnerabilities in cyber-physical systems. Through detailed analysis of the SWaT testbed, we identified 20 distinct threat consequences spanning weak authentication, insufficient data integrity, outdated firmware, and inadequate access controls. Our analysis identified six critical component-level vulnerabilities and proposed six targeted mitigation strategies aligned with NIST SP 800-53 Revision 5.

Component-level assessments excel at identifying immediate, actionable vulnerabilities in individual system components. However, the component-level findings in this paper do not capture cascading failures or complex interdependencies across stages, suggesting that component-level hardening alone is unlikely to be sufficient; system-level analysis remains necessary to address these systemic risks.

Future research should extend component-level assessment to additional CPS domains beyond water treatment, develop automated tools for component-level vulnerability scanning to reduce resource requirements, and investigate integration of component-level and system-level assessments into unified frameworks. Empirically validating the effectiveness of proposed mitigations, particularly physics-based invariant checking (VC-2) and tamper-evident audit logging (VC-4), through controlled deployment and adversarial testing on the SWaT testbed represents a critical next step. Future work should measure detection rates, response latency, and operational impact against the A6_2019 attack scenario and additional dataset attack logs.

Acknowledgements. This work was supported by the National Edge AI Hub for Real Data: Edge Intelligence for Cyber-disturbances and Data Quality, EPSRC, EP/Y028813/1.

Disclosure of Interests. The authors have no competing interests to declare that are relevant to the content of this article.

References

1. CWE - Common Weakness Enumeration: A Community-Developed List of Common Software Security Weaknesses. https://cwe.mitre.org/. Accessed 06 Aug 2024
2. Ahmed, C.M., Ochoa, M., Zhou, J., Mathur, A.P.: Scanning the cycle: timing-based authentication on PLCs. In: Proceedings of the 2021 ACM Asia Conference on Computer and Communications Security (ASIA CCS 2021), pp. 886–900. ACM, New York (2021). https://doi.org/10.1145/3433210.3453102
3. Ahmed, C.M., Gauthama Raman, M.R., Mathur, A.P.: Challenges in machine learning based approaches for real-time anomaly detection in industrial control systems. In: Proceedings of the 6th ACM on Cyber-Physical System Security Workshop (2020)
4. Alsulami, A.A., Zein-Sabatto, S.: Resilient cyber-security approach for aviation cyber-physical systems protection against sensor spoofing attacks. In: 2021 IEEE 11th Annual Computing and Communication Workshop and Conference (CCWC), NV, USA, pp. 0565–0571. IEEE (2021). https://doi.org/10.1109/CCWC51732.2021.9376158
5. Athalye, S., Ahmed, C.M., Zhou, J.: A tale of two testbeds: a comparative study of attack detection techniques in CPS. In: Rashid, A., Popov, P. (eds.) CRITIS 2020. LNCS, vol. 12332, pp. 17–30. Springer, Cham (2020). https://doi.org/10.1007/978-3-030-58295-1_2

6. Bhamare, D., Zolanvari, M., Erbad, A., Jain, R., Khan, K., Meskin, N.: Cybersecurity for industrial control systems: a survey. Comput. Secur. **89**, 101677 (2020). https://doi.org/10.1016/j.cose.2019.101677
7. Buli, N., Chestney, N., Steitz, C.: Insight: Cyberattacks on renewables: Europe power sector's dread in chaos of war (2023). https://www.reuters.com/business/energy/cyberattacks-renewables-europe-power-sectors-dread-chaos-war-2023-06-15/. Accessed 04 Aug 2024
8. Eckhart, M., Brenner, B.: Quantitative security risk assessment for industrial control systems: research opportunities and challenges. J. Internet Serv. Inf. Secur. (2019). https://doi.org/10.22667/JISIS.2019.08.31.052
9. Fung, C., Srinarasi, S., Lucas, K., Phee, H., Bauer, L.: Perspectives from a comprehensive evaluation of reconstruction-based anomaly detection in industrial control systems. In: Proceedings of the 27th European Symposium on Research in Computer Security (ESORICS 2022), pp. 3–24. Springer (2022)
10. Honkaranta, A., Leppänen, T., Costin, A.: Towards practical cybersecurity mapping of stride and CWE: a multi-perspective approach. In: Proceedings of the 29th Conference of Open Innovations Association (FRUCT), pp. 150–159. IEEE (2021). https://doi.org/10.23919/FRUCT52173.2021.9435453
11. Huseinović, A., Mrdović, S., Bicakci, K., Uludag, S.: A survey of denial-of-service attacks and solutions in the smart grid. IEEE Access **8**, 177447–177470 (2020). https://doi.org/10.1109/ACCESS.2020.3026923
12. Hutchins, E., Cloppert, M., Amin, R.: Intelligence-driven computer network defense informed by analysis of adversary campaigns and intrusion kill chains. Technical report. Lockheed Martin Corporation (2011)
13. iTrust Centre for Research in Cyber Security: SWaT security dataset (2019). Accessed via formal research request. https://itrust.sutd.edu.sg/itrust-labs_datasets/dataset_info/
14. Jacob, P.M., Jose, J., Jose, J.: An analytical approach on DFD to UML model transformation techniques. In: Proceedings of the 2016 International Conference on Information Science (ICIS), pp. 12–17 (2016)
15. Jain, A., Guin, U.: A novel tampering attack on AES cores with hardware trojans. In: 2020 IEEE International Test Conference in Asia (ITC-Asia), pp. 77–82. IEEE, Taipei, Taiwan (2020). https://doi.org/10.1109/ITC-Asia51099.2020.00025
16. Khan, R., McLaughlin, K., Laverty, D., Sezer, S.: Stride-based threat modeling for cyber-physical systems. In: 2017 IEEE 15th International Conference on Industrial Informatics (INDIN), pp. 194–200. IEEE (2017). https://doi.org/10.1109/ISGTEurope.2017.8260283
17. Kim, K., Kim, K., Kim, H.: Stride-based threat modeling and dread evaluation for the distributed control system in the oil refinery. ETRI J. **44**(6), 991–1003 (2022). https://doi.org/10.4218/etrij.2021-0181
18. Kravchik, M., Shabtai, A.: Efficient cyber attack detection in industrial control systems using lightweight neural networks and PCA, arXiv preprint arXiv:1907.01216v2 (2019)
19. Kriaa, S., Pietre-Cambacedes, L., Bouissou, M., Halgand, Y.: A survey of approaches combining safety and security for industrial control systems. Reliab. Eng. Syst. Saf. **139**, 156–178 (2015). https://doi.org/10.1016/j.ress.2015.02.008
20. Lyngaas, S.: Federal investigators confirm multiple us water utilities hit by hackers (2023). https://edition.cnn.com/2023/12/01/politics/us-water-utilities-hack/index.html. Accessed 04 Aug 2024
21. Magar, A.: State-of-the-art in cyber threat models and methodologies. DRDC-RDDC-2016-C132 DRDC Contract Report, Sphyrna Security (2016)

22. Mathur, A.P., Tippenhauer, N.O.: SWaT: a water treatment testbed for research and training on ICS security. In: Proceedings of the 2016 International Workshop on Cyber-physical Systems for Smart Water Networks (CySWater), Vienna, Austria, pp. 31–36 (2016). https://doi.org/10.1109/CySWater.2016.7469060
23. Mugarza, I., Flores, J., Montero, J.: Security issues and software updates management in the industrial internet of things (IIoT) era. Sensors **20**(24), 7160 (2020). https://doi.org/10.3390/s20247160
24. National Institute of Standards and Technology: Guide for conducting risk assessments, NIST special publication 800-30 revision 1. Technical report, JOINT TASK FORCE TRANSFORMATION INITIATIVE (2012). https://nvlpubs.nist.gov/nistpubs/legacy/sp/nistspecialpublication800-30r1.pdf
25. National Institute of Standards and Technology (NIST): Security and privacy controls for information systems and organizations. NIST Special Publication 800-53 Revision 5 (2020). https://doi.org/10.6028/NIST.SP.800-53r5
26. Nejad, H., Mathias, D.: Top-down vs. bottom-up risk assessment: Consistent, contradictory or complimentary? Technical report, Science and Technology Corporation. NASA Ames Research Center (2013)
27. Niesen, T., Houy, C., Fettke, P., Loos, P.: Towards an integrative big data analysis framework for data-driven risk management in industry 4.0. In: 49th Hawaii International Conference on System Sciences (HICSS), pp. 5065–5074. IEEE (2016). https://doi.org/10.1109/HICSS.2016.627
28. Palleti, V.R., Adepu, S., Mishra, V.K., Mathur, A.: Cascading effects of cyber-attacks on interconnected critical infrastructure. Cybersecurity **4**(1), 1–15 (2021). https://doi.org/10.1186/s42400-021-00071-z
29. Qadeer, R., Murguia, C., Ahmed, C.M., Ruths, J.: Multistage downstream attack detection in a cyber physical system. In: Katsikas, S.K., et al. (eds.) Cyber-ICPS/SECPRE -2017. LNCS, vol. 10683, pp. 177–185. Springer, Cham (2018). https://doi.org/10.1007/978-3-319-72817-9_12
30. Ralston, P., Graham, J., Hieb, J.: Cyber security risk assessment for SCADA and DCS networks. ISA Trans. **46**, 583–594 (2007). https://doi.org/10.1016/j.isatra.2007.04.003
31. Sharmeen, S., Huda, S., Abawajy, J., Ahmed, C.M., Hassan, M.M., Fortino, G.: An advanced boundary protection control for the smart water network using semisupervised and deep learning approaches. IEEE Internet Things J. **9**(10), 7298–7310 (2022). https://doi.org/10.1109/JIOT.2021.3100461
32. Shevchenko, N., Chick, T., O'Riordan, P., Scanlon, T., Woody, C.: Threat modeling: a summary of available methods. Technical report, Software Engineering Institute, Carnegie Mellon University (2018)
33. Shostack, A.: Threat Modeling: Designing for Security. Wiley (2014)
34. Singh, A., Jain, A.: Study of cyber attacks on cyber-physical system. SSRN Electron. J. (2018). https://doi.org/10.2139/ssrn.3170288
35. Sun, M., Lai, Y., Wang, Y., Liu, J., Mao, B., Gu, H.: Intrusion detection system based on in-depth understandings of industrial control logic. IEEE Trans. Industr. Inf. **19**(3), 2295–2305 (2023). https://doi.org/10.1109/TII.2022.3200363
36. Sutton, D.: Information Risk Management: A Practitioner's Guide. BCS, The Chartered Institute for IT (2014)

37. Umer, M.A., Ahmed, C.M., Jilani, M.T., Mathur, A.P.: Attack rules: an adversarial approach to generate attacks for industrial control systems using machine learning. In: Proceedings of the 2th Workshop on CPS&IoT Security and Privacy, CPSIoTSec 2021, pp. 35–40. Association for Computing Machinery, New York (2021). https://doi.org/10.1145/3462633.3483976
38. Yilmaz, E.N., Gonen, S.: Attack detection/prevention system against cyber attack in industrial control systems. Comput. Secur. **77**, 94–105 (2018)

Hiding the Heartbeat: $\mathcal{O}(1)$ Privacy-Preserving Proof of Aliveness for Critical IoT

Harishma Boyapally(✉) and Jianying Zhou

Singapore University of Technology and Design, Singapore, Singapore
{harishma_boyapally,jianying_zhou}@sutd.edu.sg

Abstract. Monitoring the operational and functional integrity of remote edge devices is paramount for decentralized Cyber-Physical Systems (CPS). While Proof-of-Aliveness (PoA) has emerged as a lightweight cryptographic primitive for secure status reporting, the existing schemes rely on static identifiers and enforced storage-computation trade-offs. Such limitations leave devices vulnerable to sophisticated traffic analysis and identity tracking, while simultaneously hindering scalability in resource-constrained environments. In this paper, we first introduce formal definitions of Aliveness-Unforgeability and Privacy, specifically targeting Identity Anonymity and Unlinkability. We propose 3PoA: a Privacy-Preserving **P**roof **o**f Aliveness protocol that replaces traditional one-way function chains with a rolling-state forwarding mechanism. 3PoA is the first to achieve simultaneous $\mathcal{O}(1)$ complexity across all metrics: setup, storage, communication, and computation. By rotating session-based pseudonyms and chaining symmetric secrets, 3PoA effectively "hides the heartbeat" of critical nodes ensuring unlinkability against external eavesdroppers and anonymity against internal verifiers. Formal security analysis and comparative evaluations demonstrate that 3PoA eliminates the performance bottlenecks of prior works while providing the robust privacy required for modern industrial IoT deployments.

Keywords: Proof of Aliveness · Cyber-Physical Systems (CPS) Security · Privacy and Anonymity · Unlinkability · Unforgeability · Constant-Time Complexity · Constant-Storage Overhead

1 Introduction

CPS and IoT have enabled unprecedented automation, but their integration has simultaneously exposed critical infrastructure to a complex array of attacks. Historically, many deployments relied on security-by-obscurity with physical isolation and network segmentation. The transition from air-gapped systems to internet-connected nodes means that a single firmware vulnerability—such as those found in the Ripple20 [7] or Amnesia:33 [16] suites—can be exploited remotely to achieve arbitrary code execution. These modern CPS connected to public networks face increased risks because edge devices often act as bridgeheads into internal networks.

S. Zonouz et al. (Eds.): DCS-CI 2026, CCIS 3109, pp. 115–135, 2027.
https://doi.org/10.1007/978-981-92-5039-4_6

A primary requirement for CPS edge nodes is the ability to prove their operational status or *aliveness* to a Control Center (CC). A straightforward approach to periodically send "heartbeat" signals may lead to replay attacks when the device is physically tampered with, powered off or destroyed. For instance, in the BlackEnergy 3 attack on the Ukrainian power grid [15], attackers suppressed legitimate status reporting while launching a Telephone Denial-of-Service (TDoS) to mask the operational state of the grid. These attacks highlight that a simple "heartbeat" is insufficient; there is a critical need for an unforgeable proof that binds a device's identity and state to a fresh, non-replayable cryptographic proof.

While traditional authentication schemes verify a device's identity, they are typically event-driven. They can only confirm a device is active at the moment of access, leaving the CC blind to failures during idle periods. Furthermore, the high computational and communication overhead of full authentication handshakes makes them expensive for use as a frequent heartbeat mechanism. The concept of Proof of Aliveness (PoA) [9] was introduced as a cryptographic primitive which enables a device to periodically and continuously prove its state of aliveness to a CC. The basic PoA protocol is setup using one-way function (OWF) chains, introduced by Lamport [12] for one-time passwords. In this protocol, the prover uses a OWF (F) and initial seed value x_N, to generate a sufficiently large OWF chain as

$$x_0 \xleftarrow{F(x_1)} x_1 \xleftarrow{F(x_2)} x_2 \ldots \xleftarrow{F(x_{N-1})} x_{N-1} \xleftarrow{F(x_N)} x_N$$

To prove aliveness, it reveals the values in the opposite direction from x_0 to x_{N-1}. Without the knowledge of the OWF trapdoor, an adversary cannot generate the valid proofs. This is a simple yet efficient solution to securely prove the aliveness of a resource-constrained device. Yet, the prover should make $(N - k)$ calls to the OWF, to generate the k-th proof and it generates a new OWF chain once all the proofs are exhausted. To support continuous proof generation, more sophisticated auto-replenishment schemes are proposed in conjunction with PoA protocols in [18]. These protocols create multiple OWFs and create a relation between two consecutive chains using commitment schemes. However, due to use of OWF chains in all the existing protocols, the storage vs computation tradeoff is not completely eliminated.

Another important requirement in CPS and IoT networks is privacy [4,14]. The notion of privacy varies depending on what should remain secret in a given system. In a decentralized power grid, if an adversary can link PoA messages to specific substations, they can identify which nodes are currently under heavy load or undergoing maintenance, increasing the possibility of large-scale physical or cyber-coordinated attacks. An external adversary monitoring network traffic should not be able to link a *proof* to its sender (identity anonymity)or, given any two proofs, determine if they originate from the same sender (unlinkability). Generally, a server node (acting as PoA Verifier) is responsible to verify the aliveness status, so PoA cannot enforce unlinkability on Verifiers. However to preserve anonymity against insider attackers, the server should not learn the identity of any uncompromised Prover. A recent attack on Mixpanel, inadvertently exposed sensitive operational data associated with OpenAI [13]). This security lapse directly elevates the risk for affected customers, increasing their susceptibility to phishing campaigns designed to deliver malware or capture further sensitive credentials. Without unlinkability, even encrypted proofs act as beacons that track a device's movements or uptime history, directly violating the privacy requirements of sensitive industrial deployments.

While the prior PoA constructions enforce strict storage-computation trade-off and broadcast static identities of Provers, our proposed 3PoA utilizes OWFs and symmetric encryption blocks to achieve $\mathcal{O}(1)$ complexity for setup, storage and computation. Furthermore, 3PoA adopts a privacy-preserving architecture that provides unlinkability against external adversaries and anonymity against internal adversaries. This makes 3PoA the first privacy-preserving aliveness protocol capable of scaling to massive IoT deployments without hardware bottlenecks.

Our Contributions. In this work, we address these limitations by introducing a new protocol and a more rigorous security framework. Our main contributions are summarized as follows:

- *Formal Security and Privacy Framework:* We provide a rigorous foundation for the PoA primitive by formalizing Aliveness-Unforgeability, Anonymity, and Unlinkability through standard cryptographic games. Our framework explicitly models the capabilities of both external eavesdroppers attempting traffic analysis and internal adversaries attempting to de-anonymize participants.
- The 3PoA Protocol: We introduce 3PoA departing from the conventional OWF chain model in favor of a state forwarding mechanism using symmetric key primitives. Through the use of unlinkable session aliases, we achieve simultaneous $\mathcal{O}(1)$ complexity across all metrics—setup, storage, communication, and computation. To our knowledge, this is the first aliveness primitive to eliminate the linear scaling bottlenecks inherent in prior designs.This design effectively breaks the storage-computation trade-off inherent in prior designs, allowing resource-constrained Provers to generate proofs indefinitely without exhausting memory or computational resources.
- *Performance Benchmarking:* We validate our claims through a comparative performance evaluation against current state-of-the-art protocols. Our results confirm that 3PoA provides superior scalability and minimal resource footprints on edge hardware while simultaneously satisfying the rigorous privacy requirements that previous $\mathcal{O}(1)$ schemes often neglect.

Organization. In Sect. 2, we discuss some relevant works. Section 3 provides the necessary building blocks and Sect. 4 introduces the formal framework and security properties 3PoA. The 3PoA Scheme is described in Sect. 5 followed by Sect. 6, to prove its security. Section 7 provides benchmarking against competing PoA protocols, confirming 3PoA's superior efficiency and $O(1)$ overhead. Finally, Sect. 8 summarizes the findings, with supporting lemma proofs detailed in Appendix A.

2 Related Works

The development of Proof of Aliveness (PoA) stems from the limitations of traditional network heartbeats and event-driven authentication in resource-constrained environments. In this section we review the evolution of these protocols and explain the technical bottlenecks, specifically the storage-computation trade-offs that 3PoA aims to eliminate.

2.1 PoA: Trade-Offs and Replenishment

Traditional heartbeat signals are the simplest form of aliveness monitoring, but it lacks cryptographic binding, making them susceptible to replay and Telephone Denial-of-Service attacks, as seen in the BlackEnergy 3 incident. While standard mutual authentication schemes provide strong security, they are typically computationally expensive for high-frequency status monitoring on edge devices. Proof of Retrievability (POR) [10] proves the existence of data on the cloud, and Remote Attestation (RA) [2] verifies the integrity of a device's memory, but they are computationally heavy for realizing PoA in lightweight CPS.

The first formal PoA primitive, Π_{OWF}, was introduced by Jin et al. [9] and later optimized by Yang et al. [18]. Similar to T/Key technique [8,11] for time-based one-time password, PoA schemes utilize one-way function (OWF) chains, a technique derived from Lamport's one-time password (OTP) system [12]. In these constructions, the Prover must either store the OWF chain of length N (high storage) or recompute every proof on the chain (high computation, $\mathcal{O}(N^2)$). To mitigate this, Π_{PRG} and Bloom Filter-based approaches $\Pi_{\mathsf{OWF}}^{\mathsf{PRG}} - \mathrm{BF}$ were proposed. This scheme uses multiple shorter chains (N)and Bloom Filters to reduce storage costs. However, these optimizations remain sublinear; the storage and computation still scale with the total number of proofs generated over the device's lifetime, eventually leading to resource exhaustion on long-lived edge devices. Unlike the existing schemes, 3PoA utilizes rolling-state forwarding mechanism: the state of the Prover and Receiver are continuously updated and the state is forwarded only if the proof generation and verification succeeds. Thus it achieves true constant overhead by replacing static chains with dynamic, symmetric-key encrypted state updates.

To prevent a device from running out of proofs, Yang et al. [18] introduced auto-replenishment schemes such as OTSRep and CommitRep. They suggest CommitRep that uses hash-based commitments, is the efficient scheme among them. However, it requires to generate fresh commitment alongside every proof. While this is a constant computation and communication overhead, the underlying PoA scheme still has storage-computation trade-off. Unlike these OWF chain based schemes, 3PoA utilizes a rolling-state forwarding mechanism and achieves true $\mathcal{O}(1)$ constant overhead across all metrics (setup, storage and computation), thus effectively eliminating the need for periodic replenishment. Furthermore, we distinguish our 3PoA from the work of Cao et al. [5], where they introduce notion of persistence in PoA. Their scheme specifically targets compromise persistence by ensuring that an adversary cannot use captured secrets to forge

Table 1. Comparative Analysis of 3PoA against Related Cryptographic Primitives

Scheme	Aliveness	Anonymity	Unlinkability	Computation	Storage
Heartbeat/Keep-Alive	✓	×	×	$\mathcal{O}(1)$	–
Remote Attestation [2]	✓	×	×	Heavy	$\mathcal{O}(1)$
DAA [3]	✓	✓	✓	Pairings	$O(1)$
Group Signatures [6]	×	✓	✓	$\mathcal{O}(N)$ or $\mathcal{O}(1)$	$\mathcal{O}(1)$
Optimized PoA [18]	✓	×	×	$\mathcal{O}(\sqrt{N})$	$\mathcal{O}(\sqrt{N})$
3PoA (Ours)	✓	✓	✓	$\mathcal{O}(1)$	$\mathcal{O}(1)$

proofs indefinitely after a device is destroyed. While 3PoA provides the O(1) efficiency required for high-frequency status checks, it can be augmented with persistent aliveness anchors to verify that the prover remains physically bound to its credentials over its entire lifespan.

2.2 Anonymity and Unlinkability

Existing PoA protocols prioritize unforgeability but largely ignore anonymity and unlinkability (Pfitzmann et al. [4,14,17]), often broadcasting static device identities. In modern CPS environments, this creates a metadata leakage vulnerability, where passive traffic analysis can reveal an external adversary, the list of alive-devices and their maintenance patterns. This effectively increases the susceptibility to coordinated physical or cyber attacks.

Providing privacy in device-reporting has traditionally relied on Direct Anonymous Attestation (DAA) (Juels et al. [3]) or Group Signatures (Chaum et al. [6]). These schemes ensure that an observer cannot identify the sender (Anonymity) or determine if two messages originate from the same sender (Unlinkability). While robust, these methods typically require complex bilinear pairings or maintain revocation lists, making them unsuitable for massive IoT deployments. 3PoA introduces these properties directly into the PoA primitive. By using unlinkable session aliases (UIDs) and hidden identities (HIDs), 3PoA ensures that even an internal verifier cannot learn the identity of an honest Prover, while external eavesdroppers in addition cannot link disparate proofs to the same source.

As discussed above, while liveness and privacy exist in the literature, no existing scheme simultaneously satisfies the stringent efficiency and privacy requirements of modern CPS. Conversely, high-privacy solutions like DAA *DAA* and Group Signatures *Group Signatures* introduce computational overheads (such as bilinear pairings). 3PoA fills this critical gap by being the first to provide formal anonymity and unlinkability with $\mathcal{O}(1)$ constant-time performance. We summarize the comparison in Table 1

3 Preliminaries

In this section, we describe PoA notion and cryptographic blocks required to introduce the security definition and protocol description. We describe the notion of anonymity and unlinkability in the context of PoA later, in Sect. 4.

3.1 Proof of Aliveness

Proof of Aliveness is a two party protocol, with one-way communication from Prover to the Verifier. The protocol is initiated using a trusted setup phase. During the online phase (on insecure communication channels), the time is divided into discrete windows of length δ_a. The protocol starts at time T_s and the prover sends unique proofs once every δ_a time, also known as the aliveness-window. The Prover device continuously sends an aliveness-proof, to the Verifier. Only when the Verifier receives a valid proof within an aliveness window, it marks the Prover device as alive. Below, we provide intuitive definition to describe PoA.

Definition 1 (PoA). *Let κ be the security parameter, T_c is the current time from a global clock, and T_s denotes the start of an online phase. The length of the aliveness-window is denoted by δ_a. A PoA protocol (Π_{PoA}) comprises the following probabilistic polynomial time (PPT) algorithms:*

1. *$\mathsf{Setup}(1^\kappa, T_s, \delta_a, \mathsf{GC})$: It is run by a trusted party to generate the credentials Γ^P and $\Gamma^{V,P}$ for Prover and Verifier. Initial states of Prover and Verifier are set as $\sigma^P = \langle \Gamma^P, T, \delta_a \rangle$ and $\sigma^{V,P} = \langle \Gamma^{V,P}, T, \delta_a \rangle$, where $T = T_s$.*
2. *$\mathsf{NextPf}(P, T_c, \sigma^P)$: It simulates the Prover. If $T < T_c < T + \delta_a$, computes the proof pf. Updates the state (σ^P) for $T = T + \delta_a$. The output is pf.*
3. *$\mathsf{Verify}(P, pf, T_c, \sigma^{V,P})$: It simulates the Verifier. If pf is a valid proof: verifier received it at $T < T_c < T + \delta_a$ and pf is correct, the state $(\sigma^{V,P})$ is updated for $T = T + \delta_a$ and output is 1. Else, output is 0.*

Correctness requires that for all honest Provers and Verifiers, the Verify algorithm deterministically outputs 1 (accept) for every proof pf generated by NextPf, provided that pf is a valid proof.

3.2 Cryptographic Building Blocks

In the context of this work, the hash function serves as the primary mechanism for both integrity and identity masking. One-Wayness (OW) is used to transform sensitive device identifiers into anonymized permanent handles, ensuring that even if a hash is intercepted, the underlying identity remains protected. Collision Resistance (CR) is leveraged to create unforgeable 'tags' that bind a proof to a specific time interval, preventing an adversary from spoofing a status report. Finally, Output Pseudorandomness (PR) allows the protocol to generate session-based pseudonyms that appear random to an observer, which is the foundation for achieving Unlinkability.

Definition 2 (Hash Function in Random Oracle Model). *A hash function H is $(\tau, \epsilon_{\mathsf{RO}})$-secure in the Random Oracle Model (RO) if, for any τ-time adversary $\mathcal{A}$, the maximum advantage $(\mathbf{Adv}^{\mathsf{H}}_{\mathcal{A}})$ in breaking any of the following three guarantees is bounded by $\epsilon_{\mathcal{RO}}$:*

1. ***One-Wayness (OW):*** *The probability that $\mathcal{A}$, given a random target output $y = \mathsf{H}(x)$, can find a preimage x' such that $\mathsf{H}(x') = y$:*

$$\mathbf{Adv}^{\mathsf{OW}}_{\mathcal{A}} = \Pr\left[\mathcal{A}(y) = x' \mid \mathsf{H}(x') = y\right] \leq \epsilon_{\mathsf{RO}}$$

2. ***Collision Resistance* (CR)*: The probability that $\mathcal{A}$ can output two distinct inputs $x_1 \neq x_2$ such that $H(x_1) = H(x_2)$:*

$$\mathbf{Adv}_{\mathcal{A}}^{CR} = \Pr[H(x_1) = H(x_2) \wedge x_1 \neq x_2] \leq \epsilon_{RO}$$

3. ***Output Pseudorandomness* (PR)*: The advantage of $\mathcal{A}$ in distinguishing the hash output $H(m)$ from a uniform random string R of the same length:*

$$\mathbf{Adv}_{\mathcal{A}}^{PR} = \left| \Pr\left[\mathcal{A}(H(m)) = 1\right] - \Pr\left[\mathcal{A}(R) = 1\right] \right| \leq \epsilon_{RO}$$

The symmetric encryption scheme is the core component that enables our rolling-state mechanism. By utilizing an IND-CPA secure scheme, the protocol can encrypt the credentials required for future proof verification. This ensures that state updates are indistinguishable from random noise to any external eavesdropper. Crucially, this primitive allows the prover to 'forward' the next secret state securely in every proof, which eliminates the need for the linear-time one-way function chains used in prior PoA constructions. This is what ultimately allows our protocol to achieve constant $\mathcal{O}(1)$ complexity.

Definition 3. *(IND-CPA **Secure Symmetric Encryption**) A symmetric encryption scheme (SKE), is indistinguishable under the chosen plaintext attack if for any $\tau-$time adversary $\mathcal{A}$, the maximum advantage ($\mathbf{Adv}_{\mathcal{A}}^{IND\text{-}CPA}$) in the following game is bounded by $\epsilon_{IND\text{-}CPA}$*

1. *$\mathcal{A}$ chooses two inputs M_0, M_1 at random and sends to the challenger $\mathcal{C}$.*
2. *$\mathcal{C}$ generates the ciphertexts (C_1, C_2) using the same key. It randomly samples the bit b to give C_b to $\mathcal{A}$.*
3. *$\mathcal{A}$ outputs bit b'.*

$$\mathbf{Adv}_{\mathcal{A}}^{IND\text{-}CPA} = \left| \Pr[\mathcal{A}(C_0) = 1] - \Pr[\mathcal{A}(C_1) = 1] \right| \leq \epsilon_{IND\text{-}CPA}$$

4 Security Framework and Threat Model

In this section, we define the formal requirements for a Privacy-Preserving Proof of Aliveness (3PoA) scheme. The primary requirement of a 3PoA scheme is to differentiate between an alive and honest device from a dead or non-operational device based on a timely response. Specifically, the device is deemed alive only if its proof satisfies the correctness property and is received within the specified time threshold. The inability of a non-operational device to successfully and timely generate a valid proof is formalized and enforced by the aliveness-unforgeability property in this section. Here, the goal of an attacker is to prove that a non-operational device as an alive device. Furthermore, a compromised Prover should not aid an adversary against honest Provers.

While the primary security concern is unforgeability, a sophisticated adversary may seek to violate the system's privacy and monitoring guarantees. These secondary objectives include determining a prover device's activity period (determine when it is alive) and learning the source identity of a proof (which node is alive).

Note: In this paper, we do not consider an adversary who can learn the secrets or actual identities of prover devices via physical access. Such attacks are classically mitigated by protocols like remote attestation or by fusing the hardware properties and software code, to achieve security against re-booting or device tampering. If an adversary succeeds to steal the secrets, then it can perpetually generate proofs without turning on the device.

4.1 Adversary Model

Our security framework considers a multi-entity environment. While the Control Center (CC) is fully trusted, the Verifier is considered semi-trusted i.e., it must be able to verify the continuous aliveness without learning the Prover's identity[1]. We consider a multi-tiered threat model as follows:

- *Malicious Eavesdropper* ($\mathcal{A}$): A network-level adversary capable of intercepting and injecting traffic. The goal of $\mathcal{A}$ is to link periodic proofs to track device movement or forge proofs for inactive devices.
- *Semi-Trusted Verifier* ($\mathcal{V}$): An intermediate entity that correctly performs verification but attempts to de-anonymize Provers. 3PoA must ensure that $\mathcal{V}$ gains no identity-revealing information.
- *Trusted Control Center* (CC): We assume the CC is a fully trusted entity that manages the root of trust and identity escrow.

It is crucial to note that the security and privacy guarantees provided by our scheme are confined to logical integrity of the protocol and the computational hardness of cryptographic problems. They do not extend the protection against breaches stemming from non-cryptographic failures. However, to mitigate the risk of compromise persistence, where an adversary might perpetually generate proofs by learning the current state of the Prover, 3PoA can be integrated with the persistent aliveness mechanisms presented in [5]. This ensures that while 3PoA maintains logical integrity and privacy during normal operations, the system remains resilient against long-term forgery resulting from physical state capture.

The following games formalize resistance against both external adversaries and curious network entities.

4.2 Aliveness-Unforgeability

The aliveness-unforgeability (AUF) property ensures that a non-operational or dead device cannot maintain a continuous illusion of being alive. A proof (pf')

[1] Note that, Verifier is expected to learn when a prover (whose identity is masked) did not send a valid proof within the aliveness threshold. Therefore, it is allowed for Verifier to learn if any two proofs originated from the same prover.

generated by an adversary ($\mathcal{A}$) is a forgery if it is either *generated without seeing the proof (pf) generated by an honest and alive device* or *different from the proof seen by the adversary i.e. $pf' \neq pf$*. It is worth noting that, network-level attack, where an adversary trivially relays the current valid proof is not a forgery since it is generated by an honest device that is alive. The security definition in [9,18] do not distinguish between a Relay Attack (network level) from a cryptographic forgery. A relay attack is not considered forgery in PoA because the event (legitimate prover being alive) actually happened, while forgery is an attempt to claim a device is alive when it is dead.

Definition 4. *(**$\mathbf{G}^{AUF}_{\Pi_{PoA}}$: Aliveness-Unforgeability Game for PoA**) A PoA protocol Π_{PoA} is (τ, ϵ_{AUF})-AUF secure if no τ-time adversary $\mathcal{A}$ has an advantage $\mathbf{Adv}^{AUF}_{\mathcal{A}}$ greater than ϵ_{AUF} in the following game $\mathbf{G}^{AUF}_{\Pi_{PoA}}$:*

1. *The challenger $\mathcal{C}$ runs Setup to simulate Verifier and a set of Provers $\mathcal{P} = \{P_1, \ldots, P_n\}$. It provides $\mathcal{A}$ with the system's public parameters.*
2. *$\mathcal{A}$ can make a polynomially bounded number of queries to the following oracles for any Prover P_i:*
   - *Proof Generation Query: $\mathcal{A}$ queries NextPf(P_i, T) to receive a valid proof for the window starting at time T.*
   - *Verification Query: $\mathcal{A}$ queries Verify(P_i, pf, T) to test the validity of any proof pf for P_i at time T.*
   - *Compromise Query [2]: $\mathcal{A}$ compromises a chosen set of Provers, $\mathcal{P}^* \subset \mathcal{P}$, and receives their current private state information.*
3. *Forgery: For the current aliveness window $[T, T + \delta_a]$ [3], $\mathcal{A}$ makes the query Verify(P, pf', T_c) corresponding to an honest, uncompromised Prover $P \notin \mathcal{P}^*$, and $T < T_c < T + \delta_a$.*

The adversary $\mathcal{A}$ wins if the challenger's Verify algorithm outputs 1 (accept) AND at least one of the following two conditions holds for the window $[T, T + \delta_a]$:

- *$\mathcal{A}$ did not query NextPf(P, T') for any T' such that $T < T' < T_c$.*
- *$\mathcal{A}$ did query NextPf(P, T') but the forged proof pf' is different from the legitimate proof pf returned by the oracle (i.e., $pf' \neq pf$).*

The advantage ($\mathbf{Adv}^{AUF}_{\mathcal{A}}$) of the adversary $\mathcal{A}$ against the AUF property of Π_{PoA} is the probability that $\mathcal{A}$ can convince the Challenger (simulating the Verifier) of a honest Prover's aliveness without a legitimate, fresh signal from that specific Prover. We define it as

$$\mathbf{Adv}^{AUF}_{\mathcal{A}} = \Pr[\mathcal{A} \text{ wins } \mathbf{G}^{AUF}_{\Pi_{PoA}}] \leq \epsilon_{AUF}.$$

[2] This query implies that, even if an attacker compromises a few devices in the system, it should not forge the proofs for uncompromised devices.

[3] δ_a is the aliveness threshold. The prover is considered un-alive, if the Verifier does not receive a valid proof during the window $[T, T + \delta_a]$.

4.3 Privacy Guarantees

PoA is said to be privacy-preserving, if an adversary cannot learn the alive/dead status of a Prover node and when it is alive/dead.

1. *Identity Anonymity:* Given a proof pf, the probability that any adversary (excluding the CC) correctly identifies the prover source is at most $\frac{1}{|\mathcal{P}|} + \epsilon$, where $\mathcal{P}$ is the set of all registered devices.
2. *Unlinkability:* Given two proofs pf_{t1} and pf_{t2} generated at different time intervals, it is *computationally hard* for an adversary to determine if they originated from the same Prover.

We define this formally below.

Definition 5. *(*$\mathbf{G}^{\mathsf{Anon}}_{\Pi_{\mathsf{PoA}}}$*: **Anonymity**) A PoA protocol is* $(\tau, \epsilon_{\mathsf{Anon}})$*-anonymous if no* τ*-time adversary* $\mathcal{A}$ *has an advantage* $\mathbf{Adv}^{\mathsf{Anon}}_{\mathcal{A}}$ *greater than* ϵ_{Anon} *in the following game* $\mathbf{G}^{\mathsf{Anon}}_{\Pi_{\mathsf{PoA}}}$*:*

1. *The challenger* $\mathcal{C}$ *runs* Setup *for an honest Prover with identity id. It runs* NextPf*,* $i-$*times using id to generate the proofs* $list.pf[i] = (pf_1, pf_2, \ldots, pf_i)$*.* $\mathcal{C}$ *randomly chooses a bit* $b \xleftarrow{R} \{0,1\}$*. It* $b = 0$*, it sets* $S = list.pf[i]$ *else* S *is set as* i *random tuples. It sends* S *to* $\mathcal{A}$*.*
2. $\mathcal{A}$ *observes the sequence* S *and outputs its guess for the identity,* b'*.*

The adversary wins if $b' = b$*. The advantage is defined as:*

$$\mathbf{Adv}^{\mathsf{Anon}}_{\mathcal{A}} = |\Pr[b' = b] - 1/2| \leq \epsilon_{\mathsf{Anon}}$$

Definition 6. *(*$\mathbf{G}^{\mathsf{UL}}_{\Pi_{\mathsf{PoA}}}$*: **Unlinkability**) A PoA protocol is* $(\tau, \epsilon_{\mathsf{UL}})$*-unlinkable if no* τ*-time adversary* $\mathcal{A}$ *has an advantage* $\mathbf{Adv}^{\mathsf{UL}}_{\mathcal{A}}$ *greater than* ϵ_{UL} *in the following game* $\mathbf{G}^{\mathsf{UL}}_{\Pi_{\mathsf{PoA}}}$*:*

1. *The challenger* $\mathcal{C}$ *prepares two distinct honest Provers,* P^a *and* P^b*, with identifiers* id^a *and* id^b*.*
2. $\mathcal{A}$ *receives the past* $i-1$ *proofs from both Provers:* S^a_{past} *and* S^b_{past}*.*
3. $\mathcal{C}$ *generates the i-th proof from both Provers:* pf^a_i *and* pf^b_i*. It tosses a random coin* $c \leftarrow \{0,1\}$*. It sends* $\mathcal{A}$ *the challenge pair* C*:* (pf^a_i, pf^b_i) *if* $c = 0$*, or* (pf^b_i, pf^a_i) *if* $c = 1$*.*
4. $\mathcal{A}$ *outputs its guess* c'*.*

The adversary wins if $c' = c$*. The advantage is defined as:*

$$\mathbf{Adv}^{\mathsf{UL}}_{\mathcal{A}} = |\Pr[c' = c] - 1/2| \leq \epsilon_{\mathsf{UL}}$$

Note: As per the privacy model, the Verifier can and is allowed to tracks the aliveness status of a specific (but anonymous) prover using session link, but an external adversary $\mathcal{A}$ should fail the Unlinkability game. This prevents semi-honest eavesdroppers from performing long-term traffic analysis or inventory mapping of critical infrastructure.

5 The 3PoA Scheme: Construction and Design

In this section, we present our Privacy-Preserving Proof of Aliveness (3PoA) scheme. 3PoA enables a Prover (remote device) to prove its aliveness status, using unlinkable proofs to a Verifier (the semi-trusted server).

5.1 System and Threat Modeling

The network consists of a trusted control center (CC), a semi-trusted server (acts as Verifier) and resource-constrained devices (act as Provers communicating over wireless network with the server). We consider a PPT adversary with full control over the communication channel, capable of interception, delay, modification, and replay of the proofs. However, it does not have physical or remote access to the Prover device. An external adversary against 3PoA has two goals: **a)** output a valid forgery on the proof or **b)** to break the privacy in the network.

5.2 Protocol Description

Let κ be the security parameter, $\mathsf{H} : \{0,1\}^* \rightarrow \{0,1\}^\kappa$ be a one-way and collision-resistant hash function modeled as Random Oracle and a symmetric encryption scheme $\mathsf{SKE} : \{0,1\}^\kappa \times \{0,1\}^\kappa \rightarrow \{0,1\}^\kappa$ be a IND-CPA secure symmetric encryption scheme. The 3PoA scheme has two phases: setup and online. In setup, CC enrols a new device by generating unlinkable identities and credentials to perform PoA. Online phase involves Prover and Verifier, without the involvement of CC.

5.3 One Time Trusted Setup and Initialization

The trusted CC runs the setup, between Prover (P), a remote device with identity id, and a Verifier (V) realized by a server, with aliveness threshold (δ_a) and online phase start time (T_s). It generates a hidden identity (HID) using a random salt for Prover anonymity. It randomly samples two strings $m_1, m_2 \xleftarrow{R} \{0,1\}^\kappa$ and hashes them with HID. $H_1 = \mathsf{H}(m_1 \| \text{HID})$ is used to verify the first proof. Then, it computes $C_2 = \mathsf{Enc}_{m_1}(H_2)$ to send the server, as a helper data to verify the next proof. To avoid linkability between two proofs, an unlinkable identity $\text{UID}_1 = \mathsf{H}(\text{HID} \| 1)$ is generated for first proof. Under the assumption that H is a one-way function, it should be noted that an adversary can learn the relation between HID and $\text{UID}_i = \mathsf{H}(\text{HID} \| i)$ in i−th session, only with negligible probability. It sets the credential for Prover as $\Gamma^P = \langle m_1, m_2 \rangle$ and Verifier as $\Gamma^{V,P} = \langle H_1, C_2 \rangle$. Their initial states are:

$$\sigma^P := \big\langle 1, \text{HID}, m_1, m_2, T_s, \delta_a \big\rangle \ , \ \sigma^{V,P} := \big\langle 1, \text{UID}_1, \text{HID}, H_1, C_2, T_s, \delta_a \big\rangle$$

Note that, the relation between id and HID is known only to the CC and the Prover node itself.

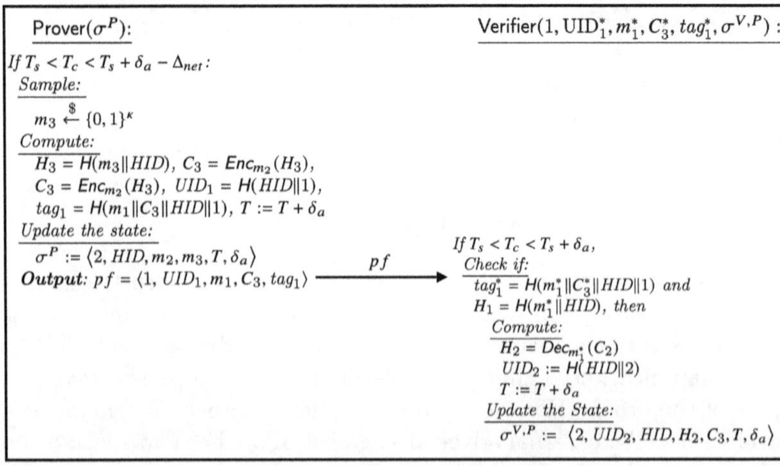

Fig. 1. Sequence of the 3PoA Online Phase. The Prover generates an unlinkable proof pf_i and updates its state σ^P at T_c, such that $T_s < T_c < T_s + \delta_a - \Delta_{net}$, where Δ_{net} is the network delay: the Verifier validates the proof within the aliveness window δ_a and evolves the session alias UID_{i+1}

5.4 Online Phase

The online phase, is described as the time period where Prover node is expected to be alive and running. It is divided into discrete intervals of length δ_a time. Figure 1, describes the generation and verification of the first proof. Here, we generalize it for i–th proof generated during $[T_s + (i-1)\delta_a,\ T_s + i\delta_a]$. The states of Prover and Verifier are $\sigma^P = \langle i, \text{HID}, m_i, m_{i+1}, T_s + (i-1)\delta_a, \delta_a \rangle$ and $\sigma^{V,P} = \langle i, \text{UID}_i, \text{HID}_i, H_i, C_{i+1}, T_s + (i-1)\delta_a, \delta_a \rangle$, where $H_i = \mathsf{H}(m_i \| \text{HID})$.

Proof Generation. It randomly samples m_{i+2} and computes credential for Verifier as $C_{i+2} = \mathsf{Enc}_{m_{i+1}}(\mathsf{H}(m_{i+1} \| \text{HID}))$. Prover computes the unlinkable identity $\text{UID}_i = \mathsf{H}(\text{HID} \| i)$. To maintain integrity against data manipulation, the Prover computes $tag_i = \mathsf{H}(m_i \| C_{i+2} \| \text{HID} \| i)$.

It sends the proof as $pf_i = \langle i, \text{UID}_1, m_i, C_{i+2}, tag_i \rangle$ before $T_s + i\delta_a - \Delta_{net}$[4]. The updated state is $\sigma^P = \langle i, \text{HID}, m_{i+1}, m_{i+2}, T_s + (i\delta_a) \rangle$.

Proof Verification. On receiving $\langle i^*, \text{UID}_i^*, m_i^*, C_{i+2}^*, tag_i^* \rangle$ at time T_c, the verifier retrieves the entry $\langle i, \text{UID}_i, \text{HID}, H_i, C_{i+1}, T, \delta_a \rangle$ from its state where $i^* = i$, $\text{UID}^* = \text{UID}$ and $T < T_c < T + \delta_a$. If $tag_i^* = \mathsf{H}(m_i^* \| C_{i+2}^* \| \text{HID} \| i^*)$ and $H_i = \mathsf{H}(m_i^* \| \text{HID})$, it accepts the proof as valid.

Then, it computes $H_{i+1} = \mathsf{Dec}_{m_i}(C_{i+1})$, $\text{UID}_2 = \mathsf{H}(\text{HID} \| (i+1))$. Finally, the Verifier's state is updated to $\sigma^{V,P} = \langle (i+1), \text{UID}_2, \text{HID},\ H_{i+1}, C_{i+2}^*, (T_s + i\delta_a), \delta_a \rangle$. If a valid i–th proof for Prover (with hidden identity HID) is not received within $T_s + i\delta_a$, the Verifier sets it as inactive/dead.

[4] We include the network delay (Δ_{net}) within aliveness-window δ_a

In the next section, we discuss the security and privacy guarantees achieved by 3PoA. Later, we present a detailed comparison of 3PoA with seminal PoA protocol in terms of design, efficiency and overheads.

6 Privacy and Security Analysis

We prove that 3PoA is aliveness unforgeable with prover anonymity (against all adversaries) and proof unlinkability (against external adversaries). For a Prover P^x ($x \in \{a,b\}$), with identity id^x, we define hidden identity HID^x and unlinkable identity UID^x. We denote the $i-$th proof of P^x as pf_i^x and the first i proofs as $list.pf^x[i]$. We denote the compromised Prover as P^*, with HID^*, UID^* and pf_i^* and $list.pf^*[i]$ corresponding notations. Similarly we define $R_i = \langle i,u,v,w,x\rangle$ as a tuple of random strings for $i-$th instance, sampled as $u,v,w,x \xleftarrow{R} \{0,1\}^\kappa$. The list $R_1,\cdots,R_i$ is denoted as $list.R[i]$.

We discuss the privacy and security of 3PoA below. The required Lemmas are provided in Appendix A. We model the security of 3PoA through a series of games, $G_0, G_1, \ldots, G_n$, where G_0 represents the real-world execution of the protocol and each subsequent game introduces a small, indistinguishable transition to a simulated environment.

Definition 7. *The 3PoA scheme described in Sect. 5.2 is anonymous (as per Definition 5), unlinkable (as per Definition 6) and secure (as per Definition 4) PoA protocol, if underlying hash function* H *is modelled as Random Oracle (as per Definition 2) and* SKE *is* $\mathsf{IND\text{-}CPA}-$*secure symmetric key encryption scheme (as per Definition 3).*

We prove security and privacy of 3PoA in the following theorems.

Theorem 1. *The 3PoA scheme is* $(\tau, \epsilon_{Anon})-$*anonymous against a* $\tau-$*time adversary in the game* $\mathbf{G}^{Anon}_{\Pi_{PoA}}$, *where* $\epsilon_{Anon} = K \cdot (2\epsilon_{RO} + \epsilon_{IND\text{-}CPA})$.

Proof. The goal of the adversary against anonymity of 3PoA is to learn the hidden identity (HID) of an honest prover. We show that the advantage is same if $\mathcal{A}$ sees $list.pf[K]$ or $list.R[K]$. From Lemma 1 (Appendix A), we can conclude

$$\mathbf{Adv}^{\mathsf{Anon}}_{\mathcal{A}} = \left| \Pr\left[\mathcal{A}(list.pf[K]) = 1\right] - \Pr\left[\mathcal{A}(list.R[K]) = 1\right]\right| \leq K \cdot (2\epsilon_{\mathsf{RO}} + \epsilon_{\mathsf{IND\text{-}CPA}})$$

Theorem 2. *The 3PoA scheme is* $(\tau, \epsilon_{UL})-$*unlinkable against a* $\tau-$*time adversary in the game* $\mathbf{G}^{UL}_{\Pi_{PoA}}$, *where* $\epsilon_{UL} = 4(2\epsilon_{RO} + \epsilon_{IND\text{-}CPA})$.

Proof. For two provers with identities id^a and id^b, the $\mathbf{Adv}^{\mathsf{UL}}_{\mathcal{A}}$ against $\mathbf{G}^{\mathsf{UL}}_{\Pi_{\mathsf{PoA}}}$ is given as

$$\mathbf{Adv}^{\mathsf{UL}}_{\mathcal{A}} = \left| \Pr\left[\mathcal{A}\binom{list.pf^a[K]}{list.pf^b[K]} = 1\right] - \Pr\left[\mathcal{A}\binom{list.pf^a[K-1], pf^b_K}{list.pf^b[K-1], pf^a_K} = 1\right]\right|$$

From Lemma 2 (Appendix A), we can conclude that $\mathbf{Adv}^{\mathsf{UL}}_{\mathcal{A}} \leq 4(2\epsilon_{\mathsf{RO}} + \epsilon_{\mathsf{IND\text{-}CPA}})$.

Theorem 3. *The 3PoA scheme is* (τ, ϵ_{AUF}) *aliveness unforgeable against a* $\tau-$ *time adversary in the game* $\mathbf{G}^{AUF}_{\Pi_{PoA}}$, *where* $\epsilon_{AUF} = (\epsilon_{ro} + i - 1) \cdot (2\epsilon_{RO} + \epsilon_{IND\text{-}CPA})$.

Proof. The adversary can a output valid forgery $pf_i^* = \langle \mathrm{UID}_i^*, m_i^*, C_{i+2}^*, tag_i^* \rangle$ on Prover (P) in two scenarios: $\mathcal{A}$ has seen $list.pf[i-1] = \langle pf_1, \cdots, pf_{i-1} \rangle$ or $\mathcal{A}$ has seen $list.pf[i]$ and $pf_i \neq pf_i^*$. The advantage of $\mathcal{A}$ is:

$$\mathbf{Adv}^{\mathsf{AUF}}_{\mathcal{A}} = \Pr\left[\mathcal{A}\ outputs\ pf_i^* \middle| list.pf[i-1]\right] + Pr\left[\mathcal{A}\ outputs\ pf_i^* \neq pf_i \middle| list.pf[i]\right]$$

- $\Pr\left[\mathcal{A}\ outputs\ pf_i^* \mid list.pf[i-1]\right]$: In Lemma 1 (Appendix. A), we computed the advantage to distinguish between $list.pf[i]$ and random $list.R[i-1]$. Therefore, the advantage $\mathcal{A}$ would have to output a valid forgery pf_i^*, when it sees $list.pf[i-1]$, instead of $list.R[i-1]$ is $(i-1)(2\epsilon_{\mathsf{RO}} + \epsilon_{\mathsf{IND\text{-}CPA}})$.
- $\Pr\left[\mathcal{A}\ outputs\ pf_i^* \neq pf_i \mid list.pf[i]\right]$: $\mathcal{A}$ sets $\mathrm{UID}_i^* = \mathrm{UID}_i$, so Verifier can retrieve the state $\sigma^{V,P} = \langle i, \mathrm{UID}_i, \mathrm{HID}, H_i, C_{i+1} \rangle$. Next, the verifiers checks if $\mathsf{H}(m_i^*) = H_i$, and $tag_i^* = \mathsf{H}(m_i^* || C_{i+2}^* || \mathrm{HID} || i)$. For this, $\mathcal{A}$ should find a collision on H_i and output tag_i^* that is valid. It can output tag_i^* only by querying H with $(m_i^* || C_{i+2}^* || \mathrm{HID} || i)$ during the game. To do this, $\mathcal{A}$ should learn HID from pf_i, whose advantage is $2\epsilon_{\mathsf{RO}} + \epsilon_{\mathsf{IND\text{-}CPA}}$ as per Lemma. 1 (Appendix. A). The total probability in this case is $\epsilon_{\mathsf{RO}} \cdot (2\epsilon_{\mathsf{RO}} + \epsilon_{\mathsf{IND\text{-}CPA}})$.

Thus, the advantage $\mathcal{A}$ to output a valid forgery is $\mathbf{Adv}^{\mathsf{AUF}}_{\mathcal{A}} \leq (\epsilon_{\mathsf{RO}} + i - 1) \cdot (2\epsilon_{\mathsf{RO}} + \epsilon_{\mathsf{IND\text{-}CPA}})$.

Corollary 1. *3-PoA is a private and secure protocol under the assumption that the underlying* H *is one-way and collision resistant hash function modelled in random-oracle model and* SKE *is a* IND-CPA *secure symmetric encryption scheme.*

7 Comparison and Performance Evaluation of 3PoA

In this section, we first discuss the various construction proposed in earlier works. Then, we present performance and efficiency of 3PoA while comparing with prior works. We use SHA3-256 as H and AES-256 as symmetric encryption scheme implemented in C for our 3PoA scheme. We run the prover on a Raspberry Pi 3 board (similar to [9,18]), while the verifier is run on a PC.

7.1 Comparison with the existing PoA schemes

The PoA schemes in [9,18] are built in (STD) and Random oracle (RO) models using one-way functions (OWF), pseudorandom generators (PRG), Bloom Filter (BF). Π_{OWF} is the seminal construction of PoA using OWF chains as discussed in Sect. 1. During the setup chooses a random x_N and generates a N-length OWF chain $x_N, \ldots, x_0$ In their protocol, the Prover stores x_N and the Verifier stores x_0 and the proofs are generated in reverse order from x_0 to x_{N-1}.

Table 2. Comparison of Overheads due to Replenishment and Setup, and Privacy Properties on Raspberry Pi 3 Board.

		Replenishment Overheard†		
Scheme	Setup Cost	Computation Complexity	Time	Privacy
Π_{OWF} RO [9] Storage Sufficient	$\mathcal{O}(N)$	10H	0.011 msec	No
Π_{OWF} RO [18] Storage Constrained				
$\Pi^{\mathsf{PRG}}_{\mathsf{OWF}}$ − BF RO [18] Storage Constrained	$\mathcal{O}(N)$	11H		
3PoA	$\mathcal{O}(1)$	0	0	Yes

† This is performed per proof generation operation.

To generate the i-th proof, the Prover makes $N - i$ calls to the OWF, leading to $\mathcal{O}(N^2)$ OWF operations, to generate N proofs. To avoid exponential overheads resulting from larger chains, $\Pi^{\mathsf{PRG}}_{\mathsf{OWF}}$ is proposed. It has multiple ($\eta = \sqrt{N}$) OWF chains, each invoking η-length Π_{OWF}. Consequently to reduce the storage corresponding to OWF chain heads, they use Bloom Filters (BF) [1], to build $\Pi^{\mathsf{PRG}}_{\mathsf{OWF}}$BF. Although, BF helps reduce the storage cost compared to earlier schemes, it is still sublinear on the number of proofs generated. Thus, the chain length should still be carefully chosen to avoid storage-computation tradeoffs.

The auto-replenishment scheme (CommitRep in RO model) based on hash-based commitments are proposed to increase the lifespan of the device. But, it requires the Prover to generate a fresh commitment during each proof generation. To reduce auto-replenishment cost, they recommend reducing its frequency, by increasing N, which ultimately leads to the original exponential and sub-linear overheads either of computation or storage. While the proposed schemes exhibit a range of trade-offs between desirable features and imposed limitations, there is no one-shot solution, shifting the choice to the user. Due to the heterogeneous nature of the existing schemes and their varying design constraints, the comparative analysis is focused on *optimality per metric*, where schemes are individually evaluated and compared on a specific feature.

The seminal works, attempt to reduce the computation and storage overheads using various techniques, and provide curated protocols depending on the available resources. However, in this paper we successfully achieve $\mathcal{O}(1)$ computation and storage, using only hash and symmetric encryption blocks. We completely eliminate the need for OWF chains, sub-chains and Bloom Filters. This proposed 3PoA protocol is suitable for network, where the prover devices are storage or computation constrained. Additionally, the verifier and prover does not have to keep additional databases to keep track of the OWF chains and perform auto-replenishment. Additionally, our protocol also supports privacy-preservation in terms of prover identity anonymization and unlinkability of the proofs.

7.2 Benchmarking 3PoA against Competing Protocols

For our evaluation with existing PoA protocols, we chose the most efficient competitors in two critical domains: Π_{OWF} RO with limited storage space (high computation time) and $\Pi^{\mathsf{PRG}}_{\mathsf{OWF}}$ – BF RO with sufficient storage capabilities. In conjunction with CommitRep, they outperform the other schemes proposed in [18].

In Table 2, we compare the schemes based on their setup cost, overheads incurred due to replenishment requires and privacy provided. Since our 3PoA scheme does not use OWF chains, it avoids the linear setup cost and replenishment schemes. The chain-based Π_{OWF} RO and $\Pi^{\mathsf{PRG}}_{\mathsf{OWF}}$ – BF RO protocols have a $\mathcal{O}(N)$ computation cost due to setup. If a prover node is compromised, the re-enrollment process (while avoiding collisions with the existing OWF-chains) takes considerable amount of time and resources. CommitRep operation is performed per proof generation operation, resulting in additional cost during proof generation.

We compare the performance of 3PoA against existing schemes in Table. 3. We evaluate 3PoA on Raspberry Pi 3, similar to [18]. To ensure a fair comparison between the proposed $\mathcal{O}(1)$ scheme and OWF chain-based protocols (which scale with the total number of proofs N), we established a standardized operational lifetime for the evaluation. Unlike 3PoA, where proof generation time remains constant, the overhead in prior works like Π_{OWF} and $\Pi^{\mathsf{PRG}}_{\mathsf{OWF}} - BF$ is intrinsically tied to the device's lifespan and frequency of proof generation. We assume a one-year device lifespan with a 10-second heartbeat interval, resulting in a total proof requirement of $N \approx 3,153,600$. This N is critical because:

- In storage constrained Π_{OWF}: The prover must recompute the chain from a stored head-node for each proof, leading to a linear $\mathcal{O}(N)$ computation time. For the assumed N, this results in an average proof generation time of 5.89 seconds, which we derived by multiplying the average number of hash operations ($N/2$) by the unit hash cost measured on the Raspberry Pi 3, which is reported as 3.74μ seconds in [18].
- In $\Pi^{\mathsf{PRG}}_{\mathsf{OWF}} - BF$: The complexity scales at $\mathcal{O}(\sqrt{N})$, which, for our assumed lifespan, results inp approximately 1,775 hash operations per proof, or 6.64 milliseconds.

For storage-sufficient Π_{OWF} schemes, proof generation is a constant memory-fetch operation taking 3.74μ seconds yet this requires prohibitive $\mathcal{O}(N)$ storage, which is approximately 100 MB on the resource-constrained IoT device. In contrast, the proposed 3PoA protocol achieves true simultaneous $\mathcal{O}(1)$ complexity for setup, storage, and computation by replacing static chains with a rolling-state forwarding mechanism that performs only 3 hashes and 1 symmetric encryption for each proof generation (0.1 milliseconds), thus breaking the storage-computation trade-off and eliminating the need for periodic replenishment cycles.

While 3PoA was directly measured on a Raspberry Pi 3, the metrics for competing schemes were analytically derived from [18]. This methodology was necessary because a single direct measurement of an OWF-chain protocol (e.g.,

Table 3. Comparison of Performance Metrics on Raspberry Pi 3 Board.

Scheme	Proof Generation Time (Avg/Worst)	Computation Complexity	Communication Cost	Storage on Prover	Bloom Filter
Π_{OWF} RO [9] Storage Sufficient	0.00374 msec	1 OWF	$\mathcal{O}(1)$	$\mathcal{O}(N)$	No
Π_{OWF} RO [18] Storage Constrained	5890 msec$^{\#}$	$\mathcal{O}(N)$·OWF‡		$\mathcal{O}(1)$	No
$\Pi^{\mathsf{PRG}}_{\mathsf{OWF}}$ − BF RO[18] Storage Constrained	6.64 msec‡	1 PRG + $\mathcal{O}(\sqrt{N})$·OWF		4κ+ $1.44\epsilon\eta$	Yes
3PoA	**0.1** msec	3H + Enc +sampling	$\mathcal{O}(1)$	$\mathcal{O}(\mathbf{1})$	No

\# Calculated for $N = 3,153,600$ (1-year lifespan, 10s interval). On average, $N/2$ hash operations are performed.
‡ Complexity scales with $\sqrt{N}$. For $N \approx 3.15 \times 10^6$, $\sqrt{N} \approx 1775$ hash operations.
† κ is the security parameter as discussed in Section. 5.2.

generating only the first few proofs) would fail to capture the progressive performance degradation that occurs as the chain is exhausted. By calculating the average overhead over a fixed N, we provide a more realistic representation of the total resource burden an edge device would face over its actual operational life.

8 Conclusion

In this work, we proposed 3PoA: a Privacy-Preserving Proof of Aliveness protocol designed to secure critical IoT and CPS infrastructure. We first established rigorous formal framework to define proof Aliveness-Unforgeability, device-identity Anonymity and proof unlinkability, thus addressing the privacy gap in existing liveness monitoring literature. By replacing traditional one-way function chains with a rolling-state forwarding mechanism, 3PoA becomes the first scheme to achieve simultaneous $\mathcal{O}(1)$ constant overhead across all critical metrics: setup, storage, computation and communication. Our comparative evaluation on Raspberry Pi 3 hardware demonstrates that 3PoA overcomes the storage-computation trade-off present in earlier schemes, maintaining a constant proof generation time of 100 μsec, regardless of the device's lifespan. Formal security proofs confirm that 3PoA effectively *hides the heartbeat* of remote nodes, ensuring that external adversaries cannot link or de-anonymize uncompromised Provers. Ultimately, 3PoA provides a robust architecture that satisfies the stringent performance and privacy requirements of modern industrial deployments.

Acknowledgments. The research is supported by the National Research Foundation, Singapore, under its National Satellite of Excellence Programme "Design Science and Technology for Secure Critical Infrastructure: Phase II" (Award No: NRF-NCR25-NSOE05-0001). Any opinions, findings and conclusions or recommendations expressed in this material are those of the author(s) and do not reflect the views of National Research Foundation, Singapore.

A Appendix

In this section, we describe and prove the Lemmas required in Section. 6. We model the security of 3PoA through a series of games, $G_0, G_1, \ldots, G_n$, where G_0 represents the real-world execution of the protocol and each subsequent game introduces a small, indistinguishable transition to a simulated environment.

Lemma 1. *Given* H^{RO}*, and* $SKE^{IND\text{-}CPA}$*, the advantage* ($Adv_{\mathcal{A}}^{(pf,R,K)}$) *of a PPT adversary* ($\mathcal{A}$) *that can distinguish between* $list.pf[K]$ *and* $list.R[K]$*, where* $K \geq 3$ *is no more than* $\epsilon_{(pf,R)} = K \cdot (2\epsilon_{RO} + \epsilon_{IND\text{-}CPA})$.

Proof. We prove this using hybrid arguments. We use a Simulator ($\mathcal{S}$), to setup P with HID.
Real Game ($G_0 = \langle pf_1, pf_2, \cdots, pf_K \rangle$)**:** The adversary $\mathcal{A}$ interacts with $\mathcal{S}$ to receive K real proofs $list.pf[K]$.
Ideal Game ($G_K = \langle R_1, R_2, \cdots, R_K \rangle$)**:** The adversary interacts $\mathcal{S}$ to receive a K simulated random proofs $listR[K]$).
For $i \geq 1$, we define a simulated proof $R_i = \langle u_i, v_i, w_i, x_i \rangle$, where the strings u_i, v_i, x_i are sampled randomly from $\{0,1\}^\kappa$. We sample w_i randomly from the ciphertext space, which is also $\{0,1\}^\kappa$.
Game ($G_i = \langle R_1, \cdots, R_i, pf_{i+1}, \cdots, pf_K \rangle$)**:** In this intermediate game, $\mathcal{A}$ receives i-simulated proofs and $K - i$ real proofs.
So, $\mathcal{A}$'s advantage $\mathbf{Adv}_{\mathcal{A}}[list(pf, R, K)]$ is the sum of the advantages gained in each transition:

$$\mathbf{Adv}_{\mathcal{A}}[pf, R, K] = \left| \Pr[\mathcal{A} \text{ wins in } G_0] - \Pr[\mathcal{A} \text{ wins in } G_K] \right|$$

Using the triangle inequality, this is bounded by the sum of individual game differences:

$$\mathbf{Adv}_{\mathcal{A}}[pf, R, K] = \sum_{i=0}^{K-1} \left| \Pr[\mathcal{A} \text{ wins in } G_i] - \Pr[\mathcal{A} \text{ wins in } G_{i+1}] \right|$$

We define ϵ_i as the advantage in i–th transition. We first compute the advantage (ϵ_1) of $\mathcal{A}$ to distinguish between G_0 and G_1. Given $\langle pf_2, \cdots, pf_K \rangle$, we find the advantage $\mathcal{A}$ would have when it sees pf_1 compared to random string (R_1). We compute it using the following hybrid arguments.

$$\langle \mathrm{UID}_1, m_1, C_3, tag_1 \rangle \xrightarrow{\epsilon_{RO}} \langle u_1, m_1, C_3, tag_1 \rangle \xrightarrow{0}$$
$$\langle u_1, v_1, C_3, tag_1 \rangle \xrightarrow{\epsilon_{IND\text{-}CPA}} \langle u_1, v_1, w_3, tag_1 \rangle \xrightarrow{\epsilon_{RO}} \langle u_1, v_1, w_3, x_1 \rangle$$

Therefore, $\epsilon_1 = 2\epsilon_{RO} + \epsilon_{IND\text{-}CPA}$.
We show the advantage (ϵ_2) of $\mathcal{A}$ to distinguish between G_1 and G_2. Since R_1 is a simulated proof, it provides zero advantage to $\mathcal{A}$. So, ϵ_2 is the advantage to distinguish between $pf_2, \cdots, pf_K$ and $R_2, pf_3, \cdots, pf_K$, which is same as ϵ_1.

Since, there are K steps, the total advantage is:

$$\mathbf{Adv}_{\mathcal{A}}[pf, R, K] = \sum_{i=1}^{K} \epsilon_1 \leq K \cdot (2\epsilon_{\mathsf{RO}} + \epsilon_{\mathsf{IND\text{-}CPA}})$$

Lemma 2. *Given* H^{RO}, *and* $\mathsf{SKE}^{\mathsf{IND\text{-}CPA}}$, *the advantage* ($Adv_{\mathcal{A}}^{pf^a,pf^b}$) *of a PPT adversary* ($\mathcal{A}$) *that can distinguish between*

$$\langle list.pf^a[K], list.pf^b[K]\rangle \text{ and } \langle (list.pf^a[K-1], pf_K^b), (list.pf^b[K-1], pf_K^a)\rangle$$

for provers P^a *and* P^b *and* $i \geq 1$, *is no more than* $\epsilon_{pf^a,pf^b} = 4(2\epsilon_{RO} + \epsilon_{cpa})$.

Proof. We use a Simulator ($\mathcal{S}$), to setup P^a and P^b with HID_A and HID_B.
Real Game $\Big(G_{real}$ **or** $G_0 = \Big\langle \big(list.pf^a[K-1], pf_K^a\big), \big(list.pf^b[K-1], pf_K^b\big)$**:** The adversary $\mathcal{A}$ interacts with $\mathcal{S}$ to receive K proofs of P^a and P^b.
Ideal Game $\Big(G_{ideal} = \langle (list.pf^a[K-1], pf_K^b), (list.pf^b[K-1], pf_K^a)\rangle$**:** The adversary $\mathcal{A}$ interacts with $\mathcal{S}$ to receive first pair: $(K-1)$ proofs of P^a, $K-$th proof of P^b and second pair: $(K-1)$ proofs of P^b, $K-$th proof of P^a.
We define R and S are the two simulated proofs. We define the intermediate games as follows:
Game $G_1 = \Big\langle \big(list.pf^a[K-1], pf_K^a\big), \big(list.pf^b[K-1], S\big)\Big\rangle$**:** Instead of pf_K^b, $\mathcal{A}$ receives a simulated proof S.
Game $G_2 = \Big\langle \big(list.pf^a[K-1], R\big), \big(pf^b[K-1], S\big)\Big\rangle$**:** Instead of pf_K^a, $\mathcal{A}$ receives a simulated proof R.
Game $G_3 = \Big\langle \big(list.pf^a[K-1], R\big), \big(pf^b[K-1], pf_K^a\big)\Big\rangle$**:** $\mathcal{S}$ sends pf_K^a instead simulated proof S.

Finally in G_{ideal} or G_4, $\mathcal{S}$ sends pf_K^a instead simulated proof S.
So, $\mathcal{A}$'s advantage $Adv_{\mathcal{A}}^{pf^a,pf^b}$ is the sum of the advantages gained in each transition:

$$Adv_{\mathcal{A}}^{pf^a,pf^b} = \big|\Pr[\mathcal{A} \text{ wins in } G_0] - \Pr[\mathcal{A} \text{ wins in } G_4]\big|$$

Using the triangle inequality, this is bounded by the sum of individual game differences:

$$Adv_{\mathcal{A}}^{pf^a,pf^b} = \sum_{i=0}^{3} \big|\Pr[\mathcal{A} \text{ wins in } G_i] - \Pr[\mathcal{A} \text{ wins in } G_{i+1}]\big|$$

From Lemma 1, we can say the advantage in $i-$th transition is bounded by $2\epsilon_{\mathsf{RO}} + \epsilon_{cpa}$. So the total advantage is: $Adv_{\mathcal{A}}^{pf^a,pf^b} \leq 4(2\epsilon_{\mathsf{RO}} + \epsilon_{cpa})$

References

1. Bloom, B.H.: Space/time trade-offs in hash coding with allowable errors. Commun. ACM **13** (1970). https://doi.org/10.1145/362686.362692
2. Brasser, F., Rasmussen, K.B., Sadeghi, A.R., Tsudik, G.: Remote attestation for low-end embedded devices: the prover's perspective. In: Proceedings of the 53rd Annual Design Automation Conference. DAC (2016). Association for Computing Machiner. https://doi.org/10.1145/2897937.2898083
3. Brickell, E.F., Camenisch, J., Chen, L.: Direct anonymous attestation. In: Proceedings of the 11th ACM Conference on Computer and Communications Security. CCS (2004). https://doi.org/10.1145/1030083.1030103
4. Brusò, M., Chatzikokolakis, K., Etalle, S., den Hartog, J.: Linking unlinkability. In: Trustworthy Global Computing - 7th International Symposium, TGC. LNCS, Springer (2012). https://doi.org/10.1007/978-3-642-41157-1_9
5. Cao, X., Yang, Z., Ning, J., Jin, C., Liu, Z., Zhou, J.: Proof of persistent aliveness. IEEE Trans. Dependable Secure Comput.,D 1–17 (2026). https://doi.org/10.1109/TDSC.2026.3656357
6. Chaum, D., van Heyst, E.: Group signatures. In: Advances in Cryptology - EUROCRYPT '91, Workshop on the Theory and Application of of Cryptographic Techniques. Springer (1991). https://doi.org/10.1007/3-540-46416-6_22
7. Cybersecurity, Agency, I.S.: ICS Advisory: Treck TCP/IP Stack (Update I) (2024). https://www.cisa.gov/news-events/ics-advisories/icsa-20-168-01TreckTCP/IPStack(UpdateI)
8. Haller, N.: RFC1760: The s/key one-time password system (1995)
9. Jin, C., Yang, Z., van Dijk, M., Zhou, J.: Proof of aliveness. In: Proceedings of the 35th Annual Computer Security Applications Conference, ACSAC. ACM (2019). https://doi.org/10.1145/3359789.3359827
10. Juels, A., Jr., B.S.K.: Pors: proofs of retrievability for large files. In: Ning, P., di Vimercati, S.D.C., Syverson, P.F. (eds.) Proceedings of the 2007 ACM Conference on Computer and Communications Security, CCS (2007). https://doi.org/10.1145/1315245.1315317
11. Kogan, D., Manohar, N., Boneh, D.: T/Key: second-factor authentication from secure hash chains. CoRR abs/1708.08424 (2017). http://arxiv.org/abs/1708.08424
12. Lamport, L.: Password authentication with insecure communication. Commun. ACM (1981). https://doi.org/10.1145/358790.358797
13. OpenAI: What to know about a recent Mixpanel security incident. https://openai.com/index/mixpanel-incident/
14. Pfitzmann, A., Köhntopp, M.: Anonymity, Unobservability, and Pseudonymity - A Proposal for Terminology. Springer (2001). https://doi.org/10.1007/3-540-44702-4_1
15. Robert M. Lee, Michael J. Assante, T.C.: TLP: White analysis of the cyber attack on the Ukrainian power grid (2016). https://nsarchive.gwu.edu/sites/default/files/documents/3891751/SANS-and-Electricity-Information-Sharing-and.pdf
16. dos Santos, D., Dashevskyi, S., Wetzels, J., Amri, A.: How embedded TCP/IP stacks breed critical vulnerabilities (2020)

17. Steinbrecher, S., Köpsell, S.: Modelling unlinkability. In: Dingledine, R. (ed.) Privacy Enhancing Technologies, Third International Workshop, PETs. Springer (2003). https://doi.org/10.1007/978-3-540-40956-4_3
18. Yang, Z., Jin, C., Cao, X., van Dijk, M., Zhou, J.: Optimizing proof of aliveness in cyber-physical systems. IEEE Trans. Dependable Secur. Comput. (2024). https://doi.org/10.1109/TDSC.2023.3335188

A Hybrid Twin Model for Evaluating Cyber-Physical Attacks Against Freight Rail Networks

Courtney Falk(✉), Dongyan Xu, Nathan Yao, Max Yu Zhen Chen, and Kira Qing Sun

CERIAS, Purdue University, West Lafayette, IN 47906, USA
{falkc,dxu,yao293,chen4724,sun1204}@purdue.edu

Abstract. Freight rail is a critical infrastructure whose size and scale make it difficult to evaluate how the multiple systems involved will interact. Validating the cybersecurity posture of a freight rail system is one such example. A hybrid twin model aids cybersecurity researchers by virtualizing some components of the rail system while building a physical simulation of other portions. Our freight rail hybrid twin implements the rail and rolling stock in model train form. Critical information technology and operational technology systems are virtualized and connected to the physical model train to help researchers understand and visualize the inter-system dynamics of cyberattacks. In our case, the researchers will use large language models to design the cyberattack with the hybrid twin model being used to validate the proposed attack.

Keywords: Critical infrastructure · Freight rail · Hybrid models · Cyberattacks

1 Introduction

Our paper describes the implementation of a hybrid physical model. This model is used to study freight rail infrastructure. We discuss the feasibility of using large language models (LLMs) for generating cyberattacks. Then we conclude by showcasing a few of the cyberattacks produced using data from our hybrid model.

Freight rail comprises the backbone of transportation in the United States. The Cybersecurity and Infrastructure Security Agency (CISA) includes freight rail as one of their sixteen critical infrastructure sectors. When it comes to security assessments, freight rail is difficult to evaluate due to its high initial cost outlay and the large amount of space required for a fully-featured rail system. A hybrid twin model addresses these problems by providing a test environment that is less expensive to implement and allows for quick and easy resets after a disaster event.

Our project implements a hybrid twin for freight rail using a combination of commercial and open source solutions designed for model train enthusiasts.

S. Zonouz et al. (Eds.): DCS-CI 2026, CCIS 3109, pp. 136–145, 2027.
https://doi.org/10.1007/978-981-92-5039-4_7

This decision serves to keep implementation costs down and to build upon the community experience that these mature tools provide. The model train components form the physical portion of the hybrid twin. These miniature parts also model the operational technology (OT) side of the model. The information technology (IT) side of the model is implemented on commercial servers hosting virtual machines (VMs).

The IT systems in our model were deliberately implemented with security weaknesses in place to allow attackers ways to either gain access to systems or otherwise interfere with communications channels. The specific weaknesses and their associated attack techniques are discussed in more detail below. Because these IT systems are virtualized, it becomes trivial to change them out with hardened versions for assessing more capable adversaries.

With our low-risk hybrid twin model we are able to evaluate cyberattacks on the IT/OT systems of a freight rail network. Our goal is to employ large language models (LLMs) in the role of the adversary. The LLMs will be provided with documentation on the model and allowed to suggest a plan of attack.

2 Prior Work

There are two main areas of prior work we considered when building our model. First were attacks (both cyber and physical) on train networks themselves. We consulted with the Surface Transportation Information Sharing and Analysis Center (ST-ISAC) for insights into the types of cybersecurity challenges that freight rail providers face. Unfortunately, the ST-ISAC has since ceased operation, leaving the Association of American Railroads as the best-suited point of contact. The closure of a dedicated organization for sharing cyber threat intelligence complicates the work of defenders and researchers alike.

2.1 Rail Attacks

Railroads have historically been a vulnerable target. They are built upon hundreds of miles of fixed track traversing a wide variety of geography. The trains themselves run at regular intervals. This allows an adversary choose the time and place of an attack. Signals and communications systems used to facilitate rail operations add another layer on top of this attack surface.

If there exists a common theme to the vulnerabilities in rail systems it would be security by obscurity. What are now understood to be secure and robust protective measures are often absent in rail systems. In łŁódź, Poland a teenager was able to build a device that manipulated the infrared signals used by the city's tram system to switch the position of rail turnouts [14]. The result was tram car collisions and passenger injuries.

Ibadah, Benavente-Peces, and Pahl assemble a widely ranging review of cyberattacks against rail networks in their 2024 paper on wholistic cybersecurity policies in rail [11]. Of note is that one third of the attacks enumerated acknowledged the role of ransomware. This demonstrates cascading risk by how attacks on backend systems can affect the delivery of rail services.

More recently, a hypothetical attack was disclosed in July, 2025 that existed in the design of the protocols for connecting trains together [20]. Because the flaw was in the design it required an entirely new protocol in order to remedy. The CVE lists the weakness as "weak authentication" (CWE-1390), which is an understatement because there is no authentication involved in the generation of the checksum value.

2.2 Hybrid Twins

Chinesta et al. defined "hybrid twin" in contrast to "digital twin" and "virtual twin" models [6]. The twins paradigm has seen applications in railroad situations already such as when Ou et al. used a digital twin model for passenger utilization of light rail in Syndey, Australia [16]. In Chinesta's article the hybrid twin is defined as a combination of a physical model as a part of a model order reduction (MOR) framework. The MOR is intended to simplify the computational demands of the overall model. This is still more a more complicated model than the work of Bai et al. where the researchers modeled a train where each car was simplified to be a particle mass [3].

As an example of the simplification, take the equation below from the work of Resende, Silva, and Tomim [18]. Equation 1, the force exerted by the curvature of rails on the train, is necessary for modeling in a purely digital environment but can be taken for granted in a hybrid environment.

$$\mathbf{F}_{curve} = -\frac{0.5b}{\rho}\, m\, g\, \hat{\mathbf{r}}_{tg} \tag{1}$$

Our project was inspired by Pacific Northwest National Laboratory's Control Environment Laboratory Resource (CELR) project [17]. PNNL's CELR platform provides the environment and tools for red and blue team activities [7]. CELR prioritizes accurate modeling of the network protocols utilized by railway equipment vendors.

3 Implementation

A hybrid model by definition combines both real, physical components and digital simulations. When building a model train layout the tracks, trains, and associated signals form the physical components. Model train hobbyists prefer to keep manual control over their layouts, performing the engineering and traffic routing duties as if they themselves were the entire freight rail enterprise. Our implementation virtualizes these functions using the same software tools that hobbyists use but in an automated manner.

3.1 The Physical Model

The physical component of our hybrid twin model was built using model train hardware. N-gauge (a 1:160 scale) was selected as the best size for building an interesting rail configuration in a four-foot by eight-foot footprint. There also exists a wide variety of engines, rolling stock, and track for N-gauge. Figure 1 below illustrates the physical train model.

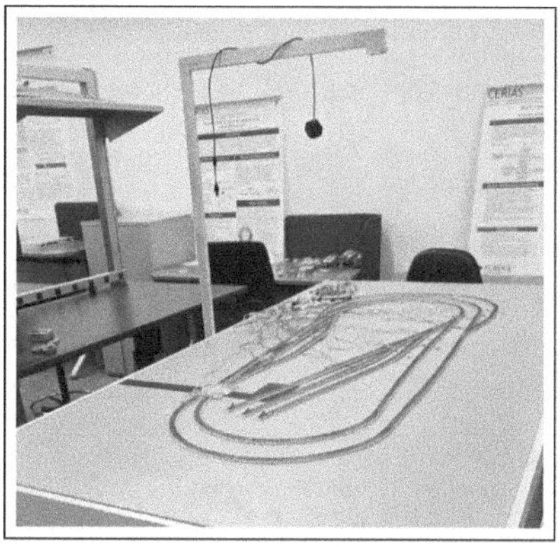

Fig. 1. The physical train layout implemented using N-gauge model railroad pieces.

Model train engines come out of the box ready to be operated by variable direct current. A higher current produces a faster speed and a lower current a slower speed. The problem is that all engines placed on the same track will run in unison. Model train enthusiasts solved the problem of operating multiple engines independently with the digital command control (DCC) protocol [4]. Instead of an analog DC power transformer as a controller, there is the need for a "command station" that can encode and issue instructions to individual engines. Our implementation used the open-source DCC-EX Command Station [9] software flashed onto a hobbyist Arduino Mega microcontroller. It is also required that each engine be equipped with a DCC-decoder board.

These DCC systems provide the mechanism for communicating with and controlling the trains individually. In order to send the control signals we utilize the Java Model Rail Interface (JMRI) [12]. JRMI is both a series of libraries implemented in the Java programming language and a series of utilities that speed up the design and deployment of model railroad layouts. Figure 2 shows the PanelPro GUI with a relatively simple layout consisting of two turnouts dividing the track into three blocks along with the accompanying lighted signals.

Detecting the location of the trains is of critical importance since the model should operate safely in its default state. In the real world this is accomplished with a combination of technologies including satellite-based systems like GPS and energized track circuits. Our hybrid model utilizes a computer vision system (OpenCV) on a cantilevered arm that can view the entirety of the layout

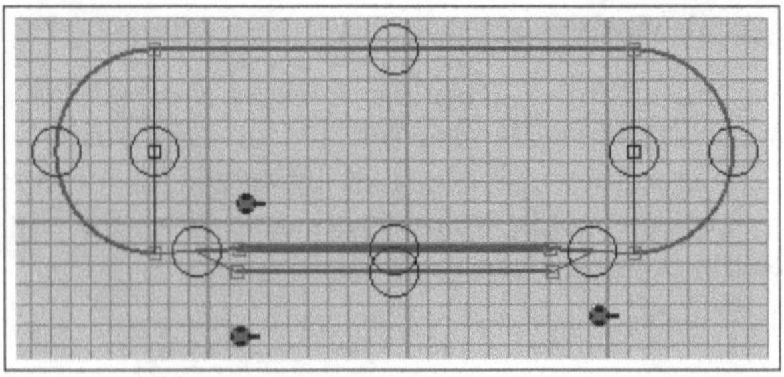

Fig. 2. Track layout in JMRI's PanelPro tool displaying two turnouts and three lighted signals.

at all times. Figure 1 illustrates the webcam placement for the OpenCV system. OpenCV communicates using the Message Queuing Telemetry Transport (MQTT) standard, which JMRI understands.

3.2 The Digital Model

The digital component of our model consists of two physical servers hosting a trio of virtualized systems and an Ethernet network as the communications backbone. One server represents the back office systems and runs a single virtual machine to host enterprise software. The other server represents the wayside components and runs two virtual machines, one for each of the engines. The wayside system also provides the USB connection to the Arduino that runs the command station software. Figure 3 below illustrates the general computer network map of our system.

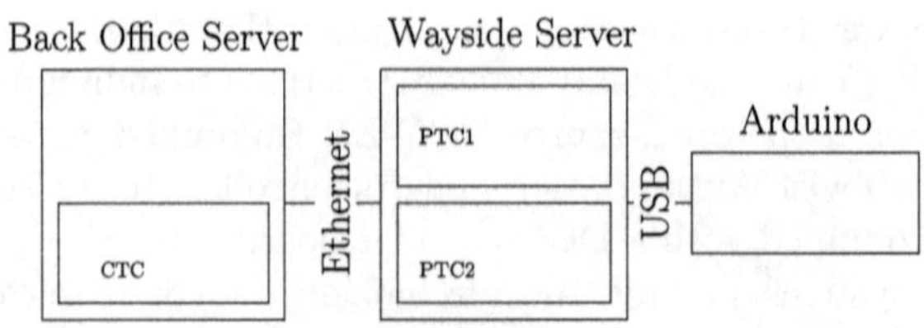

Fig. 3. Computer network map.

The back office server hosts a single virtual machine running JMRI's PanelPro application. PanelPro simulates the centralized traffic control (CTC) system whose responsibility is tracking the current state of the track, turnouts, and rolling stock.

The virtual machines on the wayside server each operate a separate train. Each train is modeled as if it includes positive train control (PTC) technology. PTCs are required on all engines in the United States as of the 2008 Rail Safety

Improvement Act [8]. If an adversary were to gain access to one of the virtualized environments in our model then they would have remote throttle control for that train.

4 Supported Attacks

The first version of our hybrid twin model implemented three IT system weaknesses that could be exploited to demonstrate the effects caused in the physical portion of the model. We classify these attacks using the MITRE ATT&CK framework [25]. Each attack consists of one or more ATT&CK techniques.

4.1 External Remote Services (T1133)

Enterprise computing environments above a certain size, or environments that are geographically distributed, rely on system administrators being able to remotely access systems without having physical access to them. Virtual Network Computing (VNC), Remote Desktop Protocol (RDP), and TeamViewer are all examples of tools that provide a remote user a graphical user interface (GUI) environment.

Adversaries can use external remote services in the exact same way as legitimate system administrators. Our CTC server has VNC installed with a weak password, a 6-character common word. When a threat actor is able to use an existing service (with user credentials) to access a machine MITRE labels this as the "External Remote Services" technique [24].

The motivation for modeling this technique was the 2015 Sandworm Team attacks against the power grid in Ivano-Frankivsk, Ukraine [21]. The Sandworm threat actors were able to use their remote access to use the human-machine interface (HMI) applications and manipulate the grid's SCADA devices.

4.2 Endpoint Denial of Service (T1499)

Denial of service (DoS) attacks target the availability of a given service. By denying access to a targeted resource, a threat actor can induce failures in the wider system. MITRE categorizes this under their "endpoint denial of service" technique [23].

In our network model we identified a bottleneck in the bridge between the IT and OT systems. Our bottleneck was the service that allowed TCP connections to communicate over a USB cable with the Arduino microcontroller that was running the DCC command station software. Using multiple scripts running in parallel we were able to saturate the service and prevent legitimate commands from making it through. The result was trains that were not correctly routed onto safe tracks, creating a collision scenario.

Cetinkaya, Ishii, and Hayakawa describe such a DoS scenario against control systems [5]. Their research focused on wireless networks, which is relevant in a freight rail network due to the fact that freight rail vendors rely on wireless backhaul links between their back office and wayside components. The train-centric Advanced Train Control System (ATCS) uses 900 MHz frequencies [2].

4.3 Adversary-in-the-Middle (T1557)

Our integrity attack involved a threat actor manipulating communications in order to alter the state of the train network. If a threat actor is able to situation themselves between two endpoints of a communication channel they may be able to insert themselves into the conversation, altering contents as they pass through. MITRE lists this as their "adversary-in-the-middle" technique [22].

Kuchar et al. investigated how to detect multiple different types of cyber-attacks to include "man-in-the-middle" attacks against wayside hardware that counts train car axles [13]. An adversary who can manipulate axle count data could either prevent data from reaching the intended recipients which may cause a collision from the undetected train, or they may create spurious axle count data that makes the track traffic seem higher than expected and causing a type of denial of service situation.

5 LLM-Designed Attacks

Large language models (LLMs) are disrupting several industries and cybersecurity is no different. LLMs can quickly explore a problem space in search of new solutions. Their approaches can seem counterintuitive to trained professionals but it is difficult to argue with the results. Researchers with Microsoft Threat Intelligence recently disclosed a campaign in which the threat actors utilized LLMs to generate code to obfuscate malicious payloads [15]. Google's Threat Intelligence Group believes that at least one threat actor they track has used AI tools to develop a complete zero-day exploit [10].

Singer et al. engineered an LLM that learned multi-step cyberattacks from the 2017 Equifax hack [19]. Chuadhry Ahmed of Newcastle University describes one such application specifically for industrial control systems (ICS) in his AttackLLM paper [1]. Ahmed's tool leverages design documentation as a part of the input. AttackLLM differs significantly from other applications of LLMs in cyberattacks such as

The challenge in using an LLM-generated cyberattack is in validating the output. Generating code or a process is fairly easy but there is still the requirement for humans to verify the results. The cost of building and maintaining a railroad is prohibitive so using a real rail network to validate is inefficient to say the least. A hybrid twin system solves the cost issue with the trade-off of requiring simplifying assumptions.

We utilized two different off-the-shelf, non-specialized LLMs in our study: OpenAI's ChatGPT 5 and Anthropic's Claude. Both performed at similar levels and discovered similar attacks. Both models were presented with the same data and the same queries. The data consisted of a document explaining the network layout in detail (services, versions, addresses, etc.) and data traces of the trains operating safely on the layout.

ChatGPT suggested rapidly sending a large number of the same turnout switching signal to a single turnout in order to desynchronize the internal state of the JMRI software with the physical state of the target turnout. PanelPro would think the turnout was meant to face in one direction when in reality it faced the other. This is similar to the real-world situation with the Therac-25 radiation therapy machine. An attacker could accomplish this attack either by having interactive access to the CTC system and using the GUI itself to send the signal barrage. Or the attacker could intercept and replay the signal as it transited the insecure Ethernet channel.

Claude caught onto the possibility of a DDoS attack identical to the situation we ourselves outlined earlier in this paper. A flood of packets set to the TCP2ser service on the Wayside Server would cause it to reject legitimate traffic, denying the CTC the ability to control turnouts. If such a DDoS attack were maintained indefinitely then it would only be a matter of the trains arranged themselves in such a way that the current layout state caused a collision.

6 Conclusions

We have shown in this paper how building a hybrid twin system allows for the modeling and analysis of cyberattack scenarios. Our specific application to freight rail saves researchers a great deal of money and time. We leveraged commercial and open source products to create a miniature scale railroad with a complement of IT/OT systems for an adversary to test against. This model allows human researchers aided by LLM-based attack generation models to attempt new attack methods.

Acknowledgments. We thank the anonymous reviewers for their timely, valuable comments and suggestions. This research was partially supported by the National Science Foundation (NSF) through Award IIS-2229876. Any opinions, findings, and conclusions in this paper are those of the authors and do not necessarily reflect the views of the sponsor.

Disclosure of Interests. The authors have no competing interests to declare that are relevant to the content of this article.

References

1. Ahmed, C.M.: AttackLLM: LLM-based attack pattern generation for an industrial control system. In: FMSys: Proceedings of the 2nd International Workshop on Foundation Models for Cyber-Physical Systems & Internet of Things, pp. 31–36. Association for Computing Machinery (2025)
2. Association of American Railroads: Manual of standards and recommended practices, section k - railway electronics. https://web.archive.org/web/20090327183011/https://www.aarpublications.com/Publications/Manual
3. Bai, W., Lin, Z., Dong, H., Ning, B.: Distributed cooperative control of a high-speed train, pp. 2716–2721 (2019)
4. Barnt, B.: NMRA standards and recommended practices. https://www.nmra.org/index-nmra-standards-and-recommended-practices

5. Cetinkaya, A., Ishii, H., Hayakawa, T.: An overview on denial-of-service attacks in control systems: attack models and security analyses. Entropy **21**(2) (2019)
6. Chinesta, F., Cueto, E.G., Abisset-Chavanne, E., Duval, J.L., Khaldi, F.E.: Virtual, digital and hybrid twins: a new paradigm in data-based engineering and engineered data. Arch. Comput. Meth. Eng. **27**, 105–134 (2019)
7. Choi, E., Edgar, T., Hofer, W., Seppala, G.: CELR rail platform: operations and maintenance guide. Tech. Rep. PNNL-32971, Pacific Northwest National Laboratory, June 2022
8. Congress: Public law 110–432 - rail safety improvement act of 2008 (2008). https://railroads.dot.gov/elibrary/public-law-110-432-rail-safety-improvement-act-2008
9. Digital Command Control Extended: DCC-EX model railroading. https://dcc-ex.com/
10. Google Threat Intelligence Group: GTIG AI threat tracker: Adversaries leverage AI for vulnerability exploitation, augmented operations, and initial access (2026). https://cloud.google.com/blog/topics/threat-intelligence/ai-vulnerability-exploitation-initial-access
11. Ibadah, N., Benavente-Peces, C., Pahl, M.O.: Securing the future of railway systems: a comprehensive cybersecurity strategy for critical on-board and track-side infrastructure. Sensors **24**(24) (2024)
12. JMRI Community: Java model rail interface. https://www.jmri.org/
13. Kuchar, K., Holasova, E., Pospisil, O., Ruotsalainen, H., Rujdiak, R., Wagner, A.: Hunting network anomalies in a railway axle counter system. Sensors **23**(6) (2023)
14. Leyden, J.: Polish teen derails tram after hacking train network, January 2008. https://www.theregister.com/2008/01/11/tram_hack/
15. Microsoft Threat Intelligence: AI vs. AI: Detecting an AI-obfuscated phishing campaign (2025). https://www.microsoft.com/en-us/security/blog/2025/09/24/ai-vs-ai-detecting-an-ai-obfuscated-phishing-campaign/
16. Ou, Y., Mihăiţă, A.S., Ellison, A., Mao, T., Lee, S., Chen, F.: Rail digital twin and deep learning for passenger flow prediction using mobile data. Electronics **14**(12) (2025)
17. Pacific Northwest National Laboratory: Freight rail platform. https://www.pnnl.gov/projects/control-environment-laboratory-resource/freight-rail-platform
18. Resende, L.P., Silva, L.T., Tomim, M.A.: Railway trajectory modeling for simulation of locomotives drive systems. In: 2017 Brazilian Power Electronics Conference (COBEP), pp. 1–8. Institute of Electrical and Electronics Engineers (2017)
19. Singer, B., Lucas, K., Adiga, L., Jain, M., Bauer, L., Sekar, V.: On the feasibility of using LLMs to autonomously execute multi-host network attacks (2025). https://arxiv.org/abs/2501.16466
20. Smith, N., Reuter, E.: End-of-train and head-of-train remote linking protocol weak authentication. Tech. Rep. CVE-2025-1727, CISA, July 2025. https://www.cve.org/CVERecord?id=CVE-2025-1727
21. Styczynski, J., Beach-Westmoreland, N.: When the lights went out: a comprehensive review of the 2015 attacks on Ukrainian critical infrastructure (2019). https://www.boozallen.com/content/dam/boozallen/documents/2016/09/ukraine-report-when-the-lights-went-out.pdf
22. The MITRE Corporation: Adversary-in-the-middle, April 2025. https://attack.mitre.org/versions/v17/techniques/T1557/

23. The MITRE Corporation: Endpoint denial of service, April 2025. https://attack.mitre.org/versions/v17/techniques/T1499//
24. The MITRE Corporation: External remote services, April 2025. https://attack.mitre.org/versions/v17/techniques/T1133/
25. The MITRE Corporation: MITRE A&TTCK® (v17), May 2025. https://attack.mitre.org/versions/v17/

Likelihood Estimation of Malicious Event Sequences in Maritime Cybersecurity

Sanjay Kumar Sharma[1], Awais Yousaf[2(✉)], Jianying Zhou[2], and Mark Voon Wei Goh[2]

[1] Information Systems Technology and Design (ISTD), Singapore University of Technology and Design (SUTD), Singapore, Singapore
sanjaykumar_sharma@alumni.sutd.edu.sg

[2] iTrust, Centre for Research in Cyber Security, Singapore University of Technology and Design (SUTD), Singapore, Singapore
awais_yousaf@sutd.edu.sg, jianying_zhou@sutd.edu.sg, mark_goh@sutd.edu.sg

Abstract. Maritime infrastructure increasingly depends on interconnected cyber-physical systems integrating IT and OT platforms. Such platforms face evolving threats that static defense models struggle to capture, particularly when adversaries operate across temporally ordered, cross-domain sequences with latent intent. To address this, we propose a probabilistic framework using Hidden Markov Model (HMM) to estimate the likelihood of event sequences being malicious, rather than the likelihood of attacks themselves. Observable attacker techniques, such as traffic anomalies, unauthorized commands, or sensor manipulation, are modeled as emissions governed by hidden tactic states, with likelihoods computed via a log-space Forward Algorithm for numerical stability. The framework aligns with the MITRE ATT & CK taxonomy and supports dynamic scenario construction across Enterprise, Industrial Control Systems (ICS), Mobile, and ATLAS domains, realistically representing threat traversal across interdependent systems. Lastly, a Flutter-based software application operationalizes the approach, enabling interactive matrix configuration, sequence visualization, and attacker parametrization. Empirical evaluations demonstrate plausible adversarial paths and differentiation of competing hypotheses. While current evaluation is simulation-based, future work will validate against real-world maritime datasets and assess resilience against operational noise and sensor drift. This framework lays the foundation for maritime-specific risk assessment and supports techniques-level reasoning tailored to maritime cyber-physical systems.

Keywords: Cyber-Physical Systems · Maritime Cybersecurity · Hidden Markov Models · MITRE ATT&CK · Probabilistic Modeling

1 Introduction

The maritime sector is rapidly digitizing, transforming vessels and ports into interconnected cyber-physical ecosystems that integrate heterogeneous layers

S. Zonouz et al. (Eds.): DCS-CI 2026, CCIS 3109, pp. 146–165, 2027.
https://doi.org/10.1007/978-981-92-5039-4_8

of Operational Technology (OT), ICS, Internet of Things (IoT) devices, and autonomous platforms. While this convergence enhances efficiency, it also introduces complex cybersecurity challenges unique to maritime environments [20]. Examples include GPS spoofing, AIS manipulation, legacy OT protocols with limited security features, intermittent satellite connectivity, and resource-constrained shipboard systems. These characteristics distinguish maritime cybersecurity from other sectors such as water, aviation or energy, where connectivity and redundancy are more robust.

Recent studies highlight the need for maritime-specific threat modeling. Yousaf et al. [17] show that frameworks like MITRE ATT & ACK, though comprehensive in terrestrial domains, lack tailored maritime support. Their work emphasizes explainable models that capture adversarial behavior in diverse contexts. Similarly, Yousaf et al. [15] advocate probabilistic approaches for autonomous cargo vessels, where deterministic models fail to capture uncertainty and temporal progression.

We propose a probabilistic framework based on HMM to infer latent adversarial tactics from observable behaviors. HMMs are particularly suited for modeling sequential dependencies and hidden states, offering tractability and interpretability compared to alternatives such as Bayesian Networks (BNs). Maritime attack sequences are formalized as stochastic processes, enabling modular representation of threat progression across shipboard and shore-based infrastructures. Observable events, such as traffic anomalies, unauthorized commands, or sensor manipulation, are modeled as emissions (techniques) governed by hidden states (tactics).

Lastly, a Flutter-based application operationalizes the framework, enabling matrix selection, dynamic graph configuration, and likelihood computation via the Forward Algorithm. Figure 1 shows the interface for selecting matrices from MITRE framework [8–11]. The application supports intuitive exploration of transitions and emissions, making it suitable for estimation of likelihood probabilities. By combining probabilistic inference with structured taxonomies, this work offers a scalable methodology for maritime threat modeling and sequence-based scenario generation. It lays the foundation for maritime-specific risk assessment where likelihood of cyber attack chain is required to be calculated quantitatively. Specifically, this paper contributes as:

- A probabilistic framework based on HMM to infer latent adversarial tactics from observable techniques.
- An interactive software application for calculation of likelihood of attack sequences with freedom to configuring MITRE matrices and simulation of adversarial tachtics and techniques.

While the current focus is on attacker modeling, future extensions will incorporate defender tactics to capture adversarial-defender interactions, further strengthening maritime cyber-defense analysis.

The rest of the paper is organized as follows. Section 2 provides related work on threat modeling frameworks and use of probabilistic modeling for maritime

Fig. 1. Flutter-based application interface for configuring MITRE matrices and adversarial sequences.

cybersecurity. Section 3 elaborates the details about proposed methodology for likelihood estimation using HMM for cyber attack sequences and methodology is applied to six conceptual attack sequences for evaluation in Sect. 4. Results are summarized and current limitations are discussed in Sect. 5. Lastly, conclusions are drawn in Sect. 6.

2 Related Work

2.1 Threat Modeling Frameworks

Structured adversarial models are central to cybersecurity analytics. Domain-specific frameworks such as MITRE ATT & ACK for ICS [10], Enterprise [9], and ATLAS [8] define tactics and techniques for OT, IT, and AI systems respectively. For example, ICS captures OT-specific actions like "Manipulation of Control" and "Impair Process Control," critical in maritime environments. ATLAS extends modeling to AI systems, relevant to autonomous platforms. Existing tools often treat domains separately, though cross-domain traversal, e.g., lateral movement between Enterprise and ICS, is vital for maritime threats.

Maritime-specific studies remain limited. One study [19] presents literature review of maritime cybersecurity with current state-of-the-art but does not provide anything related to modeling of attacker progression in cyber attack chain. Yousaf and Zhou [17] recommend adapting ATT&CK and D3FEND for threat modeling of shipboard systems like ballast water management systems and integrated bridge systems. Yousaf et al. [15] integrate Failure Mode, E ects and

Criticality Analysis (FMECA), ATT & CK, and ATLAS for quantitative risk analysis of autonomous cargo vessel. Their framework calculates likelihood based on Common Vulnerability Scoring System (CVSS). In our work, we are introducing likelihood evaluation based on probabilistic progression modeling and configurable matrix construction for threat modeling, thereby enabling quantitative analysis of adversarial sequential behavior in the maritime contexts.

2.2 Hidden Markov Models in Cybersecurity

Hidden Markov Models (HMMs) are widely applied in intrusion detection, traffic classification, and sequence analysis. One study [6] proposed use of HMMs for adaptive anomaly detection. Similarly, one study [3] presents packer detection using HMM for pattern recognition. Most implementations assume stationary environments, limiting applicability in maritime contexts characterized by intermittent connectivity and subsystem heterogeneity. This research adapts HMMs to maritime threat matrices, enabling log-space likelihood calculation and adversarial sequence simulation across hybrid systems.

2.3 Probabilistic Modeling in Maritime Cyber-Physical Systems (CPS)

Beyond maritime-specific studies, probabilistic modeling has been widely explored in autonomous cyber defense. Hammar et al. [5] propose adaptive policies via belief aggregation and rollout, while Vyas et al. [14] review simulators that treat networks as probabilistic systems. These works demonstrate the value of probabilistic reasoning in IT domains and remain limited to terrestrial infrastructures. Our contribution extends this line of research into maritime CPS, integrating IT and OT shipboard systems for cross-domain traversal analysis.

Probabilistic modeling in maritime CPS is emerging. Authors of a study published in [12] discusses CPS security for maritime assets. Some other works emphasize qualitative risk assessment [1,13,16], omitting sequence-based modeling. Tam and Raymaker [4] incorporate operational feedback but lack algorithmic implementations. This paper introduces a unified pipeline integrating matrix mapping, probabilistic simulation, and log-space likelihood evaluation within an interactive application.

The proposed framework delivers three core capabilities:

- **Matrix–TTP Mapping:** Align tactics and techniques across Enterprise, ICS, Mobile, and ATLAS domains, reflecting interdependencies in maritime CPS.
- **Likelihood Computation:** Log-space Forward Algorithm estimates sequence likelihoods, supporting analysis of temporal progression between tactics.
- **Interactive Simulation:** A Dart/Flutter based application enables matrix configuration, sequence visualization, and attacker parameter adjustment for hypothesis testing.

Together, these features advance maritime cybersecurity from static taxonomies toward adaptive simulation and probabilistic evaluation of threat sequences.

3 Likelihood Estimation Using HMM

We present a configurable sequence modeling framework for maritime CPS using HMM, where hidden states represent attacker tactics and emissions correspond to techniques. For this purpose, a software application is developed that supports both single-matrix and hybrid-matrix modes. In single-matrix mode, isolated threats are modeled using only one matrix, i.e., either of Enterprise, ICS, Mobile, or ATLAS matrices. The single-matrix mode is suitable for domain-specific campaigns. In hybrid-matrix mode, tactics are fused across multiple matrices to enable cross-domain simulation, such as maritime systems where AI navigation interfaces with OT shipboard systems. Semantic checks prevent invalid tactic–technique mappings, ensuring consistency.

3.1 Transition Matrix and Graph Construction

A directed graph over tactics q_i defines transition matrix $\mathbf{A}$ with transition probabilities a_{ij}:

$$\sum_{j=1}^{M_i} a_{ij} = 1 \tag{1}$$

The initial state vector $\boldsymbol{\pi}$ can be extracted from matrix $\mathbf{A}$ using Eq. (2):

$$\pi A = \pi \tag{2}$$

Such that:

$$\sum_{i=1}^{K} \pi_i = 1 \tag{3}$$

The software interface visualizes graphs, allows state node (tactics) and emission edges (techniques) configuration and the assignment of state transition (a_{ij}) and emission probabilities (b_{jk}). Lastly, sequences $\mathbf{O} = (o_1, \ldots, o_T)$ are generated by traversing the tactic graph. Transitions $\mathbf{A}$ guide progression, emissions are sampled from $\mathbf{B}$, and recursion or branching mimics realistic campaigns.

3.2 Emission Matrix

For each tactic q_j, techniques $\{o_k\}$ are assigned probabilities b_{jk}, forming the emission matrix $\mathbf{B}$:

$$\mathbf{B} = \begin{bmatrix} b_{11} & \cdots & b_{1N_1} \\ b_{21} & \cdots & b_{2N_2} \\ \vdots & \ddots & \vdots \\ b_{M1} & \cdots & b_{MN_M} \end{bmatrix}, \quad \sum_{k=1}^{N_j} b_{jk} = 1 \tag{4}$$

Emission matrix **B** is uniformly distributed but other distributions can also be used. Also, software interface allows users to refine probabilities to simulate nuanced behavior. The software application validates probabilities distribution in the background and flags inconsistencies, and permits contextual duplication. But it allows metadata such as latency, impact, and subsystem to be attached for annotation.

3.3 Forward Algorithm

Adversarial sequences are modeled as a first-order HMM with parameters $\lambda = (\boldsymbol{\pi}, \mathbf{A}, \mathbf{B})$. Log-probabilities are computed via the Forward Algorithm [2] as follows:

$$\alpha_1(i) = \log(\pi_i) + \log(b_{io_1}) \tag{5}$$

$$\alpha_t(j) = \log\left(\sum_{i=1}^{K} \exp\left(\alpha_{t-1}(i) + \log(a_{ij})\right)\right) + \log(b_{jo_t}) \tag{6}$$

$$\log P(\mathbf{O} \mid \lambda) = \log\left(\sum_{i=1}^{K} \exp(\alpha_T(i))\right) \tag{7}$$

3.4 Parameter Initialization and Bias

A common concern in probabilistic modeling is the manual effort required to configure emission and transition matrices. In our framework, initial probabilities are uniformly distributed due to lack of maritime specific dataset. But users may refine and update such values based on domain expertise or observed trends. This ensures that likelihood estimation reflects realistic adversarial behavior rather than arbitrary user input. To mitigate user-induced bias, the application will be enhanced in future work to validate probabilities, flag inconsistencies, and support parameter tuning.

4 Evaluation

To evaluate the proposed HMM-based attack sequence likelihood calculation, we conducted a series of simulations using a custom-built software application. The application supports two configuration modes, i.e., *Single Matrix* and *Hybrid Matrix*. Each mode enables the generation of observation sequences based on tactic-to-technique mappings derived from the MITRE ATT&CK and ATLAS frameworks, tailored to maritime cyber attack scenarios.

4.1 Configuration Modes

An application built with Flutter and Dart is developed that supports two configuration modes as follows:

Single Matrix Configuration: The configuration allows to simulate attack sequences using one of four MITRE matrices, namely Enterprise, Mobile, ICS, and ATLAS. The interface displays only tactics and techniques relevant to the chosen domain.

- **Enterprise**: IT threats such as credential access, privilege escalation, brute Force and lateral movement etc.
- **Mobile**: Mobile devices threats including app exploitation, SMS phishing, location tracking and malicious profiles etc.
- **ICS**: OT threats such as firmware manipulation, remote service exploitation, lateral tool transfer and brute force I/O etc.
- **ATLAS**: Adversarial AI threats such as evade AI model, poison training data, generate deepfakes and denial of AI service etc.

Selecting the ICS matrix, for example, enables modeling of attacks on shipboard systems (propulsion, navigation, cargo automation). Transition probabilities ensure sequences reflect realistic adversarial paths. Screenshot for single matrix configuration grabbed from our software application is depicted in Fig. 2.

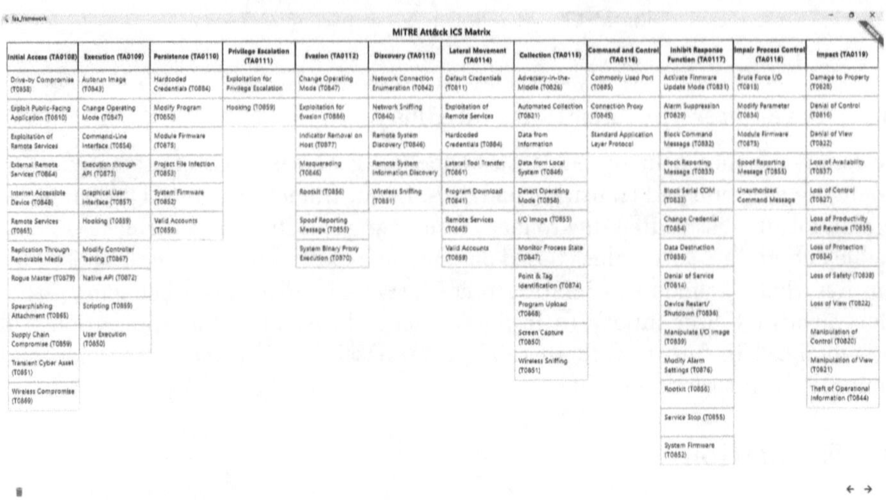

Fig. 2. Single Matrix Configuration—ICS Matrix.

Hybrid Matrix Configuration: This configuration allows tactics and techniques from any combination of Enterprise, Mobile, ICS, and ATLAS to form a unified matrix. It supports cross-domain transitions, enabling simulation of complex campaigns spanning both physical and digital maritime systems. Hybrid matrices are well-suited for APTs, such as lateral movement from ICS to enterprise IT. Figure 3 shows a hybrid matrix constructed after selective inclusion of specific tactics and techniques from all four MITRE matrices.

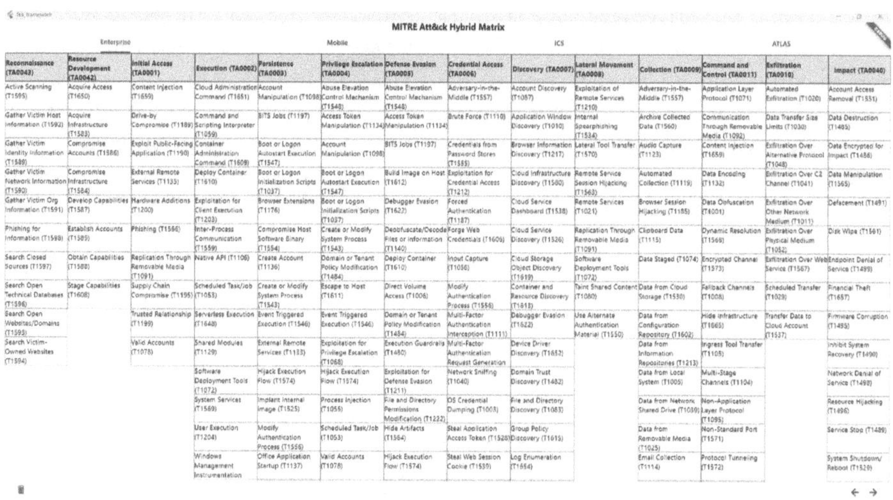

Fig. 3. Hybrid Matrix Configuration—Multi Domain Matrix.

4.2 Attack Sequence Simulations and Likelihood Evaluations

To test the tool, some of the adversarial sequences are derived from Yousaf and Zhou (2024) [17], ensuring domain relevance. Sequences are drawn in the software application as HMM after selecting suitable configuration mode. Then, state transition and emission probabilities are assigned to HMM to initialize $\lambda = (\boldsymbol{\pi}, \mathbf{A}, \mathbf{B})$. Finally, the software application provides the estimate of attack sequence likelihood using Forward Algorithm. Details of each attack sequence tested are provided in the subsequent subsections and results are given in Table 2 in Appendix A.

4.3 Sequence 1: Maritime Disruption via Remote Services

This sequence simulates a hybrid attack path that begins in the enterprise domain and escalates into ICS-specific disruption. The adversary initiates the attack using valid credentials, executes scripts to establish control, pivots into

ICS systems via remote services, and ultimately compromises safety-critical functions. The sequence reflects a realistic and structured adversarial progression across maritime CPS.

Attack Model: The attack begins with **Valid Accounts (T1078)** under *Initial Access*, indicating credential compromise, possibly via phishing or credential reuse. This is followed by **Command and Scripting Interpreter (T1059)** under *Execution*, enabling script-based control and reconnaissance on maritime workstations. The adversary then transitions into the ICS domain using **Remote Services (T0886)** under *Lateral Movement*, targeting trusted engineering interfaces to access control logic. **Connection Proxy (T0884)** under *Command and Control* facilitates covert communication, often bypassing network segmentation. The sequence culminates into **Loss of Safety (T0880)** under *Impact*, representing a direct compromise of safety-critical functions aboard the vessel. The HMM developed for this attack model is illustrated in Fig. 4.

Simulation Parameters: Depending on the attack model, transition matrix A, is defined as given in Eq. (8). Here, the restart mechanism is introduced that may be encoded in either the first or final row of the transition matrix. When present, this row assigns uniform probabilities across all tactics, enabling the adversary to reenter or restart the attack sequence from any state.

$$A = \begin{bmatrix} 0 & 1 & 0 & 0 & 0 \\ 0 & 0 & 1 & 0 & 0 \\ 0 & 0 & 0 & 1 & 0 \\ 0 & 0 & 0 & 0 & 1 \\ 0.2 & 0.2 & 0.2 & 0.2 & 0.2 \end{bmatrix} \tag{8}$$

From transition matrix A, initial state vector π is computed internally in the software application using Eq. (2) as follows:

$$\pi = \begin{bmatrix} 0.0667 \\ 0.1333 \\ 0.2000 \\ 0.2667 \\ 0.3333 \end{bmatrix} \tag{9}$$

This vector represents the initial probability distribution over the five tactics of the attack model under consideration. The edge probabilities between tactics and tactic-to-technique pairs are assigned by the user during simulation. These probabilities govern the traversal behavior of the attacker in the sequence, allowing users to explore different campaign trajectories based on the scenario. So, the emission matrix B is defined as given in Eq. (10) below:

$$B = \begin{bmatrix} 1\,0\,0\,0\,0 \\ 0\,1\,0\,0\,0 \\ 0\,0\,1\,0\,0 \\ 0\,0\,0\,1\,0 \\ 0\,0\,0\,0\,1 \end{bmatrix} \tag{10}$$

Here, each tactic only emits one technique with probability 1.0. Finally, the observation sequence O is defined as given in Eq. (11):

$$O = \begin{bmatrix} \text{Initial Access: Valid Accounts (T1078)} \\ \text{Execution: Command and Scripting Interpreter (T1059)} \\ \text{Persistence: Remote Services (T0886)} \\ \text{Defense Evasion: Connection Proxy (T0884)} \\ \text{Impact: Loss of Safety (T0880)} \end{bmatrix} \tag{11}$$

This sequence reflects a plausible adversarial path through maritime systems, starting with credential access and culminating in safety disruption. The HMM for this attack sequence is shown in Fig. 4 without edges for restart mechanism.

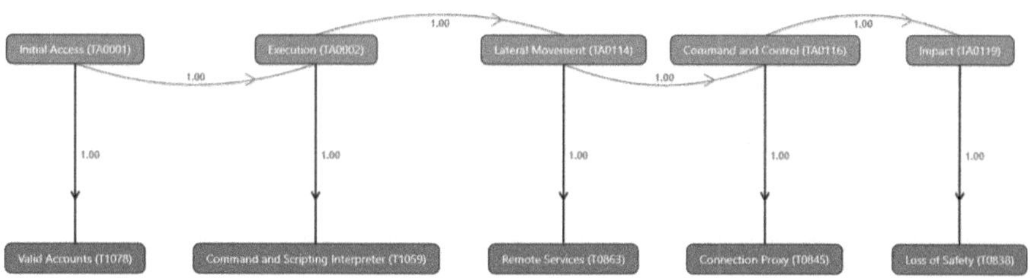

Fig. 4. HMM for Sequence 1.

And the sequence likelihood for this attack model is computed as follow:

$$P(O|\lambda) = 6.66666701436042786 \times 10^{-2}$$

Interpretation: The initial vector π indicates an adversarial bias toward the *Impact* tactic, with a probability of 0.3333, suggesting that the attacker may begin operations with disruption in mind. The transition matrix A enforces a linear progression through tactics, modeling a stepwise escalation. The final row of A introduces a restart mechanism that enables probabilistic re-entry into the attack cycle after reaching the terminal tactic. Specifically, this row assigns equal transition probabilities to all five tactics, forming a uniform distribution. This design reflects scenarios where an attacker, upon completing a campaign phase, may initiate a new phase of the attack or shifting focus to a different shipboard system. Additionally, this restart mechanism prevents premature termination and enables recursive adversarial behavior. By sampling the next tactic from a uniform distribution, it introduces variability in sequence generation and supports modeling of multi-phase campaigns and tactic repetition, which are common in persistent threat scenarios.

For simplicity, the emission matrix B maps each tactic to a unique technique. The observation sequence O aligns with the martime attack scenario, beginning with credential compromise and ending with safety loss. The computed likelihood $P(O|\lambda) = 6.67 \times 10^{-2}$ confirms that the sequence is statistically plausible under the defined HMM, validating the hybrid matrix design for modeling structured adversarial behavior.

The use of **Remote Services (T0863)** and **Connection Proxy (T0845)** reflects a stealthy pivot into ICS environments, bypassing traditional segmentation. The inclusion of **Loss of Safety (T0838)** is particularly significant, as it represents a shift from data-centric compromise to physical disruption, a critical concern in maritime cybersecurity.

Defense Recommendation: To mitigate this attack path, defenders should:

- Enforce credential hygiene and multi-factor authentication.
- Monitor script execution and command-line activity.
- Segment ICS networks and restrict remote service access.
- Detect proxy-based communication and enforce protocol whitelisting.
- Deploy safety integrity checks to detect anomalous control logic manipulation.

4.4 Sequence 2: Recursive Firmware Reinfection via Remote Access

This sequence models a recursive firmware compromise in maritime control systems, where the adversary leverages phishing and automated execution to implant persistent malware and repeatedly evade detection. The tactic flow reflects a stealthy campaign focused on long-term reinfection and operational disruption, with emphasis on embedded firmware manipulation and covert masquerading.

Attack Model: The attack initiates with **Phishing (T1566)** under *Initial Access*, targeting crew or port personnel through deceptive communication vectors. Upon successful compromise, the adversary executes malware via **Autorun Image (T0895)** under *Execution*, typically through infected bootable media or system images. Persistence is established through **Project File Infection (T0873)** under *Persistence*, where engineering configuration files are modified to embed malicious logic. This sets the foundation for a recursive reinfection loop. To evade detection, the adversary employs **Masquerading (T0849)** under *Evasion*, disguising malicious components as legitimate processes. **Remote System**

Discovery (T0888) under *Discovery* enables the adversary to identify vulnerable systems for further compromise. The attack culminates in dual-impact tactics, **Denial of Control (T0813)** and **Theft of Operational Information (T0882)** under *Impact*, disrupting control logic and exfiltrating sensitive operational data. Following impact, the adversary re-engages the reinfection loop by reactivating **Project File Infection (T0873)** and **Masquerading (T0849)**, enabling persistent compromise across firmware updates or system resets.

For sake of brevity, sequence likelihood is provided directly for this and subsequent attack sequences without explicitly mentioning of $\lambda = (\boldsymbol{\pi}, \mathbf{A}, \mathbf{B})$ and O. For this attack sequence, likelihood estimation is computed as:

$$P(O|\lambda) = 6.68462402808170971 \times 10^{-3}$$

Interpretation: The recursive loop formed by **Project File Infection (T0873)** and **Masquerading (T0849)** enables repeated reinfection without triggering conventional alerts. The absence of command-and-control or lateral movement suggests a self-contained attack designed for autonomous execution and stealth. The tactic flow is cyclic, allowing the adversary to maintain persistence even after remediation attempts.

Defense Recommendation: To mitigate this attack path, defenders should:

- Enforce strict access controls and phishing awareness training.
- Monitor autorun behaviors and validate image integrity.
- Audit engineering project files for unauthorized modifications.
- Detect masquerading via process lineage and binary validation.
- Implement safeguards against control denial and data exfiltration.
- Introduce loop detection mechanisms in firmware validation workflows.

4.5 Sequence 3: Stealthy Firmware Implant via API Execution

This sequence models a stealthy firmware-level compromise within maritime systems, initiated through API-based execution. The adversary implants persistent malware into embedded systems, evades detection using rootkit techniques, and dynamically discovers vulnerable subsystems. The attack culminates in dual-impact outcomes, i.e., loss of availability and safety, reflecting both operational disruption and critical degradation.

Attack Model: The attack begins with **Execution through API (T0871)** under *Execution*, indicating adversarial use of exposed or undocumented APIs to deploy malicious firmware components. This is followed by **Module Firmware (T1693.002)** under *Persistence*, representing a direct compromise of embedded control units. **Rootkit (T0851)** under *Evasion* enables stealthy operation and reinfection, while **Remote System Discovery (T0846)** under *Discovery* allows the attacker to identify additional targets within the ICS environment. The sequence concludes with two simultaneous impact techniques, **Loss of Availability (T0826)** and **Loss of Safety (T0880)** under *Impact*, reflecting system downtime and unsafe operational states. For this attack model, sequence likelihood is estimated as:

$$P(O|\lambda) = 6.50366480723023658 \times 10^{-4}$$

Interpretation: The use of **Execution through API (T0871)** and **Rootkit (T0851)** suggests a non-traditional entry point and persistent stealth. The recursive loop between evasion and discovery indicates adaptive targeting, with minimal reliance on lateral movement.

Defense Recommendation: To mitigate this attack path, defenders should:

- Monitor and restrict API access, especially undocumented or legacy endpoints.
- Validate firmware integrity and enforce secure boot protocols.
- Detect rootkit behavior using kernel-level integrity checks and memory forensics.
- Monitor for anomalous subsystem queries and discovery attempts.
- Implement redundancy and safety interlocks in critical control systems.
- Segment ICS networks to limit propagation and reinfection.

4.6 Sequence 4: Ballast Water Management System Attack

This sequence models a targeted attack on the ballast water management system onboard vessel. The adversary initiates access via exposed remote services, implants persistent firmware-level malware, and manipulates controller tasking to disrupt operational safety and availability. The tactic flow reflects a deliberate campaign aimed at embedded compromise and sabotage of vessel stability mechanisms.

Attack Model: The attack begins with **External Remote Services (T1133)** under *Initial Access*, indicating exploitation of exposed interfaces to infiltrate the maritime network. This is followed by **Modify Controller Tasking (T0821)** under *Execution*, suggesting adversarial manipulation of ballast control logic, potentially altering drainage or intake operations. **Modify Firmware (T1693)** under *Persistence* reflects a compromise of embedded firmware, enabling long-term control and stealth. The adversary then performs **Remote System Discovery (T0846)** under *Discovery* to map the environment and identify vulnerable subsystems. **Data from Information Repositories (T0811)** under *Collection* indicates targeted extraction of operational data, possibly to refine sabotage parameters. The sequence concludes with dual-impact tactics, i.e., **Loss of Availability (T0826)** and **Loss of Safety (T0880)** under *Impact*, representing both service disruption and critical degradation of vessel stability safeguards. For this attack model, sequence likelihood is estimated as:

$$P(O|\lambda) = 5.83571454659104379 \times 10^{-3}$$

Interpretation: The use of **Modify Controller Tasking (T0821)** highlights direct interference with task scheduling, a tactic often overlooked in traditional monitoring setups. The absence of lateral movement or evasion techniques suggests a streamlined attack, possibly executed by pre-positioned actors or through compromised remote services.

Defense Recommendation: To mitigate this attack path, defenders should:

- Restrict and monitor remote access services, especially those exposed externally.
- Validate firmware integrity and restrict update mechanisms.
- Monitor controller tasking for anomalous modifications.
- Implement safety interlocks and redundancy in ballast control systems.
- Conduct firmware audits and apply secure boot protocols.

4.7 Sequence 5: Navigation Disruption via Sensor Spoofing

This sequence models an adversary targeting vessel navigation by spoofing sensor data, leading to misinformed decisions and course deviation. The attack begins with credential-based access, followed by spoofed messages mimicking legitimate outputs. The adversary then discovers subsystems, collects data to refine spoofing, and establishes command and control using standard protocols. The impact culminates in loss of view and safety as navigation systems respond to falsified inputs.

Attack Model: The attack starts with **Valid Accounts (T1078)** under *Initial Access*. Then navigational controller is compromised by **Execution through API (T0871)** to inject spoofed sensor data via exposed interfaces. **Remote System Discovery (T0846)** enables identification of navigation-linked subsystems. **Automated Collection (T0802)** gathers environmental or system data to refine spoofing fidelity. **Standard Application Layer Protocol (T0869)** maintains communication with external infrastructure. The sequence concludes with dual impacts, i.e., **Loss of View (T0829)** and **Loss of Safety (T0880)**. For this attack model, sequence likelihood is estimated as:

$$P(O|\lambda) = 2.34 \times 10^{-3}$$

Interpretation: Automated Collection (T0802) allows to monitor and analyze navigation sensor data before launching spoofing attacks as it would lead to high impact and consequences.

Defense Recommendations: To mitigate this attack path, defenders should:

- Enforce credential hygiene and multi-factor authentication.
- Monitor API usage for anomalies.
- Segment navigation systems from external networks.
- Deploy anomaly detection for sensor streams.
- Validate responses against redundant sensor inputs.

4.8 Sequence 6: Exfiltration and Erosion of AI Model Integrity

This sequence models a cross-domain campaign beginning with firmware compromise and culminating in exfiltration and model integrity manipulation. The attacker implants persistent malware via API execution, maintains stealth through rootkit techniques, and leverages AI inference channels to ultimately degrading AI model integrity and decision-making.

Attack Model: The attack initiates with **Execution through API (T0871)** under *Execution* in the ICS matrix, where adversaries exploit exposed APIs to implant malicious firmware. This is followed by **Module Firmware (T1693.002)** under *Persistence*, indicating embedded compromise of control systems. **Rootkit (T0851)** under *Evasion* enables stealthy operation and reinfection. The attacker then transitions into the AI domain using **Exfiltration via AI Inference API (AML.T0024)** under *Exfiltration* in the **ATLAS** matrix, leveraging model behavior to extract sensitive telemetry or operational data. The sequence concludes with **Erode AI Model Integrity (AML.T0031)** under *Impact*, representing target model performance degradation and integrity. For this attack model, sequence likelihood is estimated as:

$$P(O|\lambda) = 1.11 \times 10^{-1}$$

Interpretation: AI inference as an exfiltration vector is unique, exploiting the cognitive layer of cyber-physical systems. Attacking AI systems in the vessel, attackers may exploit inference channels to subtly extract sensitive data or manipulate outputs, which can degrade trust and erode the integrity of the AI model over time, bridging classic cyber intrusion techniques with emerging AI-focused threats.

Defense Recommendations: To mitigate this attack path, defenders should:

- Monitor API execution and restrict firmware update channels.
- Detect rootkit behavior using kernel-level integrity checks.
- Audit AI model inputs and outputs for anomalies.
- Segment ICS and AI environments to prevent cross-domain leakage.

5 Discussion

Evaluation results demonstrate the efficacy of the HMM in modeling structured cyber-physical attack sequences for maritime systems. The comparative analysis summarized in Table 1 highlights diverse adversarial sequences across maritime systems. Sequence 6 shows the highest likelihood (1.11×10^{-1}), reflecting AI-layer exfiltration plausibility. Sequence 2 requires most extensive number of steps to achieve objective with relatively higher probability than sequence 6. Sequences 1, 4, and 5 represent Enterprise to ICS pivots, while Sequence 3 remains ICS matrix only. Sequence 6 uniquely involves the ATLAS matrix, introducing cognitive-layer threats. Overall, the results support the objective of modeling realistic multi-domain adversarial paths and quantifying their likelihood via probabilistic models like HMM.

Table 1. Summary of Likelihood of Attack Sequences

Sequences	MITRE Matrices	Steps	Unique Tactics	Likelihood
1. Maritime Disruption via Remote Services	Enterprise, ICS	5	5	6.67e-2
2. Recursive Firmware Reinfection via Remote Access	Enterprise, ICS	9	6	6.91e-3
3. Stealthy Firmware Implant via API Execution	ICS	6	4	6.50e-4
4. Ballast Water Management System Attack	Enterprise, ICS	7	6	5.84e-3
5. Navigation Disruption via Sensor Spoofing	Enterprise, ICS	7	6	2.34e-3
6. Exfiltration and Erosion of AI Model Integrity	ICS, ATLAS	5	5	1.11e-1

Log-space Forward Algorithm enabled robust likelihood computation, i.e., probabilistic modeling offers a powerful lens for structured threat analysis. Hybrid matrices, recursive logic, and domain-specific semantics enable clear simulation of adversarial behavior. The use of HMM with likelihoods computed via the Forward Algorithm supported systematic analysis of multi-phase attacks and it can further be used for the development of quantitative cyber risk assessment frameworks.

In this study, attack sequences were modeled using Hidden Markov Models and grounded in established adversarial knowledge bases, specifically the MITRE ATT&CK matrices (Enterprise, ICS, and Mobile) and the MITRE ATLAS framework. As part of future work, we plan to extend this modeling approach by incorporating the MariOT Matrix [7], which is tailored to maritime cybersecurity scenarios and provides domain-specific adversarial behaviors not fully captured in existing frameworks. Furthermore, to enhance the practical relevance and empirical validation of our findings, the attack sequences identified and evaluated in this paper will be deployed and tested within a MariOT testbed [18] environment. This will enable us to assess the real-world plausibility, feasibility, and impact of these modeled attack chains, thereby strengthening the robustness and applicability of our probabilistic analysis.

A key limitation of the proposed framework lies in its reliance on user-defined inputs for state transition and emission probabilities within the HMM used.

While this approach enables flexibility in modeling diverse attack scenarios, it inherently introduces subjectivity and potential bias stemming from expert judgment. This limitation is primarily driven by the absence of publicly available, high-quality datasets in maritime cybersecurity that accurately reflect real-world attack patterns and behaviors. Consequently, the probabilistic parameters used in this study may not fully capture the true statistical characteristics of adversarial activities in maritime environments. To address this limitation, future work will focus on leveraging empirically derived data as such datasets become available, allowing for the estimation of transition and emission probabilities based on observed frequencies. This shift from expert-driven to data-driven parameterization is expected to improve the accuracy, objectivity, and generalizability of the proposed modeling approach.

6 Conclusion

This study presents a probabilistic approach to modeling adversarial attack sequences by leveraging Hidden Markov Models (HMMs) and computing likelihoods through the forward algorithm. By grounding the attack chains in well-established knowledge bases, namely the MITRE ATT&CK matrices (Enterprise, ICS, and Mobile) and the MITRE ATLAS framework, we provide a structured and comprehensive representation of both traditional and AI-focused threats. The proposed methodology enables the systematic evaluation of attack progression, allowing for quantitative assessment of the likelihood of multi-stage attack scenarios.

Our findings demonstrate that integrating diverse threat matrices enhances the completeness of attack modeling, particularly when bridging conventional cyber threats with emerging AI-related risks. The probabilistic framework further adds value by enabling analysts to prioritize attack paths based on their likelihood, thereby supporting more informed decision-making in risk assessment and defensive strategy design.

As cyber-physical and AI-enabled systems continue to evolve in the maritime sector, the need for domain-specific threat modeling becomes increasingly critical. In this regard, future work will extend this research by incorporating the MariOT Matrix to better capture maritime-specific adversarial behaviors. Additionally, validating the modeled attack sequences within a MariOT testbed will allow us to assess their real-world plausibility and operational impact. These directions will further strengthen the applicability, realism, and robustness of the proposed framework, ultimately contributing to more effective cybersecurity preparedness and resilience.

Acknowledgment. The research is supported by the National Research Foundation, Singapore, under its National Satellite of Excellence Programme "Design Science and Technology for Secure Critical Infrastructure: Phase II" (Award No: NRF-NCR25-NSOE05-0001). Any opinions, findings and conclusions or recommendations expressed in this material are those of the author(s) and do not reflect the views of National Research Foundation, Singapore.

A Appendix

Table 2. Consolidated Tactic Technique Mapping, Sequences and Likelihood

Sequence	Step	Matrix	Tactic (ID)	Technique (ID)	Likelihood
1. Maritime Disruption via Remote Services	1	Enterprise	Initial Access (TA0001)	Valid Accounts (T1078)	6.67e-2
	2	Enterprise	Execution (TA0002)	Command and Scripting Interpreter (T1059)	
	3	ICS	Lateral Movement (TA0114)	Remote Services (T0886)	
	4	ICS	Command and Control (TA0116)	Connection Proxy (T0884)	
	5	ICS	Impact (TA0119)	Loss of Safety (T0880)	
2. Recursive Firmware Reinfection via Remote Access	1	Enterprise	Initial Access (TA0001)	Phishing (T1566)	6.68e-3
	2	ICS	Execution (TA0109)	Autorun Image (T0895)	
	3	ICS	Persistence (TA0110)	Project File Infection (T0873)	
	4	ICS	Evasion (TA0112)	Masquerading (T0849)	
	5	ICS	Discovery (TA0113)	Remote System Discovery (T0888)	
	6	ICS	Impact (TA0119)	Denial of Control (T0813)	
	7	ICS	Impact (TA0119)	Theft of Operational Information (T0882)	
	8	ICS	Persistence (TA0110)	Project File Infection (T0873)	
	9	ICS	Evasion (TA0112)	Masquerading (T0849)	
3. Stealthy Firmware Implant via API Execution	1	ICS	Execution (TA0109)	Execution through API (T0871)	6.5e-4
	2	ICS	Persistence (TA0110)	Module Firmware (T1693.002)	
	3	ICS	Evasion (TA0112)	Rootkit (T0851)	
	4	ICS	Discovery (TA0113)	Remote System Discovery (T0846)	
	5	ICS	Impact (TA0119)	Loss of Availability (T0826)	
	6	ICS	Impact (TA0119)	Loss of Safety (T0880)	
4. Ballast Water Management System Attack	1	Enterprise	Initial Access (TA0001)	External Remote Services (T1133)	5.84e-3
	2	ICS	Execution (TA0109)	Modify Controller Tasking (T0821)	
	3	ICS	Persistence (TA0110)	Modify Firmware (T1693)	
	4	ICS	Discovery (TA0113)	Remote System Discovery (T0846)	
	5	ICS	Collection (TA0115)	Data from Information Repositories (T0811)	
	6	ICS	Impact (TA0119)	Loss of Availability (T0826)	
	7	ICS	Impact (TA0119)	Loss of Safety (T0880)	
5. Navigation Disruption via Sensor Spoofing	1	Enterprise	Initial Access (TA0001)	Valid Accounts (T1078)	2.34e-3
	2	ICS	Execution (TA0109)	Execution through API (T0871)	
	3	ICS	Discovery (TA0113)	Remote System Discovery (T0846)	
	4	ICS	Collection (TA0115)	Automated Collection (T0802)	
	5	ICS	Command and Control (TA0116)	Standard Application Layer Protocol (T0869)	
	6	ICS	Impact (TA0119)	Loss of View (T0829)	
	7	ICS	Impact (TA0119)	Loss of Safety (T0880)	
6. Exfiltration and Erosion of AI Model Integrity	1	ICS	Execution (TA0109)	Execution through API (T0871)	1.11e-1
	2	ICS	Persistence (TA0110)	Module Firmware (T1693.002)	
	3	ICS	Evasion (TA0112)	Rootkit (T0851)	
	4	ATLAS	Exfiltration (TA0013)	Exfiltration via AI Inference API (AML.T0024)	
	5	ATLAS	Impact (TA0014)	Erode AI Model Integrity (AML.T0031)	

References

1. Basnet, S., Bolbot, V., Yousaf, A., Gunawan, S., Zhou, J., Banda, O.A.V.: An integrated STPA-STRIDE-BN framework for cybersecurity risk analysis: a case study of ship remote pilotage operations. Fut. Gener. Comput. Syst. **183**, 108515 (2026). https://doi.org/10.1016/J.FUTURE.2026.108515, https://linkinghub.elsevier.com/retrieve/pii/S0167739X26001494
2. Blunsom, P.: Hidden Markov models. Lect. Notes **15**, 18–19 (2004)
3. Hai, N.M., Tho, Q.T.: Packer identification using hidden Markov model. In: Phon-Amnuaisuk, S., Ang, S.-P., Lee, S.-Y. (eds.) MIWAI 2017. LNCS (LNAI), vol. 10607, pp. 92–105. Springer, Cham (2017). https://doi.org/10.1007/978-3-319-69456-6_8
4. Hajdu, M., Skibniewski, M., Vanhoucke, M., Horvath, A., Brilakis, I.: How many types of critical activities exist? A conjecture in need of proof. Procedia Eng. **164**, 3–11 (2016)
5. Hammar, K., Li, Y., Alpcan, T., Lupu, E.C., Bertsekas, D.: Adaptive network security policies via belief aggregation and rollout. arXiv preprint arXiv:2507.15163 (2025)
6. Lee, S., Li, L., Ni, J.: Adaptive anomaly detection using a hidden Markov model. In: International Manufacturing Science and Engineering Conference, vol. 49477, pp. 599–606 (2010)
7. Liang, A.T.W., Yousaf, A., Feng, L., Zhou, J.: Mariot matrix for maritime cybersecurity. J. Transp. Secur. **19**(1), 33 (2026). https://doi.org/10.1007/S12198-026-00354-X, https://link.springer.com/article/10.1007/s12198-026-00354-x
8. MITRE: Adversarial threat landscape for artificial-intelligence systems (ATLAS). https://atlas.mitre.org/. Accessed 23 Jun 2026
9. MITRE: Att&ck for enterprise. https://attack.mitre.org/matrices/enterprise/. Accessed 23 Jun 2026
10. MITRE: Att&ck for industrial control systems. https://attack.mitre.org/matrices/ics/. Accessed 23 Jun 2026
11. MITRE: Mitre att&ck framework. https://attack.mitre.org. Accessed 23 Jun 2026
12. Progoulakis, I., Rohmeyer, P., Nikitakos, N.: Cyber physical systems security for maritime assets. J. Mar. Sci. Eng **9**(12), 1384 (2021)
13. Sahay, R., Estay, D.A., Meng, W., Jensen, C.D., Barfod, M.B.: A comparative risk analysis on cybership system with STPA-Sec, STRIDE and CORAS. Comput. Secur. **128**, 103179 (2023). https://doi.org/10.1016/J.COSE.2023.103179
14. Vyas, S., Hannay, J., Bolton, A., Burnap, P.: Automated cyber defence: a review. arXiv preprint arXiv:2303.04926 (2023)
15. Yousaf, A., Amro, A., Kwa, P.T.H., Li, M., Zhou, J.: Cyber risk assessment of cyber-enabled autonomous cargo vessel. Int. J. Crit. Infrastruct. Prot. **46**, 100695 (2024)
16. Yousaf, A., Gunawan, S., Basnet, S., Bolbot, V., Zhou, J., Banda, O.A.V.: STPA-cyber: a semi-automated cyber risk assessment framework for maritime cybersecurity. Comput. Secur. **157**, 104559 (2025). https://doi.org/10.1016/j.cose.2025.104559, https://www.sciencedirect.com/science/article/pii/S0167404825002482
17. Yousaf, A., Zhou, J.: From sinking to saving: MITRE ATT &CK and D3FEND frameworks for maritime cybersecurity. Int. J. Inf. Secur. **23**, 1603–1618 (2024)
18. Yousaf, A., Zhou, J., Chattopadhyay, S.: Maritime operational technology shipboard testbed for cybersecurity research, training, exercises, and education. IEEE Secur. Priv. (2025). https://doi.org/10.1109/MSEC.2025.3613711, https://ieeexplore-ieee-org.sutd.idm.oclc.org/document/11204793

19. Yu, H., Meng, Q., Fang, Z., Liu, J.: Literature review on maritime cybersecurity: state-of-the-art. J. Navig. **76**(4–5), 453–466 (2023)
20. Zhou, J.: Maritime cybersecurity: a big change ahead. IEEE Secur. Priv. **23**, 4–7 (2025)

Temporal Modeling for Anomaly Detection in Industrial Control Systems

Ziyue Wang[1], Chuadhry Mujeeb Ahmed[1], Muhammad Azmi Umer[2](✉), and Rajiv Ranjan[1]

[1] Newcastle University, Newcastle, UK
{z.wang213,mujeeb.ahmed,raj.ranjan}@newcastle.ac.uk
[2] Singapore University of Technology and Design, Singapore, Singapore
azmi_umer@sutd.edu.sg

Abstract. We focus on temporal dependency modeling for anomaly detection in Industrial Control Systems (ICS) using the Secure Water Treatment (SWaT) testbed as an example. A comparative framework is designed that includes a baseline model without temporal features, a Random Forest model based on engineered temporal features, and a sequence-based LSTM model. An attack-aware temporal split is employed to ensure realistic evaluation and prevent data leakage. This framework aims to provide insights into the role of temporal modeling in enhancing ICS anomaly detection.

Keywords: ICS · Anomaly Detection · Time Series · Machine Learning · ICS Attacks

1 Introduction

Industrial control systems (ICS) are at the core of critical infrastructure in modern society, and their stable operation is crucial to multiple industries such as energy, manufacturing, and water treatment, directly impacting national security and economic development [1,6,20]. However, with the increasing integration of ICS and IT networks, these systems face growing cyber threats [9,11]. Unlike traditional IT systems, ICS security places greater emphasis on real-time availability and system integrity. Any disruption could lead to severe physical consequences, such as production interruptions, equipment damage, environmental pollution, or even public safety risks [10,11,16,17].

Current ICS attack detection methods have limitations, as they often treat data as independent observations and ignore temporal dependencies, which are central to the dynamic characteristics of ICS systems and real attack behaviour [12]. Real-world attacks often unfold gradually or exhibit delay effects in the sensor-actuator chain [8]. Therefore, effectively utilising temporal dependencies in ICS data is key to improving the robustness of attack detection [15]. This work aims to explore advanced anomaly detection methods, particularly those that utilize deep learning to enhance the cybersecurity posture of industrial

S. Zonouz et al. (Eds.): DCS-CI 2026, CCIS 3109, pp. 166–189, 2027.
https://doi.org/10.1007/978-981-92-5039-4_9

control systems. It focuses on the time-series data generated by ICS sensors and actuators and the unique challenges they present. This was done using the Secure Water Treatment (SWaT) testbed serves as the case study for this analysis.

1.1 Contributions

This paper investigates how explicitly modeling the temporal dependencies in ICS telemetry affects anomaly/attack detection. It also provides a controlled comparison between feature-based and sequence-based temporal learning paradigms. The main contributions are:

- Systematic empirical evaluation framework for temporal modeling: We establish a unified experimental setup to conduct a rigorous comparative study of temporal modeling approaches. By evaluating (i) a non-temporal baseline classifier, (ii) a feature-based temporal model (Random Forest with engineered lag/rolling features), and (iii) a sequence-based temporal model (LSTM trained on sliding-window sequences), we provide an ablation-style quantification of the explicit benefits and trade-offs of temporal modeling.
- Leakage-resistant evaluation protocol: We employ an attack-aware temporal splitting strategy that preserves temporal order while preventing event-level information leakage across train/validation/test partitions, ensuring a realistic assessment of generalization in ICS attack detection.
- Temporal feature engineering for ICS signals: We applied feature engineering strategies for multivariate ICS time series (e.g., rolling statistics, lagged observations, trend and cross-sensor dependency features) and specified how these are constructed within each data split to avoid look-ahead bias.
- Sequence construction for deep temporal learners: We formalize a sliding-window sequence representation for LSTM-based detection, including windowing and labeling choices suitable for attack intervals, and compare this representation directly against engineered-feature learning under the same data partitions.
- Empirical analysis of operational trade-offs: Through consistent evaluation within our unified benchmarking setup (including threshold optimization under class imbalance), we report empirical findings that characterize when feature-based temporal enhancement versus sequence modeling is preferable, highlighting trade-offs among detection performance, interpretability, and deployment considerations.

1.2 Organization

This paper will systematically explore the issue of temporal dependencies in ICS attack detection. Section 2 discusses about the SWaT dataset, related research, attack scenarios, background knowledge on anomaly detection, and the related work. Section 3 details the methodological approach of this study, including data preprocessing, feature engineering strategies, the selected feature-based and sequence-based models, and the experimental setup. Section 4 presents the experimental results and performance evaluation. Section 5 concludes the paper and outlines the implications of the results.

2 Background

2.1 SWaT Dataset-Related Research and Attack Scenarios

The Secure Water Treatment (SWaT) testbed is an important benchmark for industrial control system (ICS) cybersecurity research [4]. Figure 1 illustrates the simulation a real water treatment facility (six-stage process, each stage controlled by a PLC) with a production capacity of 5 gallons per hour [4]. The platform collects physical process data (sensor and actuator status) and network traffic data. The dataset comprises seven days of normal operation and four days of abnormal operation, featuring a total of 36 attack scenarios [4]. These scenarios include sensor tampering, coordinated attacks, actuator interference, and communication disruptions, providing a real-world data foundation for studying diverse ICS network threats [4].

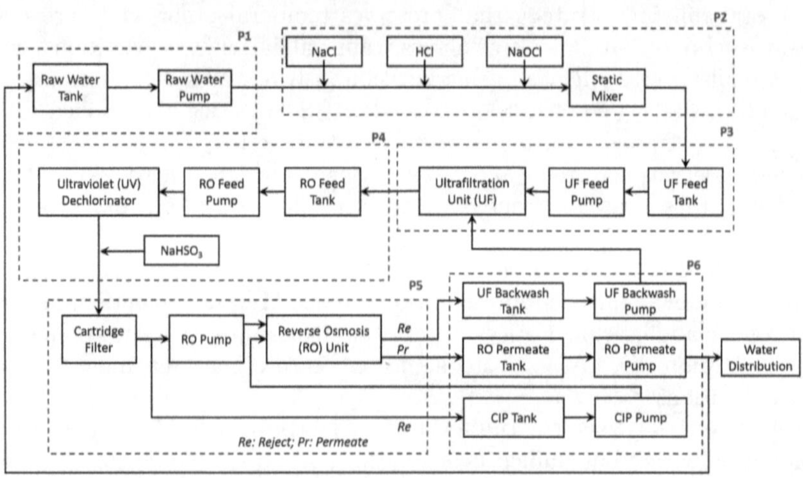

Fig. 1. SWAT testbed [4].

2.2 ICS Anomaly Detection Methods and Research Gaps

ICS anomaly detection is vital for the security of smart industrial applications, where numerous techniques have been developed to identify abnormal system behavior. Recent trends in ICS anomaly detection show a transition from manually engineered features to deep learning-based sequence modeling [3]. However, comprehensive comparisons between these diverse methodologies remain scarce.

Traditional Methods and the Lack of Temporal Information. The dynamic characteristics and attack behaviours of ICS system data are fundamentally tied to their temporal dependencies. However, existing anomaly detection methods generally treat data as independent observations, ignoring temporal information [2]. As pointed out in [2], this limits the ability to capture complex device-to-device correlations and identify complex anomalies. Furthermore, a study in [13] also emphasizes that traditional methods fail to utilize high-frequency and low-frequency temporal information in traffic. In addition, [14] find that simple sliding window approaches struggle to adequately capture underlying temporal dependencies. Therefore, establishing a baseline model that does not consider temporal context is crucial for quantifying the potential performance improvements introduced by incorporating temporal information.

Time-Series Modeling Based on Feature Engineering and Sequence-Based Deep Modeling. To address these limitations, some studies have introduced higher-order features such as lagged values and statistics through feature engineering, or shifted to deep learning sequence models that can directly learn temporal patterns. In [7], the MADICS method was proposed, which leverages the repeatability of ICS data to extract features from autocorrelation and the Discrete Fourier Transform (DFT). In [19], it is noted that traditional machine learning methods rely on labor-intensive domain-specific feature engineering; in contrast, LSTM models trained on raw sequences can achieve performance comparable to the best traditional classifiers without the need for cumbersome engineering. In [5], a deep ensemble model is introduced, which utilizes recurrent neural networks (RNN) and long short-term memory (LSTM) networks to effectively handle variable-length sequences and capture long-term relationships. However despite the widespread adoption of sequence models, many studies lack direct comparisons under unified conditions with feature engineering-based methods [19], making it challenging to accurately assess their performance improvements and applicability.

2.3 Research Gaps

Although existing studies have explored different models or architectures, the following limitations remain prevalent:

(i) Existing temporal processing strategies are often limited to static features or single-model validation, and thus lack an in-depth capture of complex, long-term temporal patterns.
(ii) Data partitioning strategies are inconsistent, and event-level information leakage may affect the reliability of results.
(iii) Most studies have not compared models' detection capabilities across multiple attack scenarios under consistent experimental conditions.

To address these research gaps, this paper compares different time-series modeling strategies under a unified configuration and quantifies their performance differences in ICS attack detection. The study compares a baseline model without time-series information, feature engineering-based time-series modeling, and sequence-based deep modeling to analyze the mechanism of time-series dependency in anomaly detection and explore the differences in effectiveness between the two modeling methods in utilizing time-series information.

3 Methodology

3.1 Research Framework

This study designs three comparative experiments to systematically evaluate the impact of time-series dependencies. We evaluate the following configurations: Model A (a baseline without temporal features), Model B (incorporating manually engineered temporal features), and Model C (a deep learning approach based on sequence modeling). The experiments aim to: 1) verify the gains of temporal feature engineering (A vs B); 2) assess the further advantages of sequence models (B vs C); 3) analyze the differences in performance and applicability between different methods.

Experiment A (Baseline Model). Experiment A provides a reference point for assessing the value of temporal dependency modeling in ICS anomaly detection. Many detection pipelines treat each timestamp as an independent observation, which can limit their ability to capture delayed effects and evolving attack behaviour. To establish a clear baseline under the same data and evaluation conditions, we train a Random Forest classifier using only the original sensor and actuator measurements, without introducing any temporal context through engineered features. The role of this experiment is to define baseline performance against which subsequent temporal approaches can be compared, allowing the performance gains from incorporating temporal information to be measured in a controlled manner.

Experiment B (Feature Based Temporal Modeling). Experiment B evaluates whether temporal dependencies can be captured effectively through feature engineering while keeping the modeling approach within a conventional machine learning setting. Prior studies often introduce limited temporal features, and the impact of those additions is not always quantified under consistent experimental conditions. In this experiment, we derive temporally informative predictors from the original time series signals, such as lagged observations and window based summary statistics, and then train a Random Forest classifier on the resulting enhanced feature set. This setup allows us to quantify the contribution of engineered temporal information and to understand how far a feature driven approach can improve detection performance before adopting sequence learning methods.

Experiment C (Sequence Based Temporal Modeling). Experiment C investigates a sequence aware approach in which temporal structure is learned directly from ordered observations rather than approximated through engineered predictors. Although recurrent neural models are widely used for time series anomaly detection, direct comparisons against feature engineered classical models under a unified evaluation protocol remain limited. To enable sequence learning, we reconstruct the multivariate ICS stream into fixed length sliding window sequences and train an LSTM classifier to learn temporal patterns within each window. The objective is to evaluate the ability of sequence based modeling to capture richer temporal dependencies than feature based modeling, and to characterize the empirical trade offs between these paradigms when applied to the same dataset partitions and evaluation metrics.

Through progressive comparison of A $\rightarrow$ B $\rightarrow$ C, this study evaluates the impact of different temporal processing strategies on ICS attack detection performance, providing empirical reference for system design.

3.2 Data Preprocessing and Stratified Attack-Aware Temporal Split

Attack-aware partitioning is a data partitioning strategy that considers the overall nature of attack events. Its core lies in treating attack events as independent partitioning units, ensuring that all samples from the same event are assigned to only one of the training set, validation set, or test set, thereby fundamentally preventing the leakage of event-level information and achieving more reliable model performance evaluation.

Data Preprocessing and Quality Control This study uses the SWaT 2015 dataset, which has a sampling frequency of 1 Hz and contains 51 sensor and actuator signals, as well as 36 labeled attack events. To ensure the reliability of model training, this study first performs data preprocessing, including missing value handling (multi-layer forward/backward filling) and time index standardization. Additionally, the initial 50,000 rows of stabilization fluctuations are removed from the normal dataset. Following the selection of 31 valid attack events, the final experimental dataset comprises 210,000 samples.

Stratified Hybrid Attack-Aware Partitioning Strategy. To address the issue of insufficient generalization in traditional attack-aware partitioning when dealing with uneven attack types, this study proposes a hierarchical hybrid partitioning strategy:

(i) Hierarchical Allocation of Attack Types: Attack types are categorised into four layers (SSSP, SSMP, MSSP, MSMP, as shown in Table 1 and Fig. 2. All attack types are evenly distributed across the training and validation sets to avoid missing certain attack patterns during the training phase [4].
(ii) Time Sequence Preservation: After hierarchically allocating attack events, normal data is still split according to time sequence (70% training, 15%

validation, and 15% testing) to preserve the natural time dependency in industrial scenarios.

(iii) Event-level independence: Partitioning is based on attack events rather than individual time points, ensuring no event overlap between the training, validation, and test sets, thereby fundamentally preventing future information leakage.

Table 1. Attack categories and their corresponding numbers.

Attack Category	Number of attacks
SSSP	26
S SMP	4
MSSP	2
MSMP	4

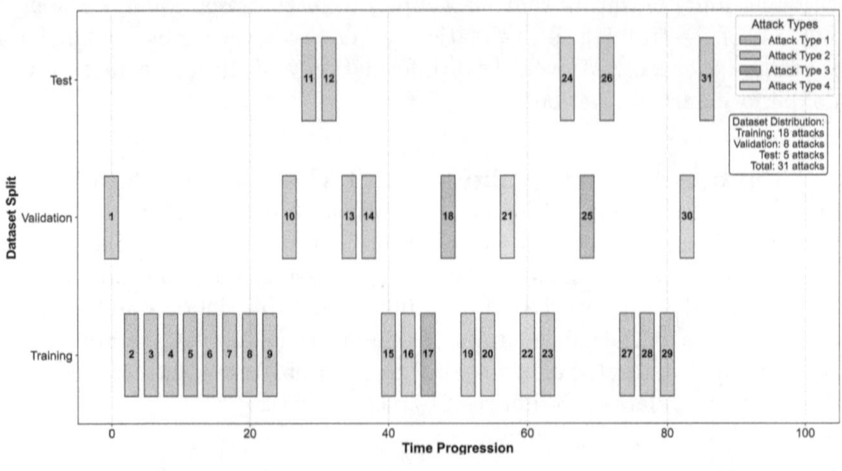

Fig. 2. Hierarchical Allocation of Attack Types.

3.3 Feature Engineering

Feature engineering is a critical step in transforming raw time-series data into learnable features for the model, and is essential for improving ICS attack detection performance. This study designed different feature engineering strategies for feature-based and sequence-based methods.

Feature-Based Approach Random Forest Time-Series Features. In the feature-based approach, this study introduces temporal information through rolling statistics, lag features, trend features, and cross-sensor correlations. To avoid data leakage, this study independently extracts temporal features within the training set, validation set, and test set, ensuring that each data partition only uses its own information for feature construction.

(i) Experiment A (Baseline): Directly uses 51 raw sensor features without temporal information.
(ii) Experiment B (Temporal Enhancement): Calculate temporal features for the top 15 important sensors, including:
   - Statistical Features (30-s sliding window): Calculate the mean, standard deviation, maximum, minimum, and change rate for each sensor.
   - Lag features: Create time lag features (10-s lag and 20-s lag) for important sensors. Lag features directly introduce historical observations, helping the model capture short-term dependencies and periodic patterns.
   - Cross-sensor features: Introduce sensor group correlation features, including correlations within pressure sensor groups, flow sensor groups, and liquid level sensor groups, as well as the frequency of binary sensor state changes. These cross-sensor features are used to model temporal coupling and physical dependencies between ICS stages, as attacks in ICS often affect multiple interconnected sensors.
   - Feature enhancement effect: The original features consisted of 51 sensor features, and the enhanced features reached 154, with an enhancement factor of approximately 3.0x (Table 2).

Table 2. Types of Feature Engineering and Mathematical Definitions.

Feature Type	Definition	Formula	Purpose
Rolling Mean	Mean within a window w	$\mu_t = \frac{1}{w}\sum_{i=1}^{w} x_{t-i}$	Capture short-term trend fluctuations
Rolling Variance	Variance within a window w	$\sigma_t^2 = \frac{1}{w}\sum_{i=1}^{w}(x_{t-i} - \mu_t)^2$	Capture short-term volatility
Lag Feature	Observation k steps earlier	x_{t-k}	Introduce historical observations to capture dependencies
Rate of Change (Slope)	Change slope within window w	$\text{slope} = \frac{x_t - x_{t-w}}{w}$	Identify temporal trends
Sensor Group Correlation	Correlation within sensor group	$\text{corr}(X, Y)$	Model interstage physical dependencies in ICS

For Sequence-Based Methods (Sequence-Based Approach) LSTM Sequence Features. Experiment C (Sequence-based Temporal modeling): To construct 3D sequence inputs for the LSTM model, this study independently constructed sliding window sequences in the training, validation, and test sets to strictly avoid data leakage. At the same time, a reasonable distribution ratio of attack categories was maintained during the sequence generation process to enhance the model's generalisation ability.

- Window Size: A 30-s sliding window is used to construct a 30-s × 51-dimensional input, meaning each sequence contains 30 time steps of data. This sliding window sequence directly captures dynamic patterns within 30 s, and the window design accounts for the delayed effects of attacks, which is critical for identifying gradually unfolding attacks in ICS systems.
- Step Size: The step size is 1-s, ensuring sufficient overlap between sequences to capture continuity.
- Labeling strategy: If the sliding window contains any attack samples, the entire window is labeled as an attack (binary label).

3.4 Model Design and Implementation Details

This section introduces the structural design, implementation tools, and key hyperparameter configurations of the two core models used in this study: random forest and LSTM. In addition, countermeasures for overfitting, performance optimization methods, and reproducibility assurance mechanisms are also provided. Through clear model settings and reasonable implementation details, the reliability and interpretability of subsequent experimental results are ensured. The relevant model configuration information is shown in Table 3.

Table 3. Comparison of Random Forest and LSTM configurations.

Configuration Item	Random Forest	LSTM
Implementation Framework	Scikit-learn	*PyTorch*
Network Architecture	Ensemble of 500 decision trees	Two-layer LSTM (64 units) + Fully Connected layer
Input Features	154 enhanced statistical features	Sliding window: 30 × 51-dimensional time-series features
Key Parameters	`max_depth=20, class_weight=balanced`	`lr=0.0005, dropout=0.3, batch_size=32`
Tuning Strategy	Feature engineering + Threshold optimization	Grid search (20 combinations) + Early stopping
Threshold Optimization	Fine-grained scan (step = 0.005)	Fine-grained scan (step = 0.005)
Application Scenario	High interpretability, suitable for low-latency environments	Suitable for scenarios requiring temporal modeling and context awareness

3.5 Evaluation Metrics

Metrics for Model Performance Assessment. To address the issue of class imbalance, this study comprehensively evaluates model performance using the following metrics:

F1-score is the primary evaluation metric in this study. It represents the harmonic mean of precision and recall, providing a balanced metric particularly suited for imbalanced data scenarios. In industrial attack detection, the F1-score is critical as it strikes the necessary balance between minimising false positives (leading to operational issues) and minimising false negatives (allowing attacks to succeed), reflecting the trade-offs in real-world operations.

$$F_1 = \frac{2 \times P \times R}{P + R} \tag{1}$$

Precision measures the proportion of correctly identified positive examples among all instances predicted as positive. In ICS, high precision is critical for minimising false positives, as false positives can lead to alert fatigue, resource waste, or even operational disruptions [18].

$$P = \frac{TP}{TP + FP} \tag{2}$$

Recall measures the proportion of true positives correctly identified among all actual positive instances. High recall is paramount in ICS environments to maximize attack detection coverage and minimize the operational risks associated with false negatives.

$$R = \frac{TP}{TP + FN} \tag{3}$$

AUC-ROC (Area Under the Receiver Operating Characteristic Curve). AUC-ROC is also used to comprehensively evaluate a model's ability to distinguish between normal and abnormal samples at different classification thresholds. It provides a robust metric for overall discriminative capability independent of specific thresholds.

Accuracy measures the proportion of correct predictions out of all predictions made by the model. It is a common overall performance metric in most classification tasks. However, in scenarios with severe class imbalance, such as ICS attack detection, accuracy can be misleading: even if the model consistently predicts 'normal,' it may still achieve a high accuracy rate. Therefore, while accuracy provides an intuitive reference for overall correctness, it is used only as an auxiliary metric in this study.

$$\text{Accuracy} = \frac{TP + TN}{TP + TN + FP + FN} \tag{4}$$

Decision Threshold Optimization. Decision threshold optimization is a key step in this study. In machine learning classification tasks, the decision threshold is the boundary point at which the model divides the predicted probability into positive (attack) or negative (normal) classes. Optimizing the decision threshold aims to find the optimal balance between precision and recall by adjusting this boundary point, particularly when dealing with class imbalance issues. This study determines the optimal threshold by analyzing the relationship curve between the F1-Score on the validation set and the threshold, which is then applied to the test set to report the final performance.

4 Results and Evaluation

4.1 Experimental Setup

This section presents and analyses the results of three comparative experiments based on the Attack-aware partitioning strategy on the SWaT dataset. Table 4 summarizes the distribution of normal and attack samples across the training, validation, and test sets, ensuring representative attack events in each subset while preserving temporal order. The F1-Score is used as the primary evaluation metric, supplemented by precision, recall, accuracy, and AUC-ROC for a comprehensive assessment.

Table 4. Dataset distribution across training, validation, and test sets.

Dataset	Total Samples	Normal Samples	Attack Samples
Training Set	154,542	111,591	42,951
Validation Set	29,831	23,887	5,944
Test Set	25,627	23,887	1,740
Total	210,000	159,290	50,710

4.2 Phase 1: Validating the Effectiveness of Temporal Feature Engineering (Experiment A vs. Experiment B)

This phase aims to quantify how temporal information contributes to model performance by comparing a baseline configuration (Experiment A) with an enhanced version incorporating time-series features (Experiment B).

Baseline Model Performance (Experiment A). In Experiment A, the random forest baseline model was trained using the original 51 sensor features. As shown in Fig. 3, by optimizing the decision threshold to 0.94, the model achieved a significant performance improvement on the test set. The F1-score improved from 0.142 at the default threshold to 0.260, while accuracy and precision also improved from 0.181 and 0.076 to 0.616 and 0.150, respectively. However, despite these improvements, the model's performance remains relatively limited. As shown in Fig. 4, although the number of false positives (FP) decreased from 20,999 to 9,851 after optimization, it remains at a high level, exposing the model's fundamental limitations when lacking temporal information.

Performance of the Time Series Feature Enhancement Model (Experiment B). In Experiment B, we introduced 154 time series features through feature engineering. As shown in Fig. 5, through decision threshold optimisation, the model performance was significantly improved, with the F1-score increasing

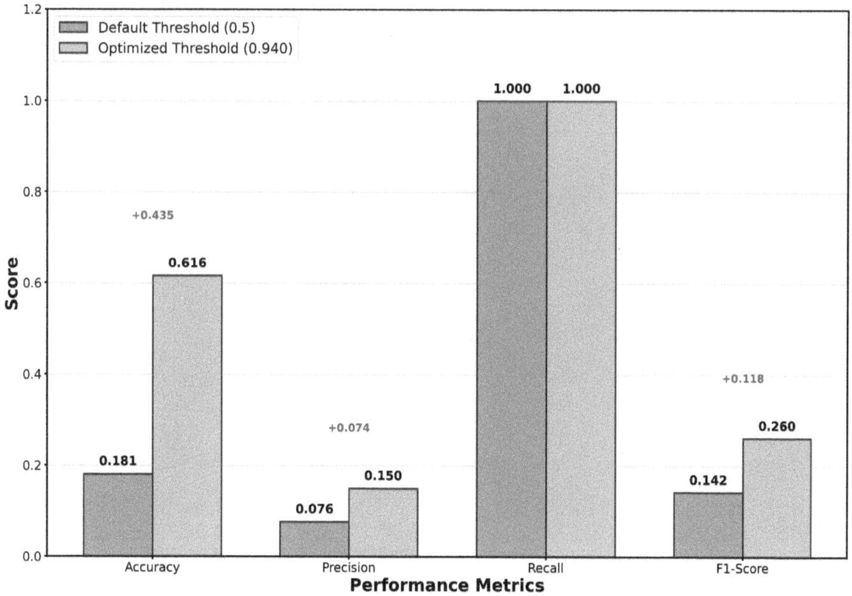

Fig. 3. Baseline Model Analysis.

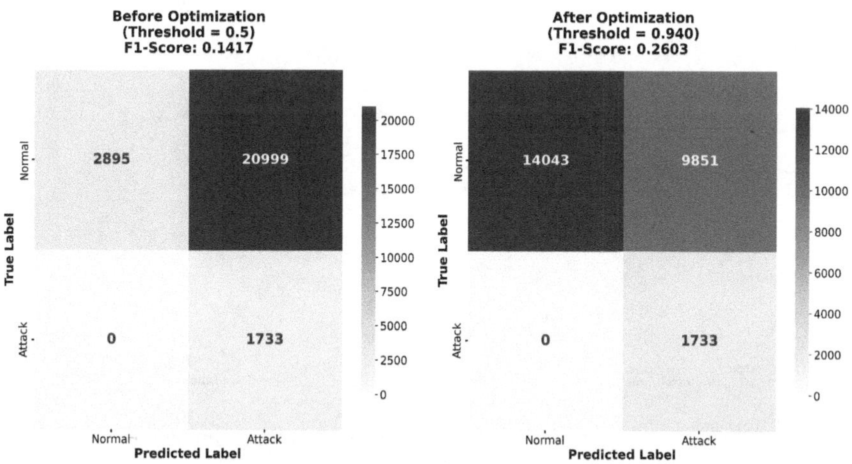

Fig. 4. Optimized confusion matrix.

from 0.190 to 0.737. The precision rate increased sharply from 0.105 to 0.974, an increase of 86.9%, while the recall rate decreased from 0.999 to 0.592. Figure 6 shows a confusion matrix that further illustrates the reasons behind the performance changes. At the default threshold, the model generated as many as 14,763 false positives (FP), but after optimization, this number dropped sharply to just 27. This directly explains why the precision rate jumped from 0.105 to

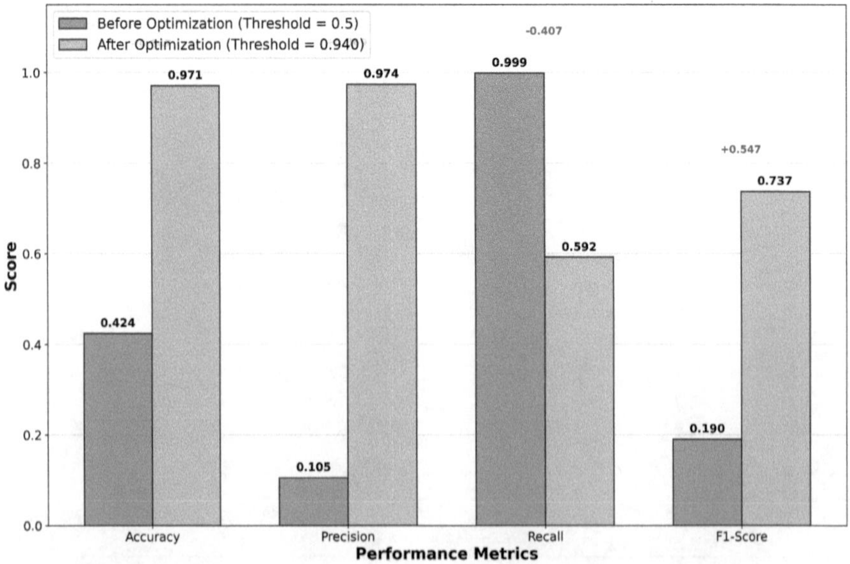

Fig. 5. Decision threshold optimization analysis of the enhanced model.

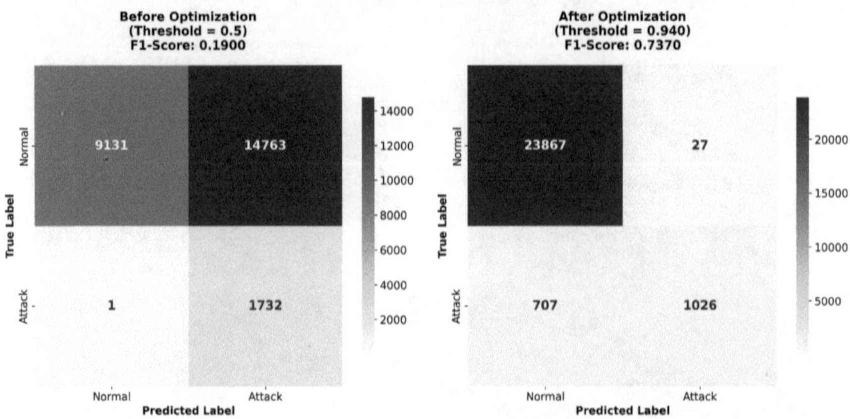

Fig. 6. Confusion matrix before and after optimization of the enhanced model.

0.974. Meanwhile, under the default threshold, the model only missed 1 attack sample (FN = 1), achieving a recall rate close to perfect (0.999). After optimization, although the recall rate decreased to 0.592, the model was able to identify attacks with extremely high precision, significantly reducing false positives, thereby achieving a balance in performance.

4.3 Comparative Analysis (Experiment A vs. Experiment B) and Feature Importance

As shown in Fig. 5, Experiment B significantly outperformed Experiment A in all metrics. The F1-score of Experiment B reached 0.737, an increase of 0.477 compared to the baseline model (0.260), representing a substantial improvement. This indicates that the introduction of temporal features has enabled the model to achieve a better balance between precision and recall. Its precision rate reached 0.974, while the baseline model only achieved 0.150, an improvement of 0.824. This indicates that temporal information significantly enhances the model's discriminative ability, enabling it to identify attacks with an extremely low false positive rate. As shown in Fig. 7, the accuracy improved from 0.616 in the baseline model to 0.971, an increase of 0.355, indicating that the enhanced model has significantly improved its overall classification ability.

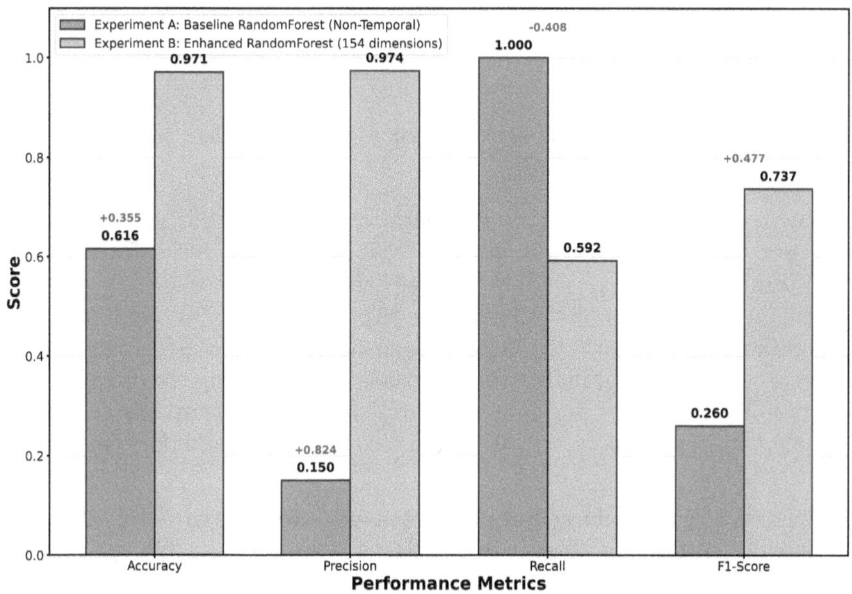

Fig. 7. Overview of Enhanced Model After Threshold Optimisation.

While the trade-off between precision and recall is expected, the decrease in the recall rate from 1.000 to 0.592 is admittedly sharp, highlighting the need for further optimization prior to real-world deployment. Although the baseline model offers a high recall rate, its extremely low precision renders it practically useless. The enhanced model sacrifices recall to achieve high precision, making it a highly practical choice for industrial environments where minimizing false positives is critical to maintaining operational reliability and preventing alarm fatigue [18]. Further analysis of the feature importance in Experiment

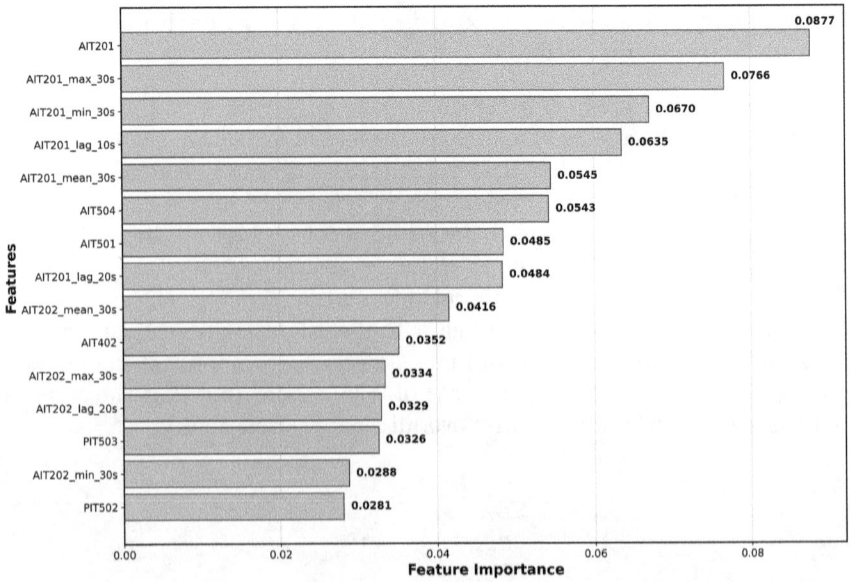

Fig. 8. The 15 most important enhanced features in Experiment B.

B reveals the core reasons behind the performance improvement. As shown in Fig. 8, over half of the top 15 important features are time-series features (e.g., `AIT201_min_30s`, where the suffix indicates a feature engineering operation such as minimum value over a 30-s window; see Appendix A for full naming conventions), rather than the original instantaneous sensor values. This validates our core hypothesis: temporal information is crucial for capturing the dynamic characteristics of ICS attacks, and domain-knowledge-based feature engineering can effectively transform this information into patterns that the model can learn.

4.4 Phase 2: Comparing Sequence-Based and Feature-Based Time Series Modelling (Experiment B vs. Experiment C)

This stage aims to compare the random forest based on feature engineering (Experiment B) with the LSTM model based on sequences (Experiment C) to evaluate the advantages and disadvantages of the two time series processing paradigms.

Feature-Based Time Series Modelling (Experiment C). In Experiment C, a two-layer LSTM model was used to directly learn from sliding window sequences to assess its ability to natively handle temporal dependencies. The final performance of the model is summarized across the subsequent figures. First, as shown in Fig. 9, the LSTM model achieved an F1-score of 0.8203 on the test set, which is higher than that of the enhanced random forest model. Its precision rate is 0.9439, recall rate is 0.7253, and accuracy rate is as high as 0.9781.

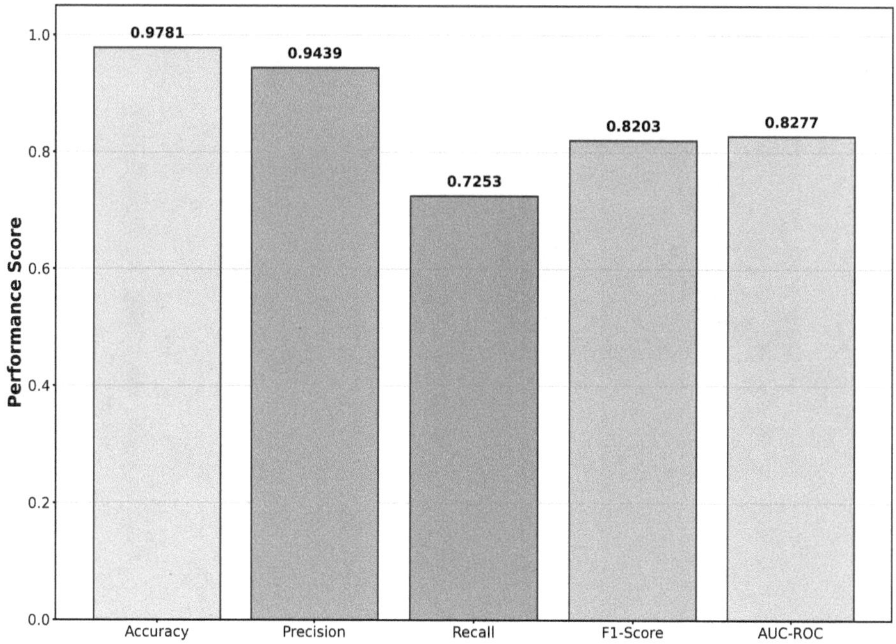

Fig. 9. Performance metrics of the LSTM model.

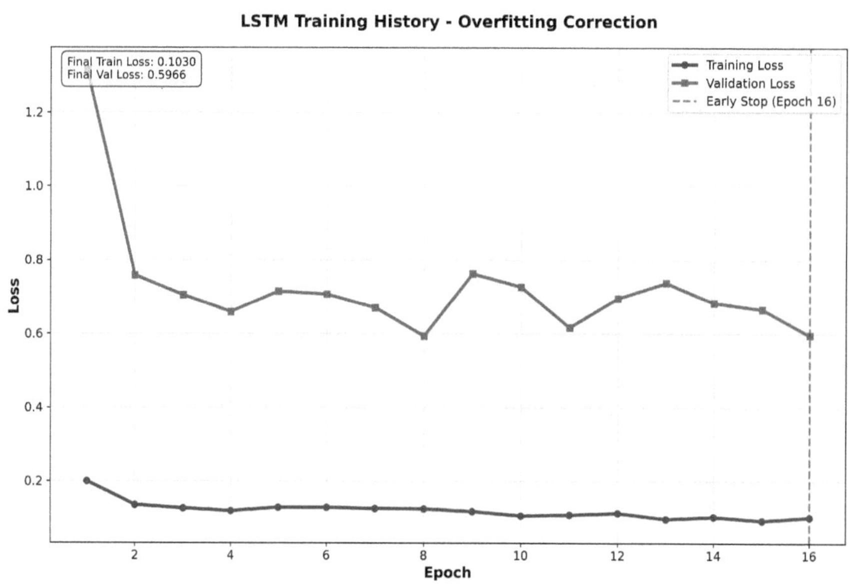

Fig. 10. Training and validation loss curves of the LSTM model.

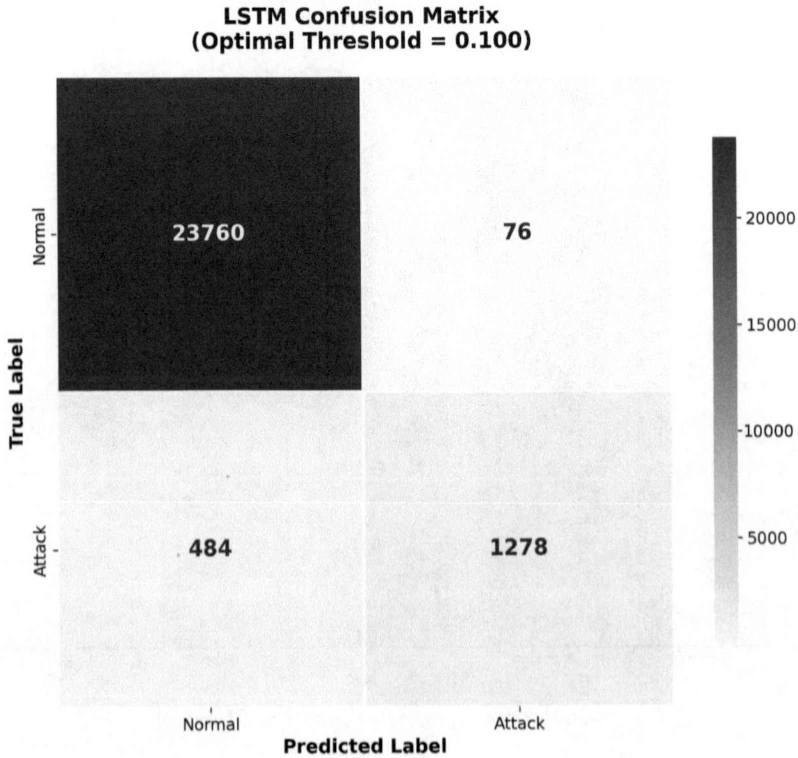

Fig. 11. Confusion Matrix of the LSTM Model at the Optimal Threshold.

The model's training process also demonstrated good convergence. As shown in Fig. 10, both training loss and validation loss exhibited a stable downward trend, with validation loss leveling off after the 12th epoch. The early stopping mechanism was triggered at the 16th epoch, effectively avoiding overfitting and ensuring the model's generalization capability. Furthermore, the model's performance at the optimal threshold was analyzed using a confusion matrix and probability distribution plot. As shown in Fig. 11, at the optimal threshold of 0.100, the model successfully classified the majority of normal samples and a portion of attack samples correctly.

Performance Comparison Between Sequence-Based and Feature-Based Models. As shown in Table 5, the comparative results of the three experiments clearly quantify the significant value of temporal modeling. The baseline model performed the worst across all metrics, with an F1-score of only 0.260. In contrast, the temporal modeling methods (Experiments B and C) achieved significant performance improvements. Among the two temporal modeling paradigms, the sequence-based LSTM model (Experiment C) outperformed the feature-based enhanced RF model (Experiment B) in terms of F1-

Table 5. Performance comparison of Baseline RF, Enhanced RF, and Optimized LSTM models.

Metric	Baseline RF	Enhanced RF	Optimized LSTM
F1-Score	0.260	0.737	0.8203
Accuracy	0.616	0.971	0.9781
Precision	0.150	0.974	0.9439
Recall	1.000	0.592	0.7253
AUC-ROC	0.750	0.865	0.8277

score, accuracy, and recall. The F1-score of the LSTM model reached 0.8203, higher than the 0.737 of the enhanced RF model. Especially in terms of recall, the LSTM model achieved 0.7253, significantly outperforming the enhanced RF model's 0.592. However, the enhanced RF model performed better in terms of accuracy (0.974) and AUC-ROC (0.865). This further supports the conclusion of this study: temporal modeling can significantly enhance performance, but the two distinct temporal modeling paradigms exhibit trade-offs in specific performance metrics.

5 Conclusions

This work aims to systematically evaluate the impact of temporal dependency processing on the performance of ICS attack detection and compares two paradigms: feature engineering-based and sequence modelling-based approaches. By designing three comparative experiments on the SWaT dataset, this study successfully validated and quantified the critical role of temporal information in ICS anomaly detection. In line with the research objectives outlined in the introduction, this study has successfully achieved the following:

- Validate the improvement in ICS attack detection performance through temporal dependency processing. The comparison results between Experiments A and B strongly support this conclusion. After introducing temporal features such as lagged features and rolling statistics, the model's F1-score significantly improved from 0.260 to 0.737.
- Compare and analyze sequence-aware methods with traditional methods enhanced through feature engineering. The comparison results of Experiments B and C provide an empirical answer to this question. The LSTM model achieved an outstanding F1-score of 0.8203, slightly outperforming the enhanced random forest, demonstrating the potential of deep sequence models in capturing complex temporal patterns. This study offers valuable references for selecting appropriate temporal modeling strategies across different industrial scenarios by comparing the performance, training costs, and interpretability of the two methods.

Additionally, this study reveals the trade-off between performance and interpretability in feature engineering and deep learning models. While LSTM outperforms in terms of performance due to its ability to model complex dynamic patterns, traditional models like random forests still hold significant advantages in training speed and feature interpretability, particularly in safety-critical industrial scenarios requiring human intervention. The study demonstrates that time-series modeling is indispensable in ICS attack detection, and different methods can be flexibly selected and combined based on actual deployment requirements, providing a technically viable path for industrial system security protection.

Acknowledgments. This work was supported by the National Edge AI Hub for Real Data: Edge Intelligence for Cyber-disturbances and Data Quality, EPSRC, EP/Y028813/1.

Disclosure of Interests. The authors have no competing interests to declare that are relevant to the content of this article.

A Feature Naming Rules

(see Table 6)

Table 6. Sensor codes, feature prefixes, and their descriptions

Prefix/Example Feature Name	Sensor Code	Description
AIT201, AIT202, AIT203, AIT402, AIT501, AIT504	AIT-xxx	Analyzer âĂŞ Instrumentation Transmitter. Measures parameters such as conductivity, pH, ORP, and hardness of water.
FIT201, FIT301, FIT401	FIT-xxx	Flow Indicator Transmitter. Measures or transmits water flow rates in different process stages.
LIT101, LIT301, LIT401	LIT-xxx	Level Indicator Transmitter. Monitors water levels in tanks at various stages.
PIT502, PIT503	PIT-xxx	Pressure Indicator Transmitter. Measures pressure in specific process points.
max_30s	—	Maximum value of the sensor reading over the past 30 s.
min_30s	—	Minimum value of the sensor reading over the past 30 s.
mean_30s	—	Mean (average) value of the sensor reading over the past 30 s.
lag_10s	—	Sensor reading value lagged by 10 s.
lag_20s	—	Sensor reading value lagged by 20 s.
lag_30s	—	Sensor reading value lagged by 30 s.

Notes

- Sensor codes (e.g., AIT201) follow the SWaT dataset naming convention, where the prefix indicates the sensor type and the number identifies its location and process stage.
- Time-window statistics (max_30s, min_30s, mean_30s) and lag features (lag_10s, etc.) are generated during feature engineering to capture temporal dynamics.

B Supplementary Experimental Results of Baseline Model:

This appendix provides supplementary charts and data for the experiment A results discussed in the main text of this paper, with the aim of providing more comprehensive analytical details and supporting evidence (Figs. 12 and 13).

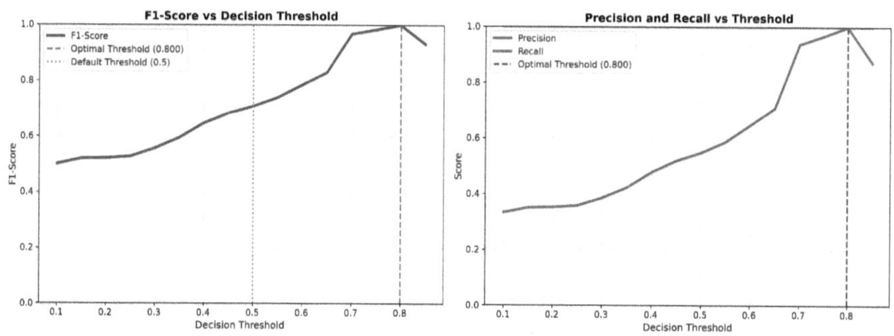

Fig. 12. Fig B.1 Performance comparison before and after baseline model optimization and threshold analysis.

C Supplementary Experimental Results of Feature-Based Temporal Modelling

This appendix provides supplementary charts and data for the experiment B results discussed in the main text of this paper, with the aim of providing more comprehensive analytical details and supporting evidence (Figs. 14 and 15).

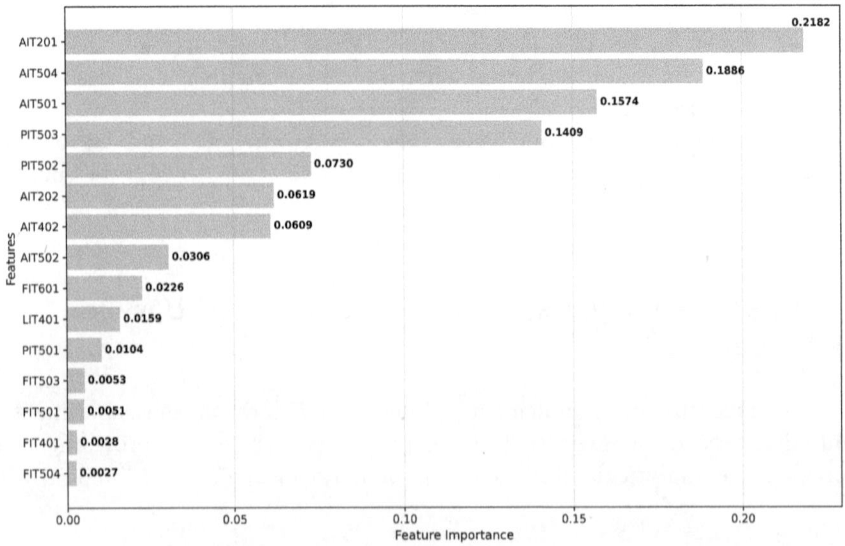

Fig. 13. Fig B.2 Top 15 Most Important Features - Random Forest Baseline Model.

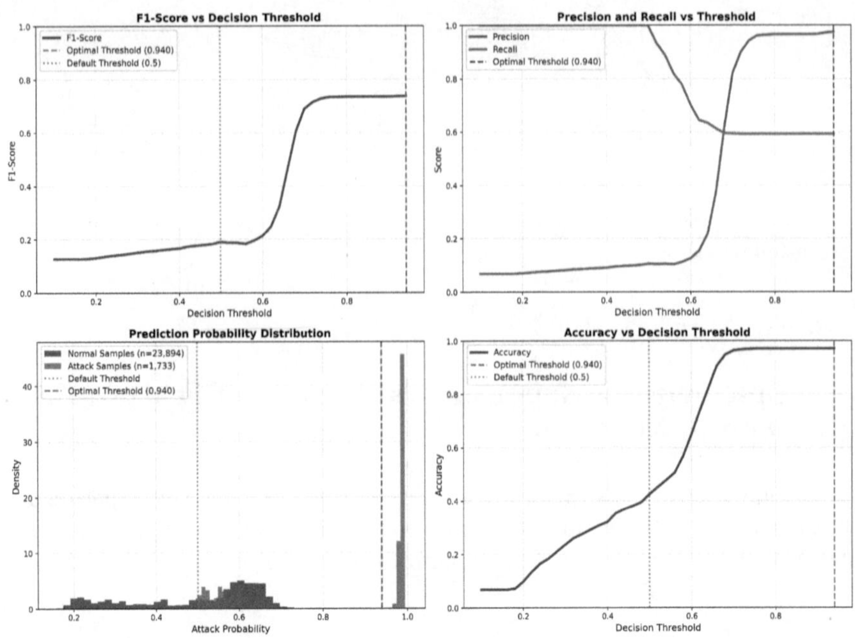

Fig. 14. Fig C.1 Performance comparison before and after Feature-based Temporal Modelling optimization and threshold analysis.

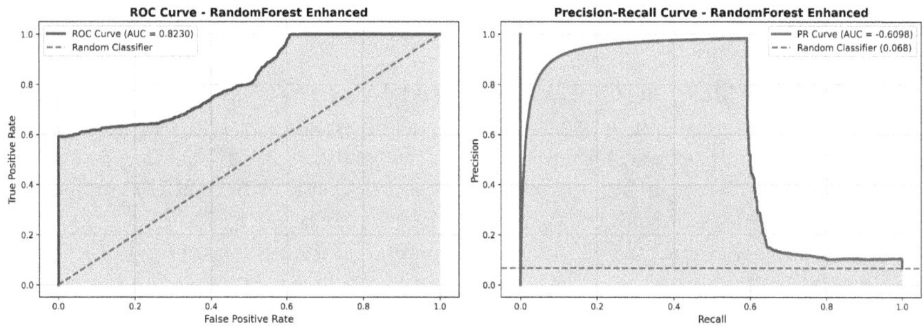

Fig. 15. Fig C.2 ROC and Precision-Recall Curves for the Random Forest Enhanced Model.

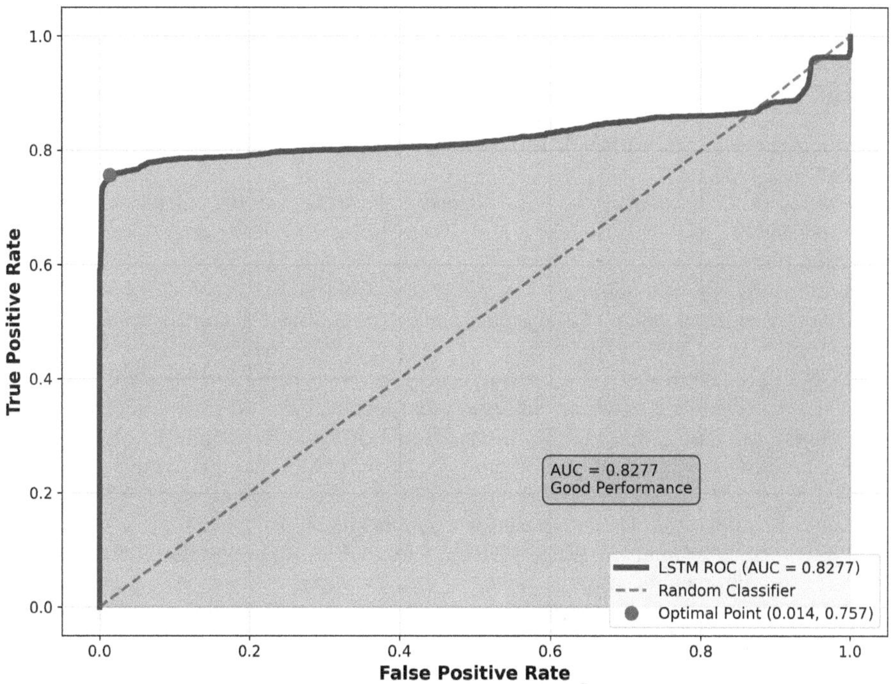

Fig. 16. Fig D.1 LSTM ROC Curve Analysis.

D Supplementary Experimental Results of Sequence-Based Temporal Modelling

This appendix provides supplementary charts and data for the experiment C results discussed in the main text of this paper, with the aim of providing more comprehensive analytical details and supporting evidence (Fig. 16).

References

1. Ahmed, C.M., Murguia, C., Ruths, J.: Model-based attack detection scheme for smart water distribution networks. In: Proceedings of the 2017 ACM on Asia Conference on Computer and Communications Security, ASIA CCS '17, pp. 101–113. Association for Computing Machinery, New York, NY, USA (2017). https://doi.org/10.1145/3052973.3053011
2. Birihanu, E., Lendák, I.: Explainable correlation-based anomaly detection for industrial control systems. Front. Artif. Intell. **7** (2025). https://doi.org/10.3389/frai.2024.1508821
3. Digital Sense AI: Detecting anomalies in time series (2025). https://www.digitalsense.ai/blog/detecting-anomalies-in-time-series. Accessed 22 Oct 2025
4. Goh, J., Adepu, S., Junejo, K.N., Mathur, A.: A dataset to support research in the design of secure water treatment systems. In: Havarneanu, G., Setola, R., Nassopoulos, H., Wolthusen, S. (eds.) CRITIS 2016. LNCS, vol. 10242, pp. 88–99. Springer, Cham (2017). https://doi.org/10.1007/978-3-319-71368-7_8
5. Iqbal, A., Amin, R.: Time series forecasting and anomaly detection using deep learning. Comput. Chem. Eng. **182**, 108560 (2024). https://doi.org/10.1016/j.compchemeng.2023.108560, https://www.sciencedirect.com/science/article/pii/S0098135423004301
6. Knapp, E.D.: Industrial Network Security: Securing Critical Infrastructure Networks for Smart Grid, SCADA, and Other Industrial Control Systems. Elsevier (2024)
7. Ángel Luis Perales Gómez, Maimó, L.F., Celdrán, A.H., Clemente, F.J.G.: MADICS: a methodology for anomaly detection in industrial control systems. Symmetry **12**(10), 1583 (2020). https://doi.org/10.3390/sym12101583
8. M. R., G.R., Ahmed, C.M., Mathur, A.: Machine learning for intrusion detection in industrial control systems: challenges and lessons from experimental evaluation. Cybersecurity **4**(1), 1–12 (2021). https://doi.org/10.1186/s42400-021-00095-5
9. Palleti, V.R., Mishra, V.K., Ahmed, C.M., Mathur, A.: Can replay attacks designed to steal water from water distribution systems remain undetected? ACM Trans. Cyber-Phys. Syst. **5**(1) (2021). https://doi.org/10.1145/3406764
10. Qadeer, R., Murguia, C., Ahmed, C.M., Ruths, J.: Multistage downstream attack detection in a cyber physical system. In: Katsikas, S.K., et al. (eds.) CyberICPS/SECPRE -2017. LNCS, vol. 10683, pp. 177–185. Springer, Cham (2018). https://doi.org/10.1007/978-3-319-72817-9_12
11. Rockwell Automation: What is ICS security? (2025). https://www.rockwellautomation.com/en-gb/company/news/blogs/what-is-ics-security.html. Accessed 22 Oct 2025
12. Sapkota, S., Mehdy, A.K.M.N., Reese, S., Mehrpouyan, H.: FALCON: framework for anomaly detection in industrial control systems. Electronics **9** (2020). https://doi.org/10.3390/electronics9081192
13. Shang, W., Qiu, J., Shi, H., Wang, S., Ding, L., Xiao, Y.: An efficient anomaly detection method for industrial control systems: deep convolutional autoencoding transformer network. Int. J. Intell. Syst. **2024**(1), 5459452 (2024)
14. Shen, L., Li, Z., Kwok, J.: Timeseries anomaly detection using temporal hierarchical one-class network. In: Advances in Neural Information Processing Systems (NeurIPS), vol. 33, pp. 13016–13026 (2020)
15. Trirat, P., et al.: Universal time-series representation learning: a survey (2024). https://doi.org/10.48550/arXiv.2401.03717

16. Umer, M.A., Ahmed, C.M., Jilani, M.T., Mathur, A.P.: Attack rules: an adversarial approach to generate attacks for industrial control systems using machine learning. In: Proceedings of the 2th Workshop on CPS&IoT Security and Privacy, pp. 35–40 (2021)
17. Umer, M.A., Ahmed, C.M., Mathur, A.P., Jilani, M.T.: Attack pattern mining to discover hidden threats to industrial control systems. Int. J. Inf. Secur. **25**(2), 70 (2026)
18. Umer, M.A., Mathur, A.P., Junejo, K.N.: Research note: the impact of absolute false positive rates on the operational reliability of anomaly detectors in critical infrastructure. Reliab. Eng. Syst. Saf., 112760 (2026)
19. Verner, A.: LSTM networks for detection and classification of anomalies in raw sensor data. Ph.D. thesis, Nova Southeastern University (2019). https://nsuworks.nova.edu/gscis_etd/1074
20. Xu, L., Shang, K., Zhang, X., Zheng, C., Pan, L.: Multi-scale feature fusion-based real-time anomaly detection in industrial control systems. Electronics **14**, 1645 (2025). https://doi.org/10.3390/electronics14081645

DCS-Water 2025

Customizing Lightweight TTnet Architecture with TFHE for Privacy-Preserving Deep Learning Time-Series Analytics in Water Infrastructures Monitoring

Alvin Wei Ze Chew[1,2](✉)

[1] Singapore University of Technology and Design, 8 Somapah Road, Singapore 487372, Singapore
[2] Bentley Systems (Singapore) Pte Ltd, 3 Harbourfront Place, #05-01 to 02 Harbourfront Tower Two, Singapore 099254, Singapore
alvin_chew@mymail.sutd.edu.sg

Abstract. This paper presents an alternative privacy-preserving deep learning framework for time-series forecasting in water infrastructure monitoring by customizing an original lightweight TTnet architecture to operate under a fully homomorphic encryption (FHE) scheme, specifically the known Fast Fully Homomorphic Encryption over the Torus (TFHE) from the present literature by adopting an honest-but-curious (semi-honest) threat model. We first introduced a modified CNN-LSTM-based TTnet model that captures complex temporal dependencies in hydraulic parameters, such as pressure, while ensuring end-to-end data confidentiality. Our method quantizes input signals and network weights into low-bit representations which can enable encrypted inference using Zama's Concrete compiler, followed by decrypting the forecasting results. We validate our proposed framework using pressure data from the BattLeDIM benchmark dataset and demonstrate that the proposed encrypted pipeline achieves an average RMSE difference of less than 0.5 mH_2O compared to plaintext predictions. Despite a latency of approximately 5.2 s per inference window, the system remains practical for low-frequency monitoring scenarios, such as data collection and sampling every 1–5 min. Additionally, the model remains competitive with traditional regressors, therefore supporting secure analytics without exposing sensitive sensor data during the entire encryption and decryption pipeline processes. Overall, our results demonstrate the feasibility of deploying secure and encrypted deep learning model pipelines for water distribution system monitoring, hence offering a promising path forward for other important operational management and monitoring applications in water infrastructure systems.

Keywords: Homomorphic Encryption · Secure Analytics · TFHE · Water Infrastructure Systems

S. Zonouz et al. (Eds.): DCS-CI 2026, CCIS 3109, pp. 193–213, 2027.
https://doi.org/10.1007/978-981-92-5039-4_10

1 Introduction

Water infrastructure systems rely on field sensors to monitor important hydraulic-related parameters such as pressure, flow rate, and water quality in the near real-time (NRT) context. In general, the sensors generate continuous, or in some cases intermittent, time-series data on defined intervals (e.g., every minute or hour) which can be leveraged to detect anomalies, predict equipment failures, and optimize water distribution operational efficiency [12]. Advanced analytics, including but not limited to numerical modeling, machine learning (ML), and deep learning models, have shown promise in extracting useful data insights for engineers and operators managing water infrastructure systems. For instance, convolutional neural networks (CNNs) and other time-series predictive models can identify subtle anomalies or patterns that are indicative of pipe bursts or contamination events in water distribution systems (WDS) [18]. In practice, these data-insight feedback mechanisms for engineers and operators can improve operational decision-making and mitigate costly infrastructure damage by enabling early interventions such as predictive maintenance.

Despite the expected operational benefits, an important data privacy challenge arises when leveraging cloud-based or third-party technologies for analyzing water-infrastructure-related data. Notably, smart water meter readings and sensor data, such as flows and pressures, may contain sensitive information about consumer water consumption patterns or operational details of critical water infrastructures, where unauthorized access or data leakage poses privacy and security risks. Strict data protection regulations and the competitive nature of utility operations often limit the sharing of raw data with external analysis platforms. For instance, recent work on smart metering systems [19] has demonstrated the use of homomorphic encryption to enable secure analysis of consumption data, hence underscoring the potential of FHE for such scenarios. Therefore, this research work motivates the need for privacy-preserving data analytics techniques that can, on one hand, extract important data values and insights while, on the other hand, not expose the raw sensor data to external analysts or third-party service providers.

Fully Homomorphic Encryption (FHE) has emerged as a promising technology to enable privacy-preserving computations on encrypted data. FHE schemes permit arbitrary computations to be performed on ciphertexts such that the results, when decrypted, match those obtained if the operations were performed on the plaintext. In the context of water infrastructures, or specifically WDS monitoring processes, FHE enables cloud or external servers to run complex analytics on encrypted sensor data without ever seeing the data in clear text. Recent advances in FHE, such as the Fast Fully Homomorphic Encryption over the Torus (TFHE) library for fast binary gate homomorphic operations [4], have since made it feasible to evaluate certain machine learning models over encrypted time-series datasets. However, applying deep learning models under FHE remains challenging due to the immense computational overhead and constraints on the types of mathematical operations supported.

This paper therefore addresses the problem statement of enabling advanced time-series analytics, i.e., deep learning methods, on water infrastructure sensor data under encryption. We modified and customized a specific deep learning model titled TTnet [3], which we have adapted for encrypted execution using the TFHE-FHE scheme [2]. In short, we modified TTnet into a customized time-series neural network architecture for modeling continuous time-series sensor datasets adopted in the popular WDS field. Our goal is to preserve the accuracy of hydraulic data, such as pressure predictions, from the modified TTnet model, which can be useful for important operational tasks such as leak detection, while ensuring that the data and intermediate computations remain confidential. In summary, this work contributes: (i) a privacy-preserving FHE analytics framework for flow and pressure time-series data in the WDS domain; (ii) a customized TTnet architecture tailored to continuous time-series data and encrypted inference; (iii) a TFHE-compatible adaptation using reduced model complexity and binary or low-bit quantized operations; and (iv) an end-to-end TFHE pipeline covering data encryption, encrypted inference, decryption, and evaluation on representative pressure time-series data from the BattLeDIM benchmark [16].

Additionally, the proposed work adopts an honest-but-curious (semi-honest) threat model. Under this model, the third-party cloud server, e.g., an analytics platform, executes all required computational steps but may attempt to infer sensitive information from any available data that it can observe, including but not limited to encrypted payloads or model metadata. In the context of water infrastructure monitoring, this is particularly relevant as hydraulic-related data, such as pressure data, can uncover operational schedules, customer usage patterns, or even signatures of possible operational vulnerabilities. Hence, it is natural and instinctive for a cybercriminal to infer this sensitive information despite being restricted to the encrypted domain. Therefore, to mitigate this risk, the above-mentioned end-to-end TFHE encryption pipeline ensures that all sensor inputs, feature maps, and output predictions remain encrypted throughout the entire inference process, where only the trusted client possesses the secret key required for decryption. The server cannot decrypt or reverse-engineer any aspect of the data, thus preserving data confidentiality even in untrusted or externally hosted computing environments. Overall, this model aligns with real-world deployment scenarios, where global utilities, such as PUB Singapore, are increasingly outsourcing computation to cloud-based AI platforms while imposing strict data protection requirements on the third-party cloud server that is assigned to process and analyze the client's data.

2 Related Studies

In general, the fear of having sensitive operational data stored and analyzed off-premises is often regarded as a major barrier to cloud service adoption. This has thus motivated research into different methods of analyzing data without exposing the original raw data measurements. For example, federated learning allows collaborative model training across multiple sites while keeping local sensor data on-premises; an approach noted for its privacy-preserving potential in water infrastructure monitoring [11]. Similarly, techniques like differential privacy inject statistical noise into time-series data to mask individual patterns. While these methods may provide partial solutions, they come with trade-offs, namely: (a) differential privacy can degrade analytics accuracy by perturbing the original data, and (b) if federated learning is adopted, it can leak information through model updates or require trust in the aggregation process [19].

In this study, we target the scenario where sensitive sensor data, e.g., pressure readings from water pipelines, must remain confidential, even as it is being processed to detect events like anomalies in a WDS. Traditional approaches to obtain privacy in data analytics often involve data anonymization, e.g., using unique identifiers. However, these known methods may not work effectively for time-series data in general since it is computationally expensive to encode and decode each temporal data point. Additionally, fine-grained temporal patterns can be complex; hence, even if personal identifiers are removed, the shape of usage curves or pressure fluctuations might reveal behaviors or identities based on domain understanding and careful engineering inspection. For instance, water pressure data with adequate time resolution can indirectly reveal water consumption routines, e.g., detecting absences or specific water-use activities, which then poses serious privacy risks [19]. Simple aggregation or removal of identifiers would not prevent such inferences. Instead, our approach leverages encryption to protect data during processing. By employing fully homomorphic encryption, in particular the TFHE scheme, we ensure that the data stays encrypted throughout the analysis.

As highlighted earlier, homomorphic encryption (HE) allows computations to be performed directly on ciphertexts, so that a cloud or third-party service can execute analytics without ever seeing plaintext values [5]. This therefore addresses the trust issue by design where the analytics platform learns nothing about the raw sensor readings, but only encrypted results. The key challenge, however, is that typical deep learning models for time-series are not designed to run on encrypted data. Operations like multiplications, convolutions, and nonlinear activations, which are trivial in plaintext, become orders of magnitude more expensive under FHE. For example, a single binary gate operation under TFHE incurs a latency on the order of milliseconds due to costly cryptographic bootstrapping, whereas an equivalent logic operation on unencrypted data is on the order of nanoseconds [2]. Moreover, certain operations, e.g., high-precision nonlinear functions like ReLU or sigmoid activations, are not directly supported or efficient in current FHE frameworks and often must be replaced or approximated by polynomial functions or lookup tables [5]. These constraints mean that naively applying a standard deep neural network to encrypted time-series data would be infeasible in practice, therefore motivating us to redesign the model architecture and computations.

Complementing the cryptographic schemes themselves, several frameworks have emerged to enable secure neural inference, combining HE with secure multiparty computation (MPC), garbled circuits, or hybrid protocols. For example, CryptoNets was among the earliest pure-HE solutions for encrypted inference, although its performance was limited. Subsequent frameworks like GAZELLE [7] improved latency dramatically by executing linear layers via HE and switching to garbled circuits for nonlinear functions such as ReLU. Chameleon [13] and SecureNN [17] further advanced the known MPC-based private inference that enables two-party collaborative inference using additive secret sharing and minimal communication overhead. These frameworks demonstrate that secure deep learning is technically feasible, especially with hybrid computation paths that reduce reliance on expensive HE operations.

To streamline adoption, libraries such as CHET [1], Microsoft EVA, and Intel's nGraph-HE offer tooling for translating standard deep learning models into FHE-compatible formats. More recently, Zama's Concrete and Concrete-ML frameworks provide high-level APIs to compile neural networks with quantized integer operations into efficient TFHE circuits [20]. These libraries support fast encrypted execution and serve as critical enablers of the privacy-preserving analytics demonstrated in our work. Hence, our implementation draws inspiration from frameworks like Concrete-ML by adopting quantized low-bit integer representations by mainly using lookup-table-based activations to minimize convolutional complexity while preserving model fidelity for extracting important information from data.

When applying the above-described secure analytics to cyber-physical systems (CPS) like water infrastructure, additional challenges need to be addressed first. Sensor streams in WDS, e.g., pressure, flow, and chlorine concentration, are generally continuous, high-volume, and time-sensitive, which make encrypted analytics demanding in terms of latency and memory footprint. Recent research in privacy-preserving smart metering systems (SMSs) has shown promise. As mentioned earlier, Xu et al. [19] proposed an FHE-enabled SMS architecture that supports data aggregation and anomaly detection over encrypted smart meter data, which ensures that even fine-grained consumption details remain private. Analogously, our proposed encrypted modified TTnet model intends to enable utilities to forecast pressure behavior over time, while keeping the original raw pressure time-series datasets confidential throughout the entire computation pipeline. It is well known that encrypted computation offers significant advantages for infrastructure operators who are outsourcing analytics to third-party cloud providers. As a rule of thumb, fully homomorphic or MPC-secured pipelines eliminate the need to trust the service provider, while preserving key compliance with regulations for the adoption and maintenance of sensitive critical infrastructure data. Although literature targeting encrypted time-series analytics for water infrastructure remains sparse, parallel work in smart grid analytics using FHE and MPC [8] provides transferable insights that we leverage in our approach.

Therefore, the technical problem in the current study can be formulated as follows: how can we redesign or modify the previously established TTnet architecture [3] for time-series analytics and modelling, while still fulfilling the computational constraints of TFHE to achieve effective analysis of complex sensor data obtained from water infrastructure applications, e.g., WDS, in general? Overall, this involves balancing two competing objectives: maintaining high computational accuracy, e.g., low root-mean-squared error (RMSE), for time-series forecasting of flow and pressure data, while reducing the computational complexity and mathematical operations so that the model can be evaluated homomorphically in a reasonable time frame. We address this through three sub-problems: adapting TTnet by preserving feasible components and simplifying layers that are unsuitable for FHE; capturing both high- or moderate-frequency transient patterns and lower-frequency diurnal or leak-related trends without excessive complexity; and designing an encrypted computation pipeline with suitable low-bit numeric representations, model parameters, key management, ciphertext sizing, and bootstrapping behavior.

In summary, by formulating the above highlighted challenges involved in our problem statement, we can thus set the stage for our proposed methodology. In the next section, we describe our approach in detail to customize/modify the TTnet model and the overall pipeline processes which optimize the privacy-preserving time-series analytics for water infrastructure monitoring, mainly pressure monitoring in WDS.

3 Methodology

As discussed, our methodology encompasses the modification of an existing TTnet neural network model architecture to design a neural network model for continuous time-series predictive modeling, followed by the development of an end-to-end encrypted analytics pipeline using the TFHE fully homomorphic encryption scheme. In the following, we describe four key components to our proposed methodology for addressing the defined problem statement, namely: (a) data modeling for continuous time-series profiles, (b) general TTnet adaptation for time-series predictive modeling with continuous data points, (c) specific architecture for pressure time-series modeling with modified TTnet, and (d) proposed FHE encryption and decryption pipeline processes.

3.1 Data Modeling for Continuous Time-Series Profiles

The first step is to define how the water sensors data can be represented and preprocessed into suitable input data structures for feeding into the modified TTnet model for continuous time-series predictive modeling. We start by considering the scenario where there is at least one time-series signal collected from water distribution systems which include, but are not limited to, pressure readings located at various junction points in the pipeline network, flow meters installed at key boundaries of the network for computing the network's flow balance, or water quality indicators, e.g., chlorine level and turbidity, which are sampled at regular intervals, e.g., every five minutes. For each sensor, we denote the time-series as $X = \{x_1, x_2, \ldots, x_T\}$ over a defined time window of interest. For example, for 5-min-interval data collected over a 24-h period, we will generate a one-dimensional (1D) time-series array of 288 data points.

To go further into the details, we employ a sliding window approach to segment the continuous time-series data. A window of length N, for instance, $N = 288$ timesteps which correspond to 24 h of data, is used as the model input. Each window captures recent readings, e.g., from the nearest previous seven days, and the proposed time-series TTnet model architecture processes the input window to perform the forecasting step. We apply standard preprocessing to each window, including normalization of sensor values, i.e., scaling readings to a fixed range or z-scores, to ensure that the modified TTnet model can learn the data patterns for optimizing the model weights. In practice, normalization parameters, average, standard deviation, or min-max range, are derived from historical data and shared between the data owner and the proposed TTnet model to consistently preprocess the required data before encryption.

Additionally, if multiple sensor modalities are required to be modeled concurrently, the input data can be structured as a multivariate time-series, where the readings of each sensor can be treated as a separate channel. Our modified TTnet architecture is flexible to accept multi-channel inputs, followed by performing 1D convolutions across time on each defined channel, i.e., parameters, and then combining the extracted information across all known channels as new inputs into the deeper layers. In our experiments, we primarily consider single-channel inputs, one parameter at a time, to simplify the encrypted computation.

Finally, before encryption, the windowed and normalized data are first quantized to an appropriate numerical precision. Since TFHE operates on binary gates, we typically represent each data point in a window as a fixed-point binary number encoded into multiple bits of ciphertext. For efficiency, our proposed approach adopts relatively low-bit precision, e.g., 8-bit or 16-bit fixed-point values, to balance accuracy and performance. Overall, this quantization ensures that when we encrypt the data, the bit-length of the plaintext values is maintained at a low rate for the subsequent benefit involved in the homomorphic operations, while not compromising the resulting model forecasting accuracy to the best possible extent.

3.2 General TTnet Adaptation for Time-Series Predictive Modeling

As mentioned above, our approach focuses on adapting the TTnet neural network architecture for efficient encrypted inference. The original TTnet is modified for time-series analytics to capture short-, medium-, and long-term temporal dependencies and behaviors. However, training or running a deep learning model on encrypted data is prohibitively slow with current FHE technology. We therefore streamline TTnet through model-complexity reduction and quantization. Convolutional layers that originally used large kernels, e.g., 5-point-wide filters, are replaced by sequences of smaller kernels, e.g., two 3-point convolutions, to approximate the same receptive field with fewer operations and within allowable bit-widths such as 16 bits. The number of filters per layer is constrained to tens rather than hundreds to limit encrypted multiply-accumulate operations. The adapted architecture also uses quantized weights, e.g., 8-bit integers, loaded as plaintext constants in the encrypted computation, and replaces nonlinear activations with FHE-friendly alternatives such as step functions or low-degree polynomial approximations.

After the above-discussed adaptations, the modified TTnet model architecture therefore becomes more compact. However, there is a trade-off, namely, while we expect a general reduction in raw accuracy compared to a large, unencrypted model, the compromise is considered necessary to achieve a level of decent computational performance with homomorphic encryption. The next subsection describes our proposed CNN-LSTM class that forms the basic building block having adapted TTnet model.

3.3 General CNN-LSTM Class with Modified TTnet for Pressure Time-Series Modeling

Pressure sensor data obtained from water distribution systems often exhibit long-term trends or periodic behaviors. For instance, daily diurnal consumption patterns, pressure changes due to changes in demand patterns, or slow-developing anomalies such as a hidden pipe leak that grows gradually over time can lead to slow and long-term changes in flow and pressure behaviors.

To model the above scenarios without incurring massive complexity, we first start off with the design of a general CNN class, as modified upon the original TTnet, that processes a down-sampled version of the input window by first applying pooling or striding on the input data, as part of the plaintext pre-processing step, for shortening the input time-series data structure. For example, from the original N points, the model will take averages or samples every m point, hence yielding a condensed series of length N/m. This process of compaction may sacrifice some important high-frequency information but still captures the overall long-term trend on the macroscale. In summary, the general CNN class involves the following key components:

1. **Convolutional Layer 1, general trend detection:** a convolution with a wider kernel, e.g., 5 or 7 points, and a moderate number of filters, e.g., 8–16, is applied to the down-sampled series. The wider kernel helps capture a trend or shape that spans a longer period, since in the original scale that kernel covers $5m$ or $7m$ points. This layer might detect patterns like "overall increase", "cyclic fluctuation", or "gradual decline" across the window.
2. **Activation function:** a simple step-function activation is adopted to indicate whether a particular trend feature is predominantly present.
3. **Convolutional Layer 2, context integration:** another convolution can combine the outputs of the first layer to detect higher-level features. For example, if the first layer detects basic upward or downward trends at different temporal positions, the second layer can combine them to recognize a complex long-term pattern. This layer can leverage a smaller kernel, 3 points, but more filters, e.g., 16 to 32, to aggregate the trends identified.
4. **Optional global aggregation:** as the CNN model is focusing on macroscale analysis, we include a form of global pooling at the end for extracting the condensed and representative features. For instance, taking the average of

the feature map across time, i.e., global average pooling, can produce a single value per filter. This yields a compact representation of the input, i.e., how strongly each macro-pattern was present in the entire window.

We then improved the CNN class by designing another variant called `CNN_LSTM`, which incorporates bidirectional LSTM (BiLSTM) layers following the convolutional feature extraction stage. This enables better modeling of temporal dependencies and back-and-forth flow dynamics in the time series. Overall, the `CNN_LSTM` architecture consists of the following components:

1. Two convolutional layers: `Conv2D(1,16,k=3,s=1,p=1)` with BatchNorm and ReLU, followed by `Conv2D(16,32,k=3,s=1,p=1)` with BatchNorm and ReLU. The above layers extract local spatial and short-term temporal features across the 288-timestep window.
2. A dropout layer with a dropout value of 0.3 to prevent overfitting.
3. Reshaping to sequence: the feature map $[B, 32, 288, 7]$ is permuted and flattened to $[B, 288, 224]$ to prepare for sequential modeling across time steps.
4. A bidirectional LSTM: a 2-layer BiLSTM with 64 hidden units per direction processes the 288-step sequence and outputs a temporal embedding of shape $[B, 288, 128]$.
5. A fully connected layer: a linear layer maps each timestep's hidden state to a single regression value, yielding a final output shape of $[B, 1, 288]$.

This revised architecture remains shallow and computationally efficient, while greatly improving the model's capacity to detect temporal variations like diurnal cycles, pressure decays, and subtle signal shifts. Importantly, the operations, namely convolutions, matrix multiplications, additions, and element-wise activations, are compatible with TFHE's binary gate logic, thus making the proposed architecture suitable for encrypted inference deployment.

3.4 General FHE Encryption and Decryption Pipeline Processes

With the modified TTnet architecture for the proposed `CNN_LSTM` class adapted for low-complexity encrypted operations, we now describe our encrypted analytics pipeline using the Concrete FHE compiler. This compiler converts fixed-shape Python functions into fully homomorphic circuits compatible with TFHE logic. Figure 1 presents a comparative overview of the two execution paradigms: (A) traditional plaintext inference where data is processed in clear form, and (B) our proposed TFHE-based secure pipeline where all model computations are conducted over encrypted inputs and outputs. The prediction components of (A) and (B) can then be compared against each other at the end of the process to check if the proposed FHE-based secure pipeline can preserve the expected model accuracy during the encryption and decryption steps as shown in Fig. 1.

As discussed earlier in the introduction section, our proposed methodology is built upon the semi-honest threat model that, however, provides avenues for adversaries to inspect and decipher ciphertexts for secret information, or attempt

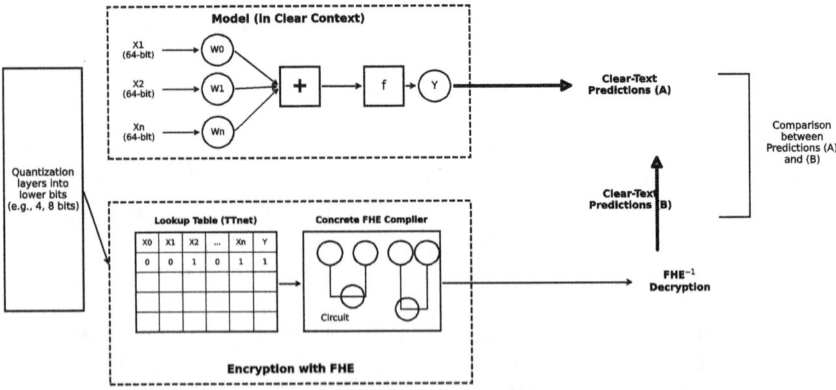

Fig. 1. Comparison between (A) plaintext inference pipeline and (B) fully homomorphic encryption (FHE) inference pipeline using the Concrete compiler. While both paths use the same model weights, the TFHE path ensures that inputs and intermediate computations remain encrypted throughout the inference process.

offline cryptanalysis on the encrypted payloads which can be detrimental to critical infrastructure monitoring. Therefore, TFHE encrypts raw sensor inputs to generate binary or low-bit ciphertexts suitable for logic circuit execution, and the encrypted inputs are transmitted to the server that then performs homomorphic inference using the compiled Concrete FHE circuit. In the entire process, the server never interacts with plaintext values, intermediate hidden states, or the model output in decoded form. The final encrypted result is returned to the client, which alone holds the private decryption key. Overall, by relying on FHE's semantic security and well-configured bootstrapping, the attack surface traditionally exposed in plaintext inference workflows is eliminated, hence making the system resilient against inference attacks.

In summary, the entire pipeline processes involved in our proposed methodology can be grouped into several key sequential steps as follows: (A) model transformation for TFHE-compatible execution; (B) quantization and client-side encryption; (C) compiling the TFHE-compatible model; (D) encrypted inference on the server; and (E) decryption of results.

Step A. Model Transformation for TFHE-Compatible Execution. The original `CNN_LSTM` model includes a bidirectional LSTM layer for modeling temporal dependencies. However, due to the recursive and dynamic nature of LSTM memory, it is incompatible with circuit-level FHE. To make the architecture homomorphically executable, we simplify the model to use: (i) two 2D convolution layers with 8-bit quantized weights, `Conv2D(1,16)` and `Conv2D(16,32)`; (ii) ReLU activations implemented as encrypted `max(0, x)` comparisons; (iii) flattened feature maps of shape $[1, 32, 288, 7] \rightarrow [1, 64512]$; and (iv) a fully connected block, FC1 ($64512 \rightarrow 512$) + ReLU + FC2 ($512 \rightarrow 288$). Therefore, this model remains compact and static, where there is no control flow, hence ensuring suitability for compilation with Zama's Concrete FHE framework.

Step B. Quantization and Client-Side Encryption. Prior to inference, each 24-h sensor window, 288 values $\times 7$ features, is normalized using precomputed historical statistics, mean and standard deviation, quantized to 8-bit fixed-point integers, and flattened into a 1D input vector of length 2,016. The client encrypts each integer using Concrete's public-key encryption to generate a set of ciphertexts which represent the binary decomposition of the input vector. These encrypted inputs are then sent to the server for homomorphic processing.

Step C. Compiling the TFHE-Compatible Model. The inference logic is manually re-implemented using only supported Python operations: integer additions and multiplications, ReLU activations via encrypted max comparison gates, and loop-based simulation of convolution and dense layers. This function is decorated with `@fhe.compiler(...)` and compiled with a dummy input to generate a Concrete circuit. All ciphertext inputs and internal states are processed with bootstrapping enabled for refreshing the ciphertexts and preserving correctness.

Step D. Encrypted Inference on the Server. The analytics server that receives only ciphertexts then runs the compiled FHE circuit. All convolution filters and FC weights are stored in plaintext. The encrypted input vector is passed through the compiled circuit. Every operation, namely convolution, activation, and FC, is performed using homomorphic logic gates. Intermediate activations and final outputs remain encrypted throughout. The above steps ensure that no plaintext information, e.g., sensor values, trends, or anomalies, is leaked to the server's end.

Step E. Decryption of Results. Upon completing the inference process, the encrypted output vector of size 288 is then returned to the client's end. Using the private decryption key, the client obtains the forecasted pressure or flow predictions for the full 24-h window. Since the compiled circuit and plaintext model share the same weights, the predictions maintain functional equivalence, while only differing in precision due to quantization.

In summary, the proposed dual-pipeline architecture, as illustrated in Fig. 1, demonstrates that it is possible to transition from traditional clear-text predictive modeling, Path A, to a fully encrypted inference pipeline, Path B, with some or minimal architectural changes. The use of Concrete's TFHE compiler ensures that there are strict data privacy guarantees in place, while also maintaining compatibility with standard neural network operations which are expressed in quantized form.

4 Experimental Results

4.1 Evaluation Dataset and Experimental Setup

To evaluate the effectiveness of our customized TTnet model with the proposed `CNN_LSTM` model, we conducted experiments using the publicly available benchmark dataset from the BattLeDIM competition [16] as highlighted above. The dataset contains simulated pressure sensor readings from a water distribution network under various operational scenarios, including normal operations and pipe leakage conditions. For our experiment, we focused on a 365-day period of continuous pressure sensor readings collected at a 5-min interval that results in 105,120 data points per sensor. In summary, our experimental setup consists of the following:

- A modified TTnet neural network model architecture with reduced parameter size, quantized weights, and simplified step-wise activation suitable for encrypted inference under the TFHE scheme.
- A single-channel time-series modeling framework, where each sensor is modeled independently using a sliding window of 2,016 points, i.e., nearest previous seven days $\times$ 24 h of pressure data as a two-dimensional input array, to predict the next 24-h 1D array of 288 pressure data points. For example, to forecast 24 h of pressure data, 288 points in total, on 8 January 2025, we leverage the 2,016 points from 1–7 January 2025 as the input data into the modified TTnet model.

The model training and validation steps were conducted in plaintext first to validate model accuracy, followed by subsequent homomorphic inference implementation tests to confirm consistency in encrypted results. In summary, we constructed a sliding window-based forecasting framework for each individual sensor. For any given day d, the model takes as input the preceding seven days of pressure data, $288 \times 7 = 2016$ points, and predicts the next 24-h window of 288 values. This results in a training dataset of 358 usable windows per sensor, spanning nearly one year. To ensure fair evaluation, the following configuration was adopted:

- The `CNN_LSTM` model was trained for 50 epochs using the Adam optimizer with learning rate 0.001.
- No random shuffling was applied to preserve sequential dependencies.
- 80% of data windows were used for training, with the remaining 20% held out for validation.

Quantization of model weights and inputs to 8-bit integers was applied prior to TFHE compilation. Encrypted inference was carried out using the Concrete library on a CPU machine. Each compiled model accepts encrypted input tensors and outputs encrypted predictions, which are decrypted using the secret key retained by the client.

4.2 Model Validation Results in Plaintext

Table 1 reports the validation Root Mean Squared Error (RMSE) for each of the 32 pressure sensors. The RMSE values were computed based on the difference between the model's predicted pressure values and the actual observed values in the assigned validation set, without any prior random shuffling for separating the original dataset pool into training and validation datasets. Table 1 also summarizes the average recorded pressures for each sensor station. By using the simple formula in Eq. (1), we leveraged the average pressure reading and the respective average RMSE loss value for each sensor station to compute its corresponding model accuracy, as also summarized in Table 1. Overall, the results demonstrate the capability of the modified TTnet using the proposed `CNN_LSTM` model architecture for generalizing unseen data and capturing relevant pressure behavior patterns. Notably, sensor n299 resulted in the highest average model accuracy of close to 100%, which highlights its strongest predictive performance, while sensor n288 recorded the lowest model accuracy of around 98.9%, which still reflects acceptable prediction accuracy in the context of encrypted inference.

$$\text{Accuracy} = \left(1 - \frac{\text{RMSE}}{\text{Avg. Pressure}}\right) \times 100\%. \tag{1}$$

4.3 TFHE-Encrypted Inference Evaluation with Concrete

In this subsection, we evaluate the feasibility and performance of executing our modified TTnet model, `CNN_LSTM`, under the TFHE-encrypted inference pipeline using the Concrete FHE compiler from Zama. The key goal is to validate whether encrypted execution preserves both predictive fidelity and computational viability in secure analytics settings. Following the architecture and flow illustrated in Fig. 1, the trained model weights were reused from the plaintext training phase and converted into an integer-only inference implementation compatible with Concrete. The model was manually rewritten using Python loops and fixed-precision arithmetic and compiled using the `@fhe.compiler` decorator. The inference pipeline for a single time window proceeds through the following stages:

1. **Preprocessing and quantization:** the $[1, 1, 288, 7]$ input tensor, 24-h pressure history, is normalized and quantized to 8-bit fixed-point integers using uniform scaling. The quantized tensor is then flattened into a 1D vector of 2,016 integers.
2. **Client-side encryption:** each quantized integer is encrypted using Concrete's public-key scheme, producing a set of ciphertexts representing the encrypted input vector. These are transmitted securely to the cloud inference server.
3. **Homomorphic evaluation:** the server runs the compiled encrypted circuit, which simulates two Conv2D layers via loop-based multiply-accumulate operations over encrypted data and plaintext weights, ReLU activations implemented as encrypted max operations using comparison gates, flattening into a 64,512-length vector, and fully connected layers FC1 and FC2 using encrypted matrix-vector multiplication.
4. **Decryption and result retrieval:** the encrypted 288-length output is returned to the client. The client decrypts the ciphertexts using the secret key to retrieve the final pressure predictions for the next 24-h window.

Table 1. Validation metrics for pressure sensors from the BattLeDIM open-source dataset.

Sensor ID	Avg. Pressure (mH_2O)	Validation Loss (RMSE)	Accuracy (%)
n229	52.1	0.02	100.0
n752	48.8	0.09	99.8
n332	56.1	0.16	99.7
n506	53.2	0.18	99.7
n726	46.8	0.19	99.6
n740	43.7	0.20	99.5
n54	36.8	0.21	99.4
n519	47.2	0.22	99.5
n722	45.8	0.24	99.5
n516	54.5	0.24	99.6
n4	33.3	0.25	99.2
n644	47.2	0.25	99.5
n31	36.6	0.25	99.3
n163	52.1	0.25	99.5
n114	53.7	0.25	99.5
n342	46.3	0.27	99.4
n495	51.5	0.28	99.5
n296	42.2	0.29	99.3
n679	47.0	0.29	99.4
n769	48.2	0.30	99.4
n636	45.2	0.33	99.3
n410	30.7	0.36	98.8
n549	54.4	0.37	99.3
n613	55.7	0.37	99.3
n429	36.4	0.41	98.9
n469	47.3	0.41	99.1
n188	54.9	0.42	99.2
n215	39.1	0.43	98.9
n105	50.2	0.50	99.0
n458	43.1	0.54	98.7
n415	45.2	0.56	98.8
n288	52.4	0.60	98.9

To evaluate the performance and fidelity of the TFHE-encrypted inference pipeline, we conducted experiments using CPU-based execution, with no GPU acceleration, with the Concrete FHE compiler. A single test window was first encoded into 8-bit quantized integers and then encrypted and processed through

Table 2. Encrypted inference runtime characteristics with the Concrete FHE compiler.

Metric	Value
Input Size (Quantized)	2,016 integers (8-bit)
Encrypted Input Size	~20.2 KB
Encryption Time	0.32 s
Encrypted Inference Time	4.81 s
Decryption Time	0.11 s
Total Inference Latency	5.24 s
Peak Memory Usage	182.7 MB
Security Level	128-bit post-quantum

the compiled homomorphic circuit. Table 2 summarizes the associated runtime metrics. Despite the cryptographic overhead of homomorphic execution, the pipeline remains feasible for low-frequency monitoring use cases, e.g., 15-min or hourly intervals.

To assess the consistency between encrypted and plaintext predictions, we compared the model outputs across 32 pressure sensors for 20 randomly selected validation windows. Table 3 reports the RMSE values between the ground truth and: (i) plaintext model predictions and (ii) decrypted TFHE predictions. Overall, the results demonstrate the following observations, namely:

- Across all sensors, the average RMSE difference between plaintext and TFHE predictions was approximately 0.47 mH_2O. Given the typical pressure range of 30–60 mH_2O in most WDS scenarios, this deviation represents a relatively small loss, below 1.5%, in predictive fidelity.
- Despite a slight increase in the RMSE values for the decrypted TFHE predictions across all sensors, the encrypted predictions consistently preserved the main structural patterns in the time series, including local maxima, diurnal peaks, slow drifts, and sudden drops. This suggests that the encrypted inference is functionally equivalent to the original model by capturing all essential features for practical WDS analytics.
- Even for sensors with higher RMSE under plaintext conditions, e.g., n415 or n288, the encrypted results remained within an acceptable upper bound, $\leq$1.2 mH_2O, hence highlighting that the TFHE inference scales well across different pressure profiles in general.

Although the encrypted inference incurs a latency of approximately 5.2 s per window, this is acceptable for WDS monitoring applications with reporting intervals of 15 min or longer. Moreover, since each sensor operates independently, encrypted inference for multiple sensors can be executed in parallel. The model structure also lends itself to future batching and GPU-accelerated homomorphic evaluation, once supported by the Concrete compiler. In summary, our encrypted TTnet inference pipeline demonstrates high fidelity and close approximation to

Table 3. Comparison of RMSE scores between plaintext predictions and decrypted TFHE predictions.

Sensor ID	Plaintext Predictions	Decrypted TFHE Predictions
n229	0.02	0.49
n752	0.09	0.69
n332	0.16	0.69
n506	0.18	0.65
n726	0.19	0.66
n740	0.20	0.60
n54	0.21	0.74
n519	0.22	0.55
n722	0.24	0.84
n516	0.24	0.84
n4	0.25	0.65
n644	0.25	0.85
n31	0.25	0.92
n163	0.25	0.92
n114	0.25	0.78
n342	0.27	0.74
n495	0.28	0.68
n296	0.29	0.62
n679	0.29	0.69
n769	0.30	0.97
n636	0.33	0.93
n410	0.36	1.03
n549	0.37	0.70
n613	0.37	0.97
n429	0.41	0.88
n469	0.41	0.81
n188	0.42	0.75
n215	0.43	1.10
n105	0.50	0.97
n458	0.54	0.94
n415	0.56	1.16
n288	0.60	1.20

plaintext inference results, acceptable latency for smart infrastructure scenarios, and strong privacy guarantees that protect sensitive time-series data at all stages of processing.

4.4 Baseline Model Comparison

To quantitatively assess the performance of the proposed modified TTnet architecture for time-series predictive analytics for the pressure sensors adopted in WDS, we further benchmarked it against three widely adopted baseline models,

Table 4. RMSE comparison between baseline models and modified TTnet model.

Sensor	RF	LR	SNN	Mod. TTnet (CNN_LSTM)
n769	0.14	0.14	0.19	0.30
n752	0.06	0.06	0.08	0.09
n740	0.14	0.14	0.18	0.20
n726	0.11	0.11	0.17	0.19
n722	0.15	0.15	0.23	0.24
n719	0.17	0.18	0.27	0.32
n679	0.15	0.16	0.24	0.29
n644	0.14	0.14	0.21	0.25
n636	0.18	0.18	0.25	0.33
n613	0.23	0.23	0.32	0.37
n549	0.18	0.19	0.26	0.37
n516	0.13	0.13	0.21	0.24
n506	0.15	0.15	0.23	0.18
n495	0.15	0.15	0.23	0.28
n469	0.19	0.19	0.28	0.41
n458	0.21	0.21	0.30	0.54
n429	0.15	0.15	0.23	0.41
n415	0.24	0.24	0.33	0.56
n410	0.18	0.18	0.26	0.36
n342	0.13	0.13	0.21	0.27
n332	0.08	0.08	0.12	0.16
n296	0.16	0.16	0.23	0.29
n288	0.24	0.24	0.35	0.60
n229	0.01	0.01	0.02	0.02
n215	0.18	0.18	0.26	0.43
n188	0.16	0.16	0.24	0.42
n163	0.12	0.12	0.18	0.25
n114	0.12	0.12	0.18	0.25
n105	0.25	0.25	0.34	0.50
n54	0.09	0.09	0.14	0.21
n4	0.09	0.09	0.13	0.25
n31	0.10	0.10	0.15	0.25

namely: Baseline Model 1, Random Forest Regressor (RF), a classical ensemble-based machine learning model known for its robustness and interpretability in regression tasks [9,14]; Baseline Model 2, Linear Regression (LR), a simple model that assumes a linear relationship between input features and the output target [10,15]; and Baseline Model 3, Shallow Neural Network (SNN), a lightweight feed-forward network trained on the same features to represent a simplified deep learning model [6].

All models were trained using identical model training configurations, e.g., training-validation splits per sensor station, to ensure fair comparison. Likewise, the primary metric used is RMSE. Table 4 presents the RMSE values for 32 pressure sensor stations under all four models. Overall, the classical models, RF and LR, yielded lower average RMSE scores due to their ability to operate directly on plaintext data without encryption constraints. In contrast, the customized TTnet, although not the best performer in terms of raw RMSE, still produces similar orders of magnitudes for the RMSE scores as compared to the other three baseline models and provides the added advantage of compatibility with TFHE encryption, hence supporting privacy-preserving analytics without revealing sensitive pipeline pressure data.

To complement the tabular results, Fig. 2 visualizes the distribution of RMSE scores for each model across all sensors. The boxplot clearly illustrates that while RF and LR have the tightest error bounds, the TTnet model, though more variable, remains competitive in terms of the orders of magnitudes for the computed RMSE scores, especially in secure, privacy-restricted contexts. The median RMSE of the TTnet falls within acceptable tolerance bounds for the current time-series analytics of the pressure sensors adopted in WDS, as explained earlier using Table 1.

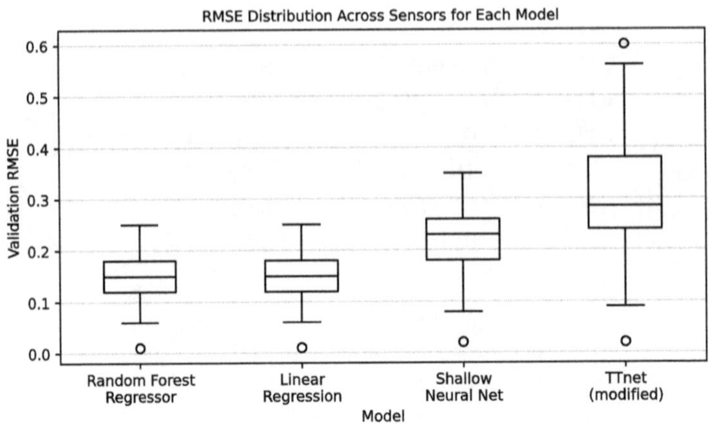

Fig. 2. RMSE distribution across sensors for each model.

5 Limitations and Future Work

At this stage, while there is some promise to the results, several limitations remain in the current study, namely:

- **Inference latency and scalability:** although encrypted inference can be reasonably feasible, approximately 5.2 s per window on CPU, real-time or high-frequency deployments may require further optimization. Currently, the Concrete FHE compiler does not support GPU acceleration or circuit batching, which therefore limits scalability for multi-sensor systems.
- **Model simplification trade-offs:** to enable FHE compatibility, we had to reduce the model complexity, e.g., removing LSTM layers and simplifying activations. While this allows encrypted execution, it may limit performance on datasets with highly non-linear or abrupt dynamics, hence posing a risk that the proposed methodology may face challenges in modelling more complex water infrastructure monitoring applications.
- **Single-channel analysis:** our encrypted experiments focused on univariate time-series data, pressure only. Multivariate encrypted modelling, e.g., combining pressure, flow, and water quality data, introduces additional complexity in both model design and ciphertext management.
- **Limited field testing:** thus far, the customized lightweight TTnet model with TFHE has only been validated using simulated BattLeDIM data. Real-world sensor noise, irregular sampling, and operational anomalies in deployed water networks will need to be addressed through more extensive field deployments and robustness testing.

Therefore, to address the above-described limitations and/or challenges identified at this stage, future research work should focus on the following: exploring the possibility of integrating encrypted GRU blocks as an alternative to LSTMs to retain memory mechanisms; leveraging hybrid privacy approaches, e.g., local differential privacy and FHE, for more efficient encrypted analytics; and exploring model distillation and pruning to further reduce circuit size and inference cost.

6 Conclusions

This study proposed a practical and scalable solution for privacy-preserving time-series analytics in water infrastructure monitoring. By customizing the TTnet deep learning architecture and adapting it to operate within the computational constraints of TFHE, and adopting an honest-but-curious (semi-honest) threat model, we demonstrated that encrypted inference over sensor data is both feasible and accurate. The proposed CNN-LSTM-based TTnet model, when compiled using Zama's Concrete compiler, preserved key predictive patterns even under encryption and maintained accuracy within acceptable bounds compared to plaintext execution. Our evaluations on the BattLeDIM dataset confirmed that encrypted predictions deviated minimally, to within 0.5 mH_2O root-mean-square-error (RMSE), from the plaintext model outputs, while providing strict confidentiality guarantees throughout the analytic pipeline. As privacy becomes a growing concern in smart city deployments, our work offers a foundational framework for enabling secure, encrypted deep learning analytics across critical infrastructure systems. We believe this approach can be extended beyond pressure forecasting to other important smart water grid applications which include, but are not limited to, secure anomaly detection, decision support, and predictive maintenance.

Acknowledgements. This study was conducted using the author's own resources and time.

Disclosure of Interests. The author declares no conflict of interest.

References

1. Bajaj, S., Cousins, D.B., Lauter, K.: HEAX: an architecture for computing on encrypted data. In: International Symposium on Computer Architecture (ISCA) (2018)
2. Benamira, A., Guérand, T., Peyrin, T., Saha, S.: TT-TFHE: a Torus fully homomorphic encryption-friendly neural network architecture. arXiv preprint arXiv:2302.01584 (2023)
3. Benamira, A., Peyrin, T., Yap, T., Guérand, T., Hooi, B.: Truth table net: scalable, compact & verifiable neural networks with a dual convolutional small Boolean circuit networks form. In: Proceedings of the 33rd International Joint Conference on Artificial Intelligence (IJCAI-24), pp. 13–21 (2024)
4. Chillotti, I., Gama, N., Georgieva, M., Izabachène, M.: TFHE: fast fully homomorphic encryption over the Torus. J. Cryptol. **33**(1), 34–91 (2020)
5. Gilad-Bachrach, R., Dowlin, N., Laine, K., Lauter, K., Naehrig, M., Wernsing, J.: CryptoNets: applying neural networks to encrypted data with high throughput and accuracy. In: International Conference on Machine Learning (ICML), pp. 201–210. PMLR (2016)
6. Judd, S.: On the complexity of loading shallow neural networks. J. Complex. **4**(3), 177–192 (1988)
7. Juvekar, C., Vaikuntanathan, V., Chandrakasan, A.: GAZELLE: a low latency framework for secure neural network inference. In: 27th USENIX Security Symposium (2018)
8. Li, F., Chen, B., Yang, H., Lin, L.: Secure time-series data aggregation in smart grids using homomorphic encryption. IEEE Trans. Smart Grid **13**(2), 1101–1113 (2022). https://doi.org/10.1109/TSG.2021.3098745
9. Liu, Y., Wang, Y., Zhang, J.: New machine learning algorithm: Random Forest. In: Liu, B., Ma, M., Chang, J. (eds.) ICICA 2012. LNCS, vol. 7473, pp. 246–252. Springer, Heidelberg (2012). https://doi.org/10.1007/978-3-642-34062-8_32
10. Montgomery, D.C., Peck, E.A., Vining, G.G.: Introduction to Linear Regression Analysis, 5 edn. Wiley (2021)
11. Moubayed, A., Sharif, M., Luccini, M., Primak, S., Shami, A.: Water leak detection survey: challenges & research opportunities using data fusion & federated learning. IEEE Access **9**, 40595–40611 (2021)

12. Ramotsoela, D., Abu-Mahfouz, A.M., Hancke, G.P.: A survey of anomaly detection in industrial wireless sensor networks with critical water system infrastructure as a case study. Sensors **18**(8), 2491 (2018)
13. Riazi, M.S., Samragh, M., Koushanfar, F., Javadi, H.H.: Chameleon: a hybrid secure computation framework for machine learning applications. In: Proceedings of the 2018 ACM Asia Conference on Computer and Communications Security (2018). https://doi.org/10.1145/3196494.3196535
14. Segal, M.R.: Machine learning benchmarks and random forest regression. Tech. rep., Technical Report (2004)
15. Su, X., Yan, X., Tsai, C.L.: Linear regression. Wiley Interdiscip. Rev. Comput. Stat. **4**(3), 275–294 (2012)
16. Vrachimis, S.G., et al.: Battle of the leakage detection and isolation methods. J. Water Resour. Plan. Manag. **148**(12), 04022068 (2022)
17. Wagh, S., Gupta, D., Chandran, N.: SecureNN: 3-party secure computation for neural network training. Proc. Priv. Enhanc. Technol. **2019**(1), 26–49 (2019). https://doi.org/10.2478/popets-2019-0003
18. Wu, Z.Y., et al.: High fidelity digital twin-based anomaly detection and localization for smart water grid operation management. Sustain. Cities Soc. **91**, 104446 (2023)
19. Xu, W., Sun, J., Cardell-Oliver, R., Mian, A., Hong, J.B.: A privacy-preserving framework using homomorphic encryption for smart metering systems. Sensors **23**(10), 4746 (2023)
20. Zama AI: Concrete and concrete-ML: fully homomorphic encryption toolchains for machine learning inference. GitHub repository (2023)

A Context-Aware Cybersecurity Assessment Methodology for Industrial Control Systems in the Water Industry

Fabian Voggensberger[1(✉)], João Henrique Schmidt[2], Victor Mattos[2], Gaurav Srivastava[1], Daniel S. Menasché[2], Amit Bhaya[2], Alan O. de Sá[3], and Alexander W. Miranda[4]

[1] Siemens Corporation, Princeton, New Jersey, USA
[2] Federal University of Rio de Janeiro (UFRJ), Rio de Janeiro, Brazil
[3] University of Lisbon, Lisbon, Portugal
[4] Georgia Institute of Technology, Atlanta, Georgia, USA

Abstract. A common approach to evaluating a system's cybersecurity maturity involves conducting assessments. Various control-based frameworks and guidelines exist for Industrial Control Systems (ICS) within the water sector. However, these frameworks provide generic security controls that are not tailored to the specific characteristics of individual plants. The relevance that individual security controls have for the cybersecurity maturity of a system is not evaluated by existing frameworks. To address these gaps, this paper introduces a novel methodology for cybersecurity assessments. A characteristics-based scoping mechanism filters relevant security controls, enabling tailored assessments aligned with the characteristics of the plant being examined. A weight-based ranking system is proposed to assess how single security controls enhance cybersecurity maturity, taking into account plant-specific characteristics, thereby enabling security prioritization under budget constraints.

Keywords: assessment · context-aware · cybersecurity · ICS · PLC · water industry

1 Introduction

Many Industrial Control Systems (ICS) are prone to cyberattacks, as cybersecurity has not always been considered during their initial design phase [33]. Such attacks can have severe consequences affecting industries, human life, and public safety, as ICS interact with the physical environment [3]. Therefore, securing ICS is essential for systems that are part of the critical infrastructure [39]. In the past, Operational Technology (OT) components within the critical infrastructure have been explicitly targeted by adversaries [7]. As modern ICS are highly interconnected, securing them against cyberattacks is even more critical [39].

The water sector showcases such highly interconnected systems. Plants are decentralized, distributed, and interactions are made over remote access protocols, which are not always sufficiently secure [31]. In fact, the water sector

S. Zonouz et al. (Eds.): DCS-CI 2026, CCIS 3109, pp. 214–235, 2027.
https://doi.org/10.1007/978-981-92-5039-4_11

has received more attention from criminals, forcing authorities to also pay more attention to security issues [7,20,21]. In 2018, the US government issued the America's Water Infrastructure Act, which makes resilience assessments mandatory [25]. Furthermore, securing water utilities has been declared a national priority [17].

However, several challenges exist. First, only limited publicly disclosed data on real-world incidents is available, making it more challenging to identify the most critical security controls. Disclosed real incidents in the literature often lack essential technical details that are crucial for further studies. Second, there is a lack of representative data on the general cybersecurity maturity within the water sector. Generally, it is known that the necessary cyber hygiene is often not present in water ICS [3,31]. Third, small water community plants typically lack the necessary investment budget and specialized cybersecurity expertise [28]. The budget is wholly used for the necessary repairs, and cybersecurity is often overlooked [17,19].

An industry-wide accepted solution for evaluating the security posture of a system is to conduct assessments, which enable the evaluation of the current maturity of an ICS [27]. We understand that a cybersecurity assessment is a systematic procedure that analyzes the compliance of the assessed system with security controls found in guidelines and standards, like [11] or [35]. Security controls are all measures that enhance the security of a system and mitigate the potential for successful attacks. Such controls can either be fully, partially, or not implemented in an ICS. An example of a security control is the use of strong passwords or multi-factor authentication. Note that we use the term security controls and controls interchangeably. Existing assessment frameworks for the water sector [11,32] contain high-level and generic security controls that cover a wide variety of systems and requirements, but these are not applicable everywhere. However, individual ICS have different requirements [37]. For example, Water Distribution Systems (WDS) are more decentralized compared to regular Waste Water Treatment Plants (WWTP). For a small WWTP without a big IT infrastructure, other security controls yield higher security improvements compared to a bigger WWTP with a complex IT infrastructure.

Our paper addresses these challenges through the following key contributions:

1. **Context-Aware Assessment Methodology**: During an initial scoping process, plant characteristics are explored to better fit the assessment framework to individual properties that differ between multiple plants. This new approach enables the conduct of context-aware cybersecurity assessments. Irrelevant security controls are filtered out during the scoping process, simplifying and accelerating the assessment process as a practical entry-level assessment method for small water utilities. The method also serves as a foundation for future research on context-aware cybersecurity.
2. **Security Control Ranking**: The implementation of security controls is crucial to a plant's cybersecurity maturity. Some security controls have higher relevance than others depending on the analyzed ICS. A novel approach to ranking security controls based on their individual relevance for a system is

presented. A formal model is introduced to quantify and rank the contribution of individual security controls to the overall cybersecurity maturity, accounting for both general cybersecurity principles and plant-specific characteristics. This enables optimal cybersecurity investments in budget-constrained OT environments. It aims to bridge an existing gap between control-based compliance checklists and sophisticated risk management approaches.

Outline. The remainder of the paper is organized as follows. Section 2 overviews guidelines for cybersecurity assessment in the water sector. In Sect. 3, recent incidents in water utilities are summarized and discussed. In Sect. 4, a system model for context-aware cybersecurity assessments is defined. We evaluate the proposed methodology and discuss our findings in Sect. 5. Finally, Sect. 6 concludes.

2 Background and State of the Art

In this section, we discuss the state-of-the-art of cybersecurity assessment frameworks. Section 2.1 introduces background, and Sect. 2.2 discusses typical ICS security threats.

2.1 Security Assessments and Frameworks

A cybersecurity assessment identifies a system's current security state to validate its security compliance [27]. It is based on an underlying standard or guideline [33]. These underlying standards and guidelines are a collection of security controls that are recommended to implement in order to improve the security posture of the system. Security controls, therefore, encompass all measures that enhance cybersecurity maturity and have the potential to prevent attacks. Such security assessments must be distinguished from risk assessments, which focus on possible consequences caused by cybersecurity incidents and rate them according to a risk model. Therefore, a security assessment can serve as a basis for a risk assessment, and the results of a risk assessment can influence specific security measures, but both methods satisfy different requirements [27]. Standard risk models account for the impact of an exploit in terms of financial or physical outcome and its likelihood [10,35]. Security assessments can be categorized along multiple dimensions. In addition to classical risk-based approaches, which use risk scenarios as a starting point, various other types of cybersecurity assessments are listed in Table 1.

In our work, we focus on the more simplistic control-based cybersecurity assessments and the impact the implementation of a security control has on the cybersecurity maturity of the system. We refer to this as the relevance weight of a security control. This paper's approach for cybersecurity assessments must not be confused with risk-based assessments. Our methodology is fundamentally control-based, but also takes high-level plant characteristics into account,

Table 1. Taxonomy of different established security assessment types and the relation of our work.

Type	Starting Point	Logic Flow	Output	Example
Threat-Based	Threat actors & tactics	Threats? → Mitigating controls?	Threat-mapped defenses	STRIDE [18], ATT&CK [36]
Asset-Based	Critical assets	Assets? → Threats? → Protection?	Asset-specific protection	Crown Jewels Analysis [23]
Control-Based	Predefined controls	Existing controls → implemented? → effective?	Compliance gap analysis	ISO 27001, NIST CSF [32], IEC 62443-2-1
Risk-Based	Risk scenarios	Risks? → Expected loss? → Risk reduction?	Risk-ranked investments	FAIR [22], OCTAVE [1], IEC 62443-3-2
This paper's approach	Plant characteristics	"What's our plant?" → "Filter controls" → "Rank by relevance"	Weighted control list	Novel (proposed)

which can be interpreted as generalized assets. While risk management through risk assessments also enhances the cybersecurity maturity, risk management is a sophisticated and labor-intensive process. In reality, not all water utilities incorporate risk management procedures requiring substantial expertise and planning. Our methodology is designed in a manner to target facilities that operate under budget and resource constraints, as are typically found in practice. Such utilities are most frequently targeted by attackers [19,21]. An easily applicable cybersecurity assessment method is urgently needed in this context.

Several guidelines and frameworks are available for questionnaire-based cybersecurity assessments in the water sector. For instance, the Environmental Protection Agency (EPA) released a cybersecurity checklist in [11], which is derived from the Cybersecurity and Infrastructure Security Agency's (CISA) Cross-Sector Cybersecurity Performance Goals (CPGs) [11]. Additionally, the National Institute of Standards and Technology (NIST) released the NIST Cybersecurity Framework (CSF) 2.0, which contains high-level and sector-independent cybersecurity outcomes [32]. CISA CPGs align with NIST CSF 2.0 [11]. In [35], NIST provides its third revision of NIST SP 800-82, a comprehensive guideline on security measures for general OT security. Table 2 lists some exemplary security controls out of the EPA's cybersecurity checklist [11]. These security controls, which we also denote as f_j, will be used in Sect. 5 to evaluate our proposed methodology.

Based on a particular framework or guideline, the scope of an assessment remains constant. There is currently no possibility of automatically adapting a framework based on individual plant characteristics and other analysis outcomes,

Table 2. Selected security controls f_j out of EPA's cybersecurity checklist [11].

j	ID	Description
1	1.E.	Patch/Mitigate known vulnerabilities within the recommended timeframe
2	2.A.	Change default passwords
3	2.D.	Immediately disable access when access is no longer required due to retirement, change of role, termination, or other factors
4	2.E.	Separate user and privileged (e.g., Admin, Operator) accounts
5	2.F.	Segment OT and IT networks
6	2.G.	Detect and block repeated unsuccessful login attempts
7	2.H.	Require multi-factor authentication (MFA) wherever possible, but at a minimum to remotely access utility OT/IT networks
8	2.K.	Use effective encryption to maintain the confidentiality of data in transit
9	2.L.	Use encryption to maintain the confidentiality of stored sensitive data
10	2.M.	Use email security controls to reduce common email-based threats, such as spoofing, phishing, and interception
11	2.V.	Prohibit the connection of unauthorized hardware (e.g., USB devices, removable media, laptops brought in by others) to OT and IT assets
12	2.W.	Ensure that assets connected to the public Internet expose no unnecessary exploitable services (e.g., remote desktop protocol)
13	2.X.	Eliminate connections between OT assets and the Internet

like attacker models, that are part of threat modeling [18,43]. Currently, certain parameters are implicitly assumed within the underlying framework, while others are missing. The cybersecurity checklist [11] provided by the EPA does not include any security controls to physically secure a utility, for instance. Many security controls in assessment frameworks are only relevant for specific plant characteristics. Especially smaller ICS often face the issue of having only limited budget and capacities for additional security measures [17,19]. Therefore, specific prioritization mechanisms are needed to maximize security gains within limited investment budgets, based on individual plant characteristics. To the best of the authors' knowledge, no simple and easy-to-use cybersecurity assessment method has been published to address this problem so far. Therefore, we present a novel cybersecurity assessment methodology in Sect. 4, which can be easily applied in practice without the need for complex impact analysis and risk modeling. Note that cybersecurity decision making is a risk-centric procedure [22]. Our method cannot replace advanced risk methodologies. Instead, it aims to connect cybersecurity decision-making and control-based compliance, ensuring a baseline level of cybersecurity.

Our approach of proposing context-aware cybersecurity assessments is also connected to some stages in the process of threat modeling. Threat modeling in terms of system evaluation is commonly defined as the analysis of a system's architecture to identify potential security threats and select mitigation techniques [43]. Certain complex methodologies, such as HAZOP, STPA-sec,

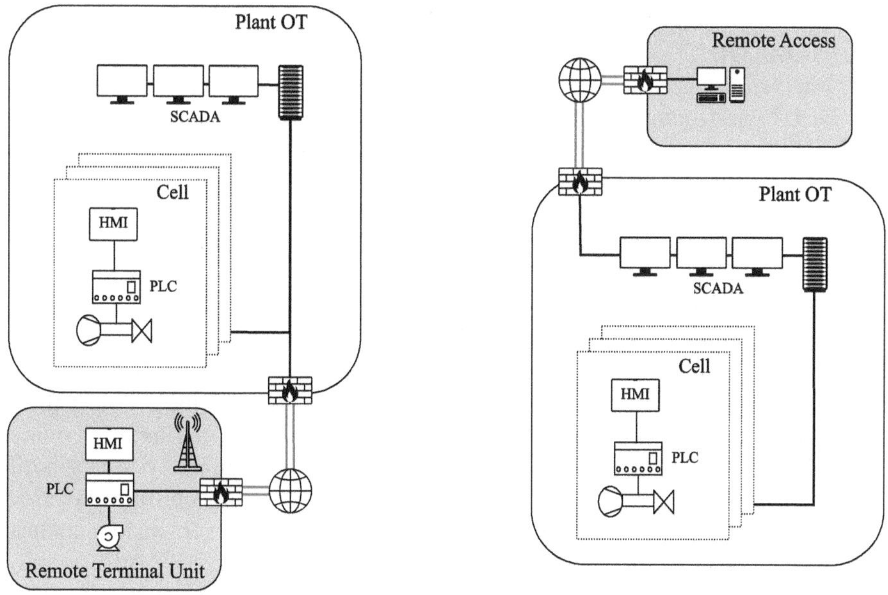

(a) **Remote Terminal Units (RTUs)** are production components not located at the central facility.

(b) **Remote Access** is often used for engineering and/or process control without physical presence.

Fig. 1. Typical remote connections in ICS.

OCTAVE, and PASTA, primarily focus on system safety and risks [24], which is not within the scope of our work on cybersecurity assessments. However, threat taxonomies, such as STRIDE [18], which consider threats from an attacker's perspective, can be beneficial for an initial scoping process. This is because attacker models and a priori knowledge about potential threats in a system aid in cybersecurity decision-making. While our current scoping process only covers what we define as plant characteristics, the systematic mapping of attacker capabilities, threats, and the system architecture, taken from the broader context of threat modeling, will be considered in future work to extend plant characteristics in the initial scoping.

2.2 ICS Security Threats and Mitigation in Water Industry

The FBI, NSA, CISA, and EPA analyze the current threats to ICS in critical infrastructure and the water sector in [19–21]. They also provide recommendations to mitigate the realization of possible threats. The importance of secure remote connections, which are established for various reasons, is outlined. On one side, water systems are highly distributed. PLCs control processes in remote locations, not directly at the central facility. These Remote Terminal Units (RTUs) are connected over a communication channel to the central SCADA system [17]. This is depicted in Fig. 1a. On the other hand, operators and engineers may be permitted to connect to the SCADA system for remote maintenance or process control [37]. This is depicted in Fig. 1b. Both communication channels for RTUs and remote access are critical, as they open the doors for remote attackers into the OT environment if they are not adequately secured [17]. NIST recommends using Virtual Private Networks (VPNs) [35,37], where the VPN servers are located within a Demilitarized Zone (DMZ). To simplify presentation, Fig. 1 does not depict the recommended layered network architecture.

In addition to insecure remote connections, the importance of network segmentation is frequently highlighted [11,17,19,21,35,37]. IT and OT components should be put into separate networks divided by a DMZ, providing jump hosts, firewalls, and communication diodes to regulate communication between both networks. According to [17], typical water systems have a flat network structure with IT and OT components in the same network segment. Other water utilities employ layered network architectures that separate IT and OT networks via a DMZ and firewall policies [17]. A layered network architecture increases the difficulty of attacking OT systems, as an attacker must bypass multiple layers of security to reach an asset in the production network. However, insider attackers have generally direct access to the DMZ or OT network [17]. Additionally, attacks like Stuxnet [12] have successfully overcome multiple defense layers in the past by chaining various exploits together. This tactic was initially described as the ICS cyber kill chain in [4].

3 Real Incidents in Water ICS

Only minimal data is available about real-world attacks on ICS in the water sector. Publicly disclosed attacks often lack essential technical details [19], making it hard to evaluate the relevance of specific existing security measures. The real incidents in water ICS, reviewed by Hassanzadeh et al. in [17], do not always target the OT environment specifically. Over the course of 19 years, only 15 incidents are presented, highlighting the scarcity of data. However, the authors outline that the estimated number of unreported cases is much higher. Some other incidents in the water sector between 2019 and 2021 are shortly described in [19]. Except for one incident caused by an insider, the other incidents were all caused by general IT ransomware.

We researched recent incidents involving water ICS through Internet searches and present them in Table 3. Only clearnet sources were considered, while the dark web was not part of our scope. By manually linking the incidents to several security controls from the cybersecurity checklist for water ICS [11], it is demonstrated how the incident could have been potentially mitigated. The corresponding controls are listed in Table 2 in Sect. 2.1. We do not claim that single security controls could have prevented those incidents, but use them to motivate assessors to implement security controls found in current guidelines and frameworks. It also highlights the relevance of the security controls. To get a better picture of incidents in the water sector over time, we plotted the number of disclosed incidents that are documented in [17] and [19] together with the

Table 3. Relation of security controls to real-world incidents. The linked security controls can be found by their ID in Table 2.

Incident	Year	Location	Controls	Key Lessons
Bowman Avenue Dam [40,41]	2013	New York	2.F., 2.H., 2.W., 2.X.,	First documented nation-state attack on U.S. water infrastructure; cellular modem vulnerability
San Francisco WTP [8]	2021	California	2.D., 2.E., 2.W.	Stolen credentials from former employee; privilege separation critical
Oldsmar WTP [6,13]	2021	Florida	1.E., 2.A., 2.W., 2.X.	Attribution contested; Windows 7, TeamViewer vulnerabilities; $30K+ remediation
Aliquippa Municipal Water [26,28]	2023	Pennsylvania	1.E., 2.A., 2.W., 2.X.	Supply chain attack targeting Israeli equipment; default passwords exploited; part of broader campaign (4+ utilities)
County Mayo WTP [30]	2023	Ireland	1.E., 2.A.	CVE-2023-6448 exploited with default credentials; 2-day outage affecting 160–180 households
Veolia Water [15]	2024	Multiple U.S. states	2.D., 2.E., 2.K., 2.L.	PII theft; affects 8,500 facilities globally; demonstrates scale of exposure
Muleshoe, Hale Center, Abernathy [5,14]	2024	c Texas	2.F., 2.G., 2.H., 2.X.	37,000 brute-force attempts over 4 d; tank overflow in Muleshoe; VPN breach in Abernathy
Tipton (W)WTP [2,29]	2024	Indiana	2.F., 2.X.	Russian "Cyber Army" actors; amateur-level; minimal disruption but demonstrates targeting
Arkansas City [16]	2024	Kansas	2.F., 2.W., 2.X.	Manual operations switchover; demonstrates incident frequency
American Water [34]	2024	14 states in the U.S.	2.D., 2.E., 2.F.	Largest U.S. utility (14M customers); IT/OT boundary protected water operations; billing down 1+ week

data from our work in Fig. 2. There is a clear trend of an increasing number of attacks against water utilities in recent years.

New York's Bowman Avenue Dam incident [17,40,41] represents one of the earliest documented nation-state attacks on U.S. water infrastructure. The incident highlights a 12-year history of threats from a nation-state actor with sophisticated capabilities, targeting a major metropolitan area. Between August 28 and September 18, 2013, an Iranian Revolutionary Guard Corps-affiliated actor operating from Iran repeatedly gained unauthorized remote access to the Bowman Dam SCADA system in Rye, New York. He obtained operational data, including water levels, temperature, and the status of the sluice gate. Although the sluice gate was offline for maintenance and inoperable, the intrusion incurred over $30,000 in remediation costs. The incident validates the high importance of security controls 2.W, 2.X, and 2.H, as no firewalls or authenticated access controls were in place. Instead, the SCADA system was directly internet-accessible. A cellular modem vulnerability was exploited to gain access.

On January 15, 2021, during the San Francisco Bay Area incident [8], a water treatment plant was attacked via stolen TeamViewer credentials from a former employee. The intruder remotely accessed a facility computer and deleted programs used for water treatment. The breach, discovered the following day, prompted the facility to reset all passwords and reinstall software. A separation of privileges according to security control 2.E. might have mitigated the attack or at least prevented loss of critical project files. Control 2.D. also relates to that attack - disabling the TeamViewer account of a former employee, which was used for the attack, would have reduced the threat potential.

The incident in Oldsmar, Florida [6,13] might have been a misattribution or an insider incident instead of an external cyberattack [42]. However, reports clearly revealed critical weaknesses: the outdated operating system Windows 7 and the software TeamViewer, which is used for remote connections, were installed. No network segmentation was present.

On November 25, 2023, the Cyber Av3ngers, an Iranian-backed, Islamic Revolutionary Guard Corps-affiliated threat actor, exploited default passwords in Unitronics PLCs at the water booster station in Raccoon Township, Pennsylvania [26,28]. The attackers disabled automated water pressure controls and switched to manual operations. They left a message stating "Every equipment'made in Israel' is Cyber Av3ngers legal target". This recent incident demonstrates supply chain vulnerabilities, where exploited default passwords in Unitronics PLCs were used to disable automated water pressure controls and switch to manual operations. The incident is a direct validation of control 2.A. - the change of default passwords. Additionally, PLCs were internet-accessible. The incident is part of a coordinated campaign related to the Israel-Hamas conflict. A technical in-depth analysis has been published by Claroty's Team 82 in [38].

In December 2023, at the County Mayo Water Treatment Plant (WTP), vulnerability CVE-2023-6448 was exploited using default admin credentials [30]. The attackers defaced the HMI with "You have been hacked. Down with Israel". Patching of vulnerabilities - security control 1.E. - prevents attackers from gain-

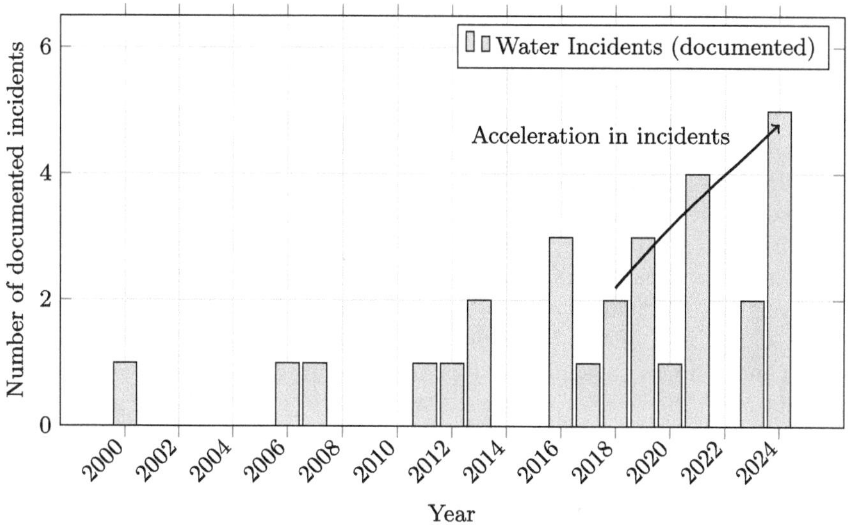

Fig. 2. Frequency analysis on documented incidents in water utilities based on the summarized incidents in this work, [17], and [19].

ing access by crafting exploits. Additionally, the use of admin credentials is addressed by security control 2.A., which involves changing default passwords. The manufacturer released a patch that enforces the change of the default password, highlighting the relevance of patching for other utilities that have not been affected yet. The attack resulted in a two-day outage, as operations were shifted to manual control.

As another example, in 2024, the incident at Hale Center, Abernathy, could have been detected earlier by following security control 2.G., which involves detecting and blocking unsuccessful repeated login attempts [5,14]. Indeed, firewall logs revealed 37,000 brute-force login attempts over a four-day period. It is also worth noting that stolen credentials may have contributed to the breach of the VPN connection in this incident. Security control 2.H. - using multi-factor authentication - mitigates the issue of leaked or stolen credentials. In Abernathy, where the attackers breached the VPN, they accessed the HMI and attempted to change passwords. A 10-second video shows them randomly clicking through menus and switching the lead well alternator. The same attackers accessed the HMI in nearby Muleshoe, causing a tank overflow.

On April 14, 2024, the water and wastewater treatment plant in Tipton, Indiana, had been under attack by the "Cyber Army of Russia" [2,29]. Officials stated that there were only minimal disruptions, and the plant stayed operational. A video showing manipulations on an HMI was released by the hackers on Telegram. The facility remained operational with minimal disruption and no impact on the process. According to the CEO of the consulting firm Infinity Squared Group, the threat actor was operating at an amateur level and only caused some minor, annoying disturbances [29].

The cyber incident on September 22, 2024, in Arkansas City [16] forced the water treatment facility to switch to manual operations. Although operations were not impacted, the incident highlights the ongoing frequency of threats.

The incident at American Water [34] on October 3, 2024 is the largest U.S. water utility breach. The company has 14 million customers across 14 states and 18 military installations. Although the water treatment process remained unaffected by the attack, the customer portal and billing systems were shut down for over a week. The incident demonstrates that even well-resourced utilities are vulnerable to attacks. Network segmentation prevented an operational impact, highlighting its importance.

We did not find any publications that included a detailed analysis of the provided incidents with sufficient technical details to analyze the root cause of the attacks. This makes a more detailed discussion of the incidents and a systematic mapping of incidents and security controls hardly possible. While the presented data does not provide sufficient insights into the root causes of the incidents, it reinforces the relevance of conducting assessments. Additionally, the fact sheets and technical reports [19–21] released by the US authorities also highlight the relevance of the security controls that can be found in Table 2 in Sect. 2.1.

4 Assessment Methodology

This section introduces a formal model for context-aware cybersecurity assessments that adapts to the individual characteristics of each plant. Our assessment methodology primarily employs a control-based approach, which also incorporates high-level plant characteristics, similar to asset-centric assessments. It cannot be used as a replacement for other security assessment approaches, such as those listed in Table 1 in Sect. 2.1. There is a tradeoff between complexity and usability of assessments. We discuss this for our approach in Sect. 5.3. Generally, our method is designed as an entry point to initiate cybersecurity assessments within the overall cybersecurity maturity journey of a utility. Table 4 summarizes the notation used in our formal model. Note that a higher relevance weight for a security control also indicates a greater impact on improvements in the cybersecurity maturity of an ICS. We refer to this using the term relevance weight.

Each real-world ICS has a unique setup. Some plants have flat network architectures and utilize RTUs, while others might have a layered network architecture with remote engineering capabilities. Some ICS can have network storage for file uploads, while others do not [37]. Additionally, the water sector itself has a diverse set of requirements. Regulatory aspects differ even within one sector, like between water and wastewater treatment. Furthermore, WWTPs have distinct architectures and processes on the physical layer compared to a WDS. Some processes are centralized at a single facility, while others are highly decentralized. Even within the same sub-sector, such as water treatment, certain distinguishing factors exist between different ICS architectures, including

Table 4. Notation for our formal model for context-aware cybersecurity assessments.

Symbol	Meaning
Θ	Static set of all possible ICS characteristics.
θ_i	i-th characteristic in Θ.
F	Assessment framework.
f_k	k-th security control.
Θ_j	Static subset of Θ relevant to control f_j.
$\Theta^{(A)}$	Dynamically scoped plant characteristics for assessment A.
$\tilde{F}$	Customized framework filtered by $\Theta^{(A)}$.
$\delta_{i,j}$	Relevance of θ_i for f_j.
ω_j^{base}	Baseline relevance weight of control f_j.
ω_j	Relevance weight of control f_j.

the underlying network topology. Certain characteristics, like the use of remote access or the network design, influence the cybersecurity decision-making about the implementation and relevance of security controls in the assessment. Applying one uniform assessment framework to different ICS covering a variety of setups requires granular security controls. However, certain security controls will then be irrelevant for the assessment in a specific plant, which can increase complexity for scenarios where not all controls are required. This makes the conduct of such assessments complicated for small utilities lacking the knowledge, staff, and budget to select important security controls.

We define all of these differing properties, which are relevant as key factors for cybersecurity decisions, as plant characteristics. The static and finite set Θ contains all plant characteristics for all ICS and is time independent. Table 5 lists some exemplary characteristics that influence cybersecurity decision-making, and are found in many ICS. Note that Θ has a higher dimensionality in reality. We only listed several characteristics that can be derived from security controls in current security guidelines [11,19,21] and mainly refer to the network architecture of the system.

Unlike full threat modeling approaches, we do not explicitly enumerate threats or derive attack paths, but only consider system properties that influence the relevance of security controls in this work. In future work, the plant characteristics may be further extended to incorporate attacker models or explicit

Table 5. Exemplary set $\Theta = \{\theta_1, \ldots, \theta_6\}$ of possible plant characteristics that influence cybersecurity decisions. The dimensionality of Θ is greater in practice.

i	θ_i	i	θ_i	i	θ_i
1	RemoteAccess	3	FlatNetwork	5	FileUpload
2	RemoteTerminalUnit	4	SegmentedNetwork	6	SensitiveData

threat information, which would more closely align with an approach that incorporates full threat modeling.

An assessment framework $F = \{f_0, f_1, \ldots, f_{k-1}\}$ consists of k security controls. First, we determine to what extent the implementation of control f affects cybersecurity maturity aspects regarding each characteristic $\theta_i \in \Theta$ by surveying subject matter experts (SMEs). Each survey participant selects all relevant characteristics for each control f_j using a relevance factor $\delta_{i,j} \in [0, 10]$ indicating the extent of relevance. If the SME assigns $\delta_{i,j} = 0$ for f_j, f_j is irrelevant for θ_i. If the SME assigns a value greater than zero, the control's relevance for the characteristic is proportional to the assigned value. A value of $\delta_{1,9} = 10$ for f_9 means that implementing f_9 has the highest relevance for the characteristic of remote access (θ_1), while $\delta_{6,9} = 1$ means that implementing f_9 has only limited relevance for the characteristic of sensitive data (θ_6). We also ask for the assignment of a general characteristics independent rank ω_j^{base} in our SME survey to rank the overall relevance of each f_j. In the model, ω_j^{base} reflects the general, plant-independent relevance weight of implementing security control f_j on the overall cybersecurity maturity, where $\omega_j^{\text{base}} \in [1, 10]$.

Overall and characteristics-dependent relevance are used to rank different security controls later. First, we initiate the scoping process to select only relevant security controls based on the plant's characteristics. The subset of relevant characteristics $\Theta_j \subset \Theta$ for each control f_j includes all θ_i where the relevance factors $\delta_{i,j}$ are not zero,

$$\Theta_j = \{\theta_i \in \Theta \mid \delta_{i,j} \neq 0\}. \tag{1}$$

A customized assessment framework $\tilde{F} = \{\tilde{f}_0, \tilde{f}_1, \ldots, \tilde{f}_{n-1}\} \subseteq F$ consists of n controls. It is derived by filtering F based on an assessment's characteristics $\Theta^{(A)}$,

$$\tilde{F} = \{f_j \in F \mid \Theta_j \cap \Theta^{(A)} \neq \emptyset\}, \quad \tilde{F} \subseteq F, \tag{2}$$

where the set $\Theta^{(A)} \subseteq \Theta$ of relevant characteristics for the specific assessment A is determined prior to conducting A. $\Theta^{(A)}$ includes all relevant characteristics that need to be assessed for a concrete plant and can dynamically vary over time. This way, only relevant security controls are included in the assessment.

Although the characteristics in $\Theta^{(A)}$ may appear to be outcomes of a complete assessment, they are derived from a lightweight preliminary scoping process. This scoping step requires only basic information and is not intended to replace the assessment, but rather to simplify its complexity. In practice, it is essential to ensure that all relevant characteristics are in scope for every assessment. Otherwise, important controls might be omitted.

Finally, an assessment

$$A|_{\tilde{F}} = \{(f_j, \omega_j) | f_j \in \tilde{F}\} \tag{3}$$

for a plant P based on a customized assessment framework $\tilde{F}$, is a set of ordered pairs of security control and individual weight, (f_j, ω_j). Weights are used to rank the individual security controls. The value of ω_j is proportional to the relevance weight that the implementation of a security control has on the cybersecurity maturity of an individual plant. It is based on two factors: On one hand, different security controls have a general relevance that is identical for every assessment and independent of characteristics. On the other hand, individual characteristics contribute differently to the relevance of one security control. Therefore, we calculate ω_j for security control f_j as

$$\omega_j = \omega_j^{\text{base}} + \sum_{i|\theta_i \in \{\Theta^{(A)} \cap \Theta_j\}} \delta_{i,j}, \tag{4}$$

where the sum accounts for all relevance factors $\delta_{i,j}$ for all scoped characteristics that apply to the corresponding security control. Note that this is not equivalent to the sum of all relevance factors for a control. A relevance factor, which is not zero, is only added in the sum if the initial scoping process included the corresponding characteristic. ω_j ranks the same security control differently depending on the scope of characteristics that concretely apply to the individual ICS.

5 Evaluation and Discussion

In Sect. 5.1, we present and discuss the results from our conducted expert survey, which was used to parametrize the weights linking security controls to plant characteristics. In Sect. 5.2, a case study is presented that demonstrates how the proposed methodology can be applied to rank security controls. Section 5.3 discusses current shortcomings and opens new directions for future research.

While our defined plant characteristics have a higher dimensionality in reality, we limit our evaluation to the six exemplary dimensions out of Table 5 in Sect. 4. We also limit the evaluation to the exemplary security controls found in Table 2 in Sect. 2.1. The security controls we use in our evaluation are most frequently mentioned in cybersecurity fact sheets for critical infrastructure and the water sector released by US authorities [19,21], highlighting their importance in reality. The importance of our selected security controls can also be seen through our summary on recent real-world incidents in Sect. 3.

5.1 Survey Results

We conduct an internal SME survey with three experienced cybersecurity assessors for the characteristics listed in Table 5 in Sect. 4 and the security controls provided in Table 2 in Sect. 2.1. The results are depicted in Table 6. We calculated the arithmetic mean for all ω_f^{base} and relevance factors δ from the SME questionnaires and rounded the results to zero decimal places.

The SME survey results can be interpreted using the intuitive security control 2.F. for network segmentation. 2.F. has a high base score, as network segmentation is overall one of the most critical security controls. Additionally, it has a high score of 10 for flat network architectures, while it has a small score of only 3 for plants that already have segmented network architectures. For a plant

Table 6. SME survey results for the security controls out of Table 2. Characteristics θ_i according to Table 5 were assigned to each security control where $\delta_{i,j} \neq 0$. If $\delta_{i,j} = 0$ the control is not relevant for θ_i. Labeling and relevance factor assignment were evaluated independently.

Security control f_j by ID	Relevance $\delta_{i,j}$ of control f_j for plant characteristic θ_i						
	Baseline Relevance	Remote Access θ_1	Remote Terminal Unit θ_2	Flat Network θ_3	Segmented Network θ_4	File Upload θ_5	Sensitive Dataθ_6
ID	ω_f^{base}	$\delta_{1,j}$	$\delta_{2,j}$	$\delta_{3,j}$	$\delta_{4,j}$	$\delta_{5,j}$	$\delta_{6,j}$
1.E.	10	10	3	3	3	0	2
2.A.	9	10	9	3	3	0	3
2.D.	7	8	2	2	2	1	3
2.E.	8	5	2	6	3	2	6
2.F.	9	7	3	10	3	0	0
2.G.	6	8	5	0	2	2	0
2.H.	9	9	5	3	2	0	0
2.K.	6	3	3	0	0	2	7
2.L.	6	0	0	0	0	0	8
2.M.	7	3	0	0	0	8	3
2.V.	8	0	0	6	3	3	2
2.W.	10	10	3	7	5	0	0
2.X.	8	7	0	10	5	0	0

only scoping a flat network architecture before the actual assessment, 2.F. will be included in the assessment, and its rank will be 19. A different plant that already has a segmented network architecture will still include the security control, but with a lower individual relevance of 3 instead of 10. The overall rank is 12, not 19. An assessment with scoped sensitive data only will not include 2.F. at all.

The base scores for each security control also reflect the relevance of measures found in currently recommended mitigation strategies [19–21]. Generally, it can be seen that all security controls have a high overall relevance greater than 5, highlighting the importance of all controls. The most critical controls are 1.E. and 2.W., with the highest possible base score of 10. According to our data, unnecessary exposure of OT assets and vulnerability mitigation are more critical for ICS environments than the use of internal encryption and blocking of unsuccessful login attempts. This aligns with current standards and practices. The real-world incidents that we listed in Table 3 also support this. The San Francisco Bay Area incident was caused by stolen TeamViewer credentials [8]. The recommended mitigation strategy is security control 2.W., as openly available remote connections are easier to exploit. The County Mayo incident was caused by the use of default credentials [30]. As the component was exposed to the internet, an exploitation was feasible.

Further comparison between the SME survey results in Table 6 and the incident-related controls in Table 3 in Sect. 3 reveals that access controls, such as 2.W. and 2.X., are implicated in real incidents and were likewise highlighted by experts as especially critical in flat networks. This alignment reflects the reality that many water plants still operate with flat network architectures, as noted in [17], highlighting the importance of prioritizing access control measures.

We asked our survey participants to name the three most important security controls. Interestingly, we got very different responses. For control 2.X., for example, two assessors stated that eliminating connections between OT assets and the internet is most critical, while the third assessor explained that such connections are fine as long as they are done securely through the DMZ. It highlights that our SME survey is biased by the personal experience of each expert. This might affect the reproducibility of our survey results with different survey participants and differently formulated security controls. All experts explained having difficulties in ranking the security controls, as they were too generic. For example, Control 1.E. - mitigating and patching of a vulnerability - should be separated into two controls.

5.2 Illustrative Case Study

We conduct an illustrative case study to further evaluate how our proposed methodology facilitates the ranking and filtering of security controls. Assume an assessment for two different ICS in the water sector that have the following properties:

- **Facility A** is a Water Distribution System (WDS) consisting of various RTUs. Remote access is primarily used due to the plant's distributed nature. The SCADA system at the central facility is separated from the IT network using a DMZ. The IT network integrates an automated billing system for the customers. The initial scope using plant characteristics out of Table 5 is dynamically determined before the assessment as $\Theta^{(A)} = \{\theta_1, \theta_2, \theta_4, \theta_6\}$.
- **Facility B** is a small, rural Waste Water Treatment Plant (WWTP), where the treatment is done at a central facility. The plant does not have RTUs, and access to HMIs can only be established onsite, as no network ports are externally exposed. The plant does not have a segmented network architecture. The initial scope using plant characteristics out of Table 5 is $\Theta^{(A)} = \{\theta_3\}$.

Table 7 shows all scoped security controls for both plants and the resulting weights according to Equation (4) using the data out of Table 6.

For Facility A, security control 2.A. - changing default passwords - with a relevance of 34 is the most important, whereas for Facility B, security control 2.F. - network segmentation - has the highest relevance of 18. Note that the relevance only refers to concrete assessments. Therefore, the relevance between different facilities can only be compared when properly normalized. In fact, changing default passwords is most important for Facility A, as many remote connections and RTUs are utilized. For Facility B, network segmentation is indeed the most

Table 7. Ranks of filtered security controls for facilities A/B after initial scoping.

f_j	1.E.	2.A.	2.D.	2.E.	2.F.	2.G.	2.H.	2.K.	2.L.	2.M.	2.V.	2.W.	2.X.
Facility A	✓	✓	✓	✓	✓	✓	✓	✓	✓	✓	✓	✓	✓
$\omega_f\|_A$	28	34	22	24	22	21	25	19	14	13	13	28	20
Facility B	✓	✓	✓	✓	✓	✗	✓	✗	✗	✗	✓	✓	✓
$\omega_f\|_B$	13	12	9	14	19	-	12	-	-	-	14	17	18

critical aspect. This is also one of the most mentioned threat mitigation strategies in the literature. Note that further network segmentation is less critical for Facility A, as it is already established. Some controls are ranked low in both scenarios: For example, control 2.V. - prohibiting unauthorized hardware - ranks low in both cases. However, specific threat scenarios, such as insider attacks, would make the control more relevant. This is a limitation, as the current plant characteristics are not sufficient for proper threat modeling.

When linking the results of our case study back to recent real-world incidents in the water sector, as summarized in Table 3 in Sect. 3, we make the following observations: It is not easily possible to link our case-study scenario with the real-world incidents, as only little empirical data about the described incidents can be found. Our fictional Facility A mainly uses remote access. While it is not clear how attackers got access to the HMI during the real-world incident in Indiana [2,29], the exploited weakness in the County Mayo incident [30] was the use of remote access in combination with default credentials. Changing default passwords is also the security control in our evaluation for Facility A, which has the highest relevance factor of 34. While we cannot state that our assessment methodology would have prevented the exact incident for the specific scenario in County Mayo, the ranking of our security controls is in accordance with real-world observations. Our methodology could have only prevented the incident in County Mayo if it had been applied proactively before the attack.

We demonstrated in our illustrative case study that our context-aware assessment methodology tailors assessment frameworks to different ICS by accounting for plant characteristics. The same security controls are represented by different relevance weights in both assessments, reflecting their individual contribution to the overall security posture. We note that some security controls are omitted for the assessment of a concrete plant if they are not applicable. Confidentiality and email security are not relevant for the small WWTP without an IT network and customer data. While this exclusion is sensible in our scenario, it needs to be further examined whether important security controls are being omitted due to incorrect or biased SME survey results.

5.3 Threats to Validity and Future Work

Conceptual: The method in its current form comes at some cost. First, there is no economic justification for cybersecurity investments to the management, as there is no risk quantification that calculates the magnitude of loss from incidents based on their impact and analyzes the return on security investments over time. Second, static weights are derived from a one-time survey of SMEs. However, the threat landscape changes rapidly, making the approach less adaptable if weights are not frequently reevaluated. Third, interactions between different controls through synergistic combinations are currently not captured in our approach. Future work will therefore discuss and compare different assessment approaches with our proposed methodology and further improve our context-aware cybersecurity approach to address these conceptual limitations.

Note that the scope of an assessment is time-dependent and may vary between two different assessments for the same plant, as plant characteristics can change over time. While we do not explicitly analyze time dependence in this paper, a new assessment will automatically adapt to new circumstances, as the initial scoping will be different if plant characteristics change that impact cybersecurity decision-making. We will further investigate time-dependence in future work. Additionally, our current work does not consider the implementation complexity of a security control. If a high-ranked security control is complex to implement, the implementation of other controls that are less important may still improve cybersecurity maturity as an effective and easy-to-implement measure with higher practical feasibility.

Methodological: While we note that the number of survey participants and the dimensionalities of controls and characteristics are small for representative findings that are transferable to real-world scenarios, our findings demonstrate the intended outcome of our methodology. Due to the small number of survey participants and the limited characteristics and security controls, our results in Table 6 should not be used for a real-world assessment at this time. Instead, the results motivate a systematic, large-scale study in future work, which will also analyze the reliability of the survey results by incorporating expert elicitation. While we did not address such expert calibration and potential bias in the conducted survey, there exist standardized ways, like Cooke's Classical Model [9] to ensure optimal weighting using calibration questions and information scoring. Such systematic expert elicitation approaches ensure reproducibility and strengthen the empirical foundation of future data.

There are more characteristics in reality that are not covered here. For example, attacker modeling, which is commonly part of threat modeling, should be considered before an assessment instead of focusing on the currently defined plant characteristics only. Security controls for insider attack prevention are highly different than the ones used against remote attacks, and our analyzed set of controls is not representative of the complete guidelines found in reality. Future work needs to analyze whether remote access is overemphasized at the expense of other aspects, such as file uploads, sensitive data, or other characteristics, that are currently not represented.

6 Conclusion

Control-based assessments are a key approach to analyzing the cybersecurity maturity of an industrial control system. In this paper, we introduced a context-aware assessment methodology that enables ranking of security controls to prioritize the most critical measures for budget-constrained water system environments. The first SME survey results indicate that some controls are consistently rated as more relevant, while their relevance weights vary across different plant characteristics. We demonstrated how our context-aware assessment methodology is used to tailor assessments to individual plant characteristics and how security controls are ranked for the most actionable improvements in cybersecurity maturity.

This work opens up several avenues for future research. In particular, the next step involves conducting a systematic expert elicitation with more security controls and detailed plant characteristics to make accurate and reproducible statements. Another promising direction involves extending our initial scoping mechanism with methods used in threat modeling, including different attacker models and potential threats. Additionally, several conceptual aspects, such as temporal scoring of cybersecurity assessments, implementation complexity, and economic justification, require further attention in future work. Finally, we also envision a web-based assessment tool to validate the methodology in practice and gather broader data on the water-sector cyber maturity.

Disclosure of Interests. The authors have no competing interests to declare that are relevant to the content of this article.

References

1. Alberts, C.J., Behrens, S.G., Pethia, R.D., Wilson, W.R.: Operationally Critical Threat, Asset, and Vulnerability Evaluation (OCTAVE) Framework, Version 1.0: Tech. rep., Defense Technical Information Center, Fort Belvoir, VA (1999). https://doi.org/10.21236/ADA367718
2. Antoniuk, D.: Russian hackers claim cyberattack on Indiana water plant (2024/04/23). https://therecord.media/russia-hackers-cyberattack-tipton-indiana, Accessed 04 Aug 2025
3. Asghar, M.R., Hu, Q., Zeadally, S.: Cybersecurity in industrial control systems: issues, technologies, and challenges. Comput. Netw. **165**, 106946 (2019). https://doi.org/10.1016/j.comnet.2019.106946
4. Assante, M.J., Lee, R.M.: The Industrial Control System Cyber Kill Chain. Whitepaper, SANS (Oct 2015). https://www.sans.org/white-papers/36297
5. Bleiberg, J.: Tiny Texas City Repels Russia-Tied Hackers Eyeing Water System (2024-04-22). https://www.insurancejournal.com/news/southcentral/2024/04/22/770918.htm, Accessed 18 April 2025
6. CISA: Compromise of U.S. Water Treatment Facility (Feb 2021). https://www.cisa.gov/news-events/cybersecurity-advisories/aa21-042a, Accessed 30 Sep 2025
7. Defending OT Operations Against Ongoing Pro-Russia Hacktivist Activity (May 2024). https://www.cisa.gov/resources-tools/resources/defending-ot-operations-against-ongoing-pro-russia-hacktivist-activity, Accessed 09 April 2025

8. Collier, K.: 50,000 security disasters waiting to happen: The problem of america's water supplies (2021/06/17). https://www.nbcnews.com/tech/security/hacker-tried-poison-calif-water-supply-was-easy-entering-password-rcna1206, Accessed 24 April 2025
9. Cooke, R.M.: Experts In Uncertainty: Opinion and Subjective Probability in Science. Oxford University Press (Oct 1991)
10. Cárdenas, A.A., Amin, S., Lin, Z.S., Huang, Y.L., Huang, C.Y., Sastry, S.: Attacks against process control systems: risk assessment, detection, and response. In: Proceedings of the 6th ACM Symposium on Information, Computer and Communications Security, ASIACCS 2011, pp. 355–366. Association for Computing Machinery, New York (Mar 2011). https://doi.org/10.1145/1966913.1966959
11. EPA: Guidance on Improving Cybersecurity at Drinking Water and Wastewater Systems. Tech. rep., U.S. Environmental Protection Agency (Sep 2024). https://www.cisa.gov/resources-tools/resources/epa-guidance-improving-cybersecurity-drinking-water-and-wastewater-systems, Accessed 09 April 2025
12. Falliere, N., O Murchu, L., Chien, E.: W32. Stuxnet Dossier. White paper, Symantec Corp (2011). https://www.security.com/threat-intelligence/stuxnet-dossier-espionage, Accessed 21 April 2025
13. Greenberg, A.: A Hacker Tried to Poison a Florida City's Water Supply (Feb 2021). https://www.wired.com/story/oldsmar-florida-water-utility-hack/, Accessed 30 Sep 2025
14. Greenberg, A.: Hackers Linked to Russia's Military Claim Credit for Sabotaging US Water Utilities (Apr 2024). https://www.wired.com/story/cyber-army-of-russia-reborn-sandworm-us-cyberattacks/, Accessed 30 Sep 2025
15. Greig, J.: Water facility operator says ransomware attack affected north america division (2024/01/23). https://therecord.media/water-operator-ransomware-attack-north-america, (Accessed 03 Aug 2025
16. Greig, J.: CISA warns of continuing attacks on water systems after kansas town reports incident (2024/09/25). https://therecord.media/cisa-warns-of-attacks-aginst-water-systems-arkansas-city, Accessed 24 April 2025
17. Hassanzadeh, A., Rasekh, A., Galelli, S., Aghashahi, M., Taormina, R., Ostfeld, A., Banks, M.K.: A review of cybersecurity incidents in the water sector. J. Environ. Eng. **146**(5), 03120003 (2020). https://doi.org/10.1061/(ASCE)EE.1943-7870.0001686
18. Howard, M., Lipner, S.: The Security Development Lifecycle. Microsoft Press, Redmond, WA, USA (May (2006)
19. Joint Advisory: Ongoing Cyber Threats to U.S. Water and Wastewater Systems. Tech. Rep. AA21-287A, CISA (Oct 2021), https://media.defense.gov/2021/oct/14/2002873650/-1/-1/0csa_ongoing_cyber_threats_to_u.s._water_and_wastewater_systems_20211014.pdf, Accessed 07 Aug 2025
20. Joint Advisory: IRGC-Affiliated Cyber Actors Exploit PLCs in Multiple Sectors, Including US Water and Wastewater Systems Facilities. Tech. Rep. AA23-335A, CISA (Dec 2024). https://www.cisa.gov/news-events/cybersecurity-advisories/aa23-335a
21. Joint Advisory: Iranian Cyber Actors May Target Vulnerable US Networks and Entities of Interest. Fact Sheet, CISA (Jun 2025). https://www.cisa.gov/resources-tools/resources/iranian-cyber-actors-may-target-vulnerable-us-networks-and-entities-interest, Accessed 07 July 2025
22. Jones, J.: An Introduction to Factor Analysis of Information Risk (FAIR). Whitepaper, Risk Management Insight LLC (2005)

23. Kertzner, P., Carter, C., Hahn, A.: Crown Jewels Analysis (CJA) for Industrial Control Systems (ICS). Whitepaper, MITRE (Dec 2022). https://www.mitre.org/news-insights/publication/crown-jewels-analysis-industrial-control-systems
24. Khan, R., McLaughlin, K., Laverty, D., Sezer, S.: STRIDE-based threat modeling for cyber-physical systems. In: 2017 IEEE PES Innovative Smart Grid Technologies Conference Europe (ISGT-Europe), pp. 1–6 (Sep 2017). https://doi.org/10.1109/ISGTEurope.2017.8260283
25. Klobuchar, A.: 115th Congress (2017-2018): America's Water Infrastructure Act of 2018 (Oct 2018). https://www.congress.gov/bill/115th-congress/senate-bill/3021/text, archive Location: 2018-06-07
26. Kovacs, E.: Hackers Hijack Industrial Control System at US Water Utility (Nov 2023). https://www.securityweek.com/hackers-hijack-industrial-control-system-at-us-water-utility/, Accessed 30 Sep 2025
27. Leszczyna, R.: Review of cybersecurity assessment methods: applicability perspective. Comput. Sec. **108**, 102376 (2021). https://doi.org/10.1016/j.cose.2021.102376
28. Levy, M.: Cybersecurity at water utilities a national concern after Pa. water authority hacked (Jan 2024). https://whyy.org/articles/cybersecurity-water-utilities-hacking/, Accessed 30 Sep 2025
29. Lyngaas, S.: Russia-linked hacking group claims to have targeted Indiana water plant (2024/04/22). https://www.cnn.com/2024/04/22/politics/russia-linked-hacking-group-targets-indiana-water-plant, Accessed 04 Aug 2025
30. Martin, A.: Two-day water outage in remote Irish region caused by pro-Iran hackers. https://therecord.media/water-outage-in-ireland-county-mayo, Accessed 04 Aug 2025
31. Mattos, V., Schmidt, J.H., Bhaya, A., Sá, A.O.D., Menasché, D.S., Srivastava, G.: Design and Detection of Covert Man-in-the-Middle Cyberattacks on Water Treatment Plants. In: Proceedings of the 2025 Workshop on Re-design Industrial Control Systems with Security (RICSS 2025), p. 9, ACM, Taipei, Taiwan (2025). https://doi.org/10.1145/3733823.3764516
32. National Institute of Standards and Technology: The NIST Cybersecurity Framework (CSF) 2.0. Tech. Rep. NIST CSWP 29, National Institute of Standards and Technology, Gaithersburg, MD (Feb 2024).https://doi.org/10.6028/NIST.CSWP.29
33. Qassim, Q.S., Jamil, N., Daud, M., Patel, A., Ja'affar, N.: A review of security assessment methodologies in industrial control systems. Inform. Comput. Sec. **27**(1), 47–61 (2019)
34. Sears, D.: American Water's Security Incident: ransomware or Something Else? (Oct 2024). https://www.forescout.com/blog/american-water-security-incident-ransomware/, Accessed 30 Sep 2025
35. Stouffer, K., et al.: Guide to Operational Technology (OT) Security. Tech. Rep. NIST Special Publication (SP) 800-82 Rev. 3, National Institute of Standards and Technology (Sep 2023). https://doi.org/10.6028/NIST.SP.800-82r3
36. Strom, B.E., Applebaum, A., Miller, D.P., Nickels, K.C., Pennington, A.G., Thomas, C.B.: MITRE ATT&CK: Design and Philosophy. Whitepaper 19-01075-28, MITRE (Mar 2020). https://www.mitre.org/news-insights/publication/mitre-attck-design-and-philosophy
37. Tang, C., Faatz, D.: Cybersecurity for the Water and Wastewater Sector: Build Architecture. Operational Technology Remote Access, Tech. Rep. NIST TN 2283 ipd, National Institute of Standards and Technology, Gaithersburg, MD (2024). https://doi.org/10.6028/NIST.TN.2283.ipd

38. Team82: From Exploits to Forensics: Unraveling the Unitronics Attack (Aug 2024). https://claroty.com/team82/research/from-exploits-to-forensics-unraveling-the-unitronics-attack, Accessed 01 Oct 2025
39. Tuptuk, N., Hazell, P., Watson, J., Hailes, S.: A systematic review of the state of cyber-security in water systems. Water **13**(1), 81 (2021)
40. U.S. Department of Justice, Office of Public Affairs: Seven Iranians Working for Islamic Revolutionary Guard Corps-Affiliated Entities Charged for Conducting Coordinated Campaign of Cyber Attacks Against U.S. Financial Sector — United States Department of Justice (Mar 2016). https://www.justice.gov/archives/opa/pr/seven-iranians-working-islamic-revolutionary-guard-corps-affiliated-entities-charged, Accessed 30 Sep 2025
41. US District Court Southern District of New York: Indictment, United States v. Fathi. Federal indictment 16 CRIM 48, U.S. Department of Justice, Washington, D.C. (Mar 2016). https://www.justice.gov/archives/opa/file/834996/dl
42. Vasquez, C.: Did someone really hack into the Oldsmar, Florida, water treatment plant? New details suggest maybe not. (Apr 2023). https://cyberscoop.com/water-oldsmar-incident-cyberattack/, Accessed 30 Sep 2025
43. Xiong, W., Lagerström, R.: Threat modeling - a systematic literature review. Comput. Sec. **84**, 53–69 (2019). https://doi.org/10.1016/j.cose.2019.03.010

A Group-Aware Temporal Autoencoder for Anomaly Detection in Water Treatment Systems

Babangida Abdullahi(✉) and Sunil Vadera

School of Science, Engineering and Environment, University of Salford, Manchester, UK
b.abdullahi1@edu.salford.ac.uk, s.vadera@salford.ac.uk

Abstract. The increasing integration of cyber-physical systems in critical infrastructure like water treatment plants has heightened their vulnerability to sophisticated attacks, necessitating robust anomaly detection mechanisms. This paper presents a Temporal-Enhanced LSTM Autoencoder (TE-LSTM-AE), a novel two-stage deep learning model designed to detect cyber-physical attacks by learning complex temporal patterns and cross-sensor correlations. Our group-aware architecture processes sensor data based on functional characteristics (e.g., hydraulic performance, water quality) in dedicated autoencoders before a fusion stage learns critical inter-group relationships. Evaluated on the Secure Water Treatment (SWaT) testbed, the model achieved state-of-the-art performance with a precision of 0.9910, recall of 0.9244, F1-score of 0.9565, and a near-perfect ROC-AUC of 0.9971. Detailed analysis demonstrates the model's high sensitivity to disruptions in hydraulic correlations and its unique capability to identify a concentrated multi-stage attack campaign. The results establish that a physics-informed, group-based approach is superior for achieving high-fidelity, interpretable, and robust anomaly detection in complex industrial environments.

Keyword: Critical Infrastructure Security · Industrial Control Systems (ICS) · Anomaly Detection · Cyber-Physical Systems (CPS) · Deep Learning · LSTM Autoencoder · Secure Water Treatment (SWaT) · Time Series Analysis, Sensor Data Correlation · Attack Detection

1 Introduction

Critical Infrastructure (CI) systems, such as water treatment, water distribution, and power generation, are fundamental to societal welfare and economic stability [1–6]. Modern CI often operate as Cyber-Physical Systems (CPS), where physical processes are meticulously monitored by sensors and controlled by actuators through Programmable Logic Controllers (PLCs) and Supervisory Control and Data Acquisition (SCADA) systems connected via intricate communication networks [1, 2, 7]. This sophisticated integration, while enabling efficient automation, concurrently introduces significant vulnerabilities to cyber threats [3, 7–9].

S. Zonouz et al. (Eds.): DCS-CI 2026, CCIS 3109, pp. 236–255, 2027.
https://doi.org/10.1007/978-981-92-5039-4_12

Cyber-attacks on Industrial Control Systems (ICS) within water utilities can precipitate severe consequences, ranging from physical plant damage and service disruption to environmental contamination and public safety hazards [3, 7, 10]. Notable incidents, such as the sewage spill at Maroochy water services in Australia attributed to an insider attack [11], underscore the imperative for robust defence mechanisms. Therefore, early and accurate detection of anomalous behaviours is paramount to ensure the continuous and reliable operation of these vital systems [3].

Traditional signature-based intrusion detection systems, reliant on predefined attack patterns, prove inadequate against novel, or "zero-day," attacks [3, 7]. This limitation has propelled the adoption of anomaly-based detection approaches, which seek to identify deviations from established normal system behaviour[4]. The advent of Machine Learning (ML) and Deep Learning (DL) techniques has profoundly transformed anomaly detection in ICS. These data-driven methods possess an inherent capability to learn complex, non-linear patterns from vast quantities of historical operational data, thereby significantly reducing the extensive manual effort typically required to model intricate plant designs [7, 9].

However, the landscape of anomalies in water treatment plants is diverse and complex. It encompasses both fast-acting cyber-attacks that aim for rapid disruption, and slowly developing issues, such as sensor drift. Sensor drift, characterized by a gradual and often imperceptible deviation in sensor readings, can subtly compromise operational reliability and water quality over extended periods without immediate alerting. Existing anomaly detection methods frequently struggle with the dual challenge of discerning between these distinct types of anomalies and dynamically adapting to the evolving operational conditions of ICS without necessitating frequent and costly retraining [3].

This work aims to addresses these critical gaps by proposing a 2-stage enhanced Long Short-Term Memory (LSTM) autoencoder [7]. The novel architecture is designed to leverage latent vectors from the first stage encoder as input to the second LSTM encoder [12]. The primary objective is to develop a sophisticated mechanism capable of detecting both fast and slow anomalies in water treatment plants. The approach is predicated on enhancing the robustness and accuracy of anomaly detection in dynamic ICS environments by refining the learning of complex, multi-temporal patterns.

Key Contributions

The principal contributions of this work are threefold:

1. A Novel Group-Aware, Two-Stage Deep Learning Architecture: We propose the Temporal-Enhanced LSTM Autoencoder (TE-LSTM-AE), a novel model that moves beyond monolithic processing. It employs dedicated autoencoders for functionally distinct sensor groups (e.g., hydraulic, water quality) to learn specialized temporal features, followed by a second fusion stage that explicitly learns critical cross-system correlations, a capability lacking in many existing data-driven methods.
2. Unprecedented Detection Performance with Minimal False Alarms: Our model achieves state-of-the-art results on the SWaT dataset, demonstrating an exceptional balance between high precision (0.9910) and high recall (0.9244), yielding an F1-score of 0.9565. Crucially, the near-perfect precision signifies the model generates minimal false positives (only 1 FP), a critical advancement for operational deployment where alarm fatigue undermines security systems.

3. In-Depth Analytical Validation and Explainability: We provide a comprehensive, multi-faceted analysis that moves beyond aggregate metrics. This includes a temporal analysis revealing the model's ability to reconstruct complex multi-stage attack campaigns, a sensitivity analysis quantifying the importance of cross-group correlations, and a sensor-level investigation that links model detections to specific physical manifestations of attacks, enhancing practical interpretability for operators.

The remainder of this paper is structured as follows. Section 2 reviews related work in ICS anomaly detection, highlighting the limitations of existing methods. Section 3 details the proposed TE-LSTM-AE methodology, including data preprocessing, the group-based architecture, and the anomaly detection process. Section 4 presents a comprehensive analysis and discussion of the experimental results, covering overall performance, temporal patterns, cross-group correlations, and training dynamics. Finally, Sect. 5 concludes the paper by summarizing the findings and outlining directions for future work.

2 Related Work

The increasing sophistication of Critical Infrastructure (CI) systems [2, 13, 14], often operating as Cyber-Physical Systems (CPS) [2, 4, 14], with interconnected sensors, actuators, and control systems like Programmable Logic Controllers (PLCs) and Supervisory Control and Data Acquisition (SCADA) [2, 4, 9, 15], has concurrently introduced significant vulnerabilities to cyber threats [3, 15]. Ensuring the integrity and reliable operation of these systems, such as water treatment and distribution plants, is paramount [3, 4, 7, 16]. Anomaly detection research for Industrial Control Systems (ICS) in water utilities is broadly classified into design-centric (or physics-aware) and data-driven (or physics-unaware) approaches [11, 16, 17]. While design-centric methods leverage plant design knowledge for explainable outcomes and anomaly localization [4, 11, 16], they suffer from considerable manual effort and maintenance challenges, particularly with complex or modified processes [11, 13]. Data-driven methods, conversely, learn process dynamics from historical datasets [4, 16], offering faster deployment but often failing to capture intricate temporal dependencies and complex interrelationships among ICS components, which can lead to high false positive rates and hinder detection of unknown attacks due to the impracticality of acquiring labelled data for all threats [3, 4, 11, 18–20].

In recent years, deep learning (DL) models have gained significant attention for time series anomaly detection (TSAD) in CPS, excelling at learning complex behavioural patterns [4, 21, 22]. These models include forecasting-based approaches, such as Long Short-Term Memory (LSTM) networks and Gated Recurrent Units (GRUs), which are well-suited for capturing temporal information and long-term dependencies in sequential data [8, 21, 23–26]. Reconstruction-based models, often employing Autoencoders (AEs), learn compact representations of normal data, detecting anomalies based on high reconstruction error [3, 12, 22, 23, 25, 27, 28]. However, existing deep learning models face several critical challenges. They can be vulnerable to adversarial sample attacks, where malicious perturbations evade detection [2, 3, 12, 28]. Mode collapse is another issue for deep one-class classification and autoencoder-like structures, where learned features converge to a single point, limiting model adaptability; this can be mitigated by regularization techniques like noise injection or penalizing minibatch variance [6, 22]. Furthermore, achieving complete feature extraction is challenging; direct application of attention mechanisms can be computationally intensive, and models may not fully utilize temporal position information or differentiate periodic changes, leading to incomplete representation [17]. Research shows that smaller convolution kernels (e.g., 1×3) are more effective at capturing local jump features and sudden changes in ICS data than larger ones [18]. Computational complexity and resource constraints, particularly for small and medium-sized industrial environments, also necessitate lightweight solutions [4, 14, 18, 19, 26].

To address these limitations, more sophisticated and hybrid approaches have emerged. Time-frequency fusion feature attention encoding (TFANet), for instance, extracts and fuses features from both time and frequency domains, leveraging attention mechanisms to calculate feature weights and capture long-distance correlations, enhancing anomaly classification [29]. Multi-model or ensemble approaches are also employed, where multiple independent models monitor specific system stages or components, or multiple classifiers are combined to improve detection coverage, handle heterogeneous data, and detect diverse, previously unseen attacks [17–20, 30–33]. This includes Global-Local Integration (GLIN) in Graph Neural Networks (GNNs) to overcome limitations of purely local aggregation by incorporating global perspectives on system semantics [34]. Importantly, the integration of design knowledge (physics-informed machine learning) into data-driven models, as demonstrated by frameworks like AICrit and DAE, significantly improves detection performance, reduces false positives, and aids in the localization and explanation of anomalies, especially in heterogeneous and dynamic ICS environments [2, 6, 11, 13, 16, 30]. This is vital as purely data-driven methods often struggle with localization and interpretability [3, 11, 16]. Dimensionality reduction techniques like Principal Component Analysis (PCA) and Kernel PCA (KPCA) are also used to reduce computational burden and address high-dimensional data, though their effectiveness can vary depending on whether design knowledge is considered [11, 19, 21].

The proposed two-stage enhanced Long Short-Term Memory (LSTM) autoencoder (TE-LSTM-AE) builds upon these advancements, specifically leveraging the strengths of reconstruction-based deep learning [3, 22, 23, 25, 35] and modularity [27]. By employing a Group LSTM Encoder to generate Group Latent Vectors that feed into a Cross-Group LSTM Encoder and Decoder, the model aims to refine the learning of complex, multi-temporal patterns and enhance the robustness and accuracy of anomaly detection [1, 5, 21, 23–25, 34]. This architecture addresses the challenge of extracting features that represent dynamic temporal changes and long-distance correlations [26, 27], similar to the motivations behind time-frequency fusion [21] or multi-faceted feature extraction approaches [18]. The focus on Threshold Optimization and Anomaly Decision components within the model highlights the critical need for effective anomaly scoring and thresholding to minimize false alarms and improve overall detection efficacy, a persistent challenge in ICS anomaly detection [2–4, 6, 11, 12, 17–19, 26, 34] [34]. The effectiveness of such models is rigorously evaluated using standard metrics like Accuracy, Precision, Recall, F1-Score, and Area Under the Receiver Operating Characteristic Curve (AUC-ROC) [6, 7, 11, 12, 17, 18, 20, 21, 23, 36, 37], often validated on realistic testbeds such as the Secure Water Treatment (SWaT) and Water Distribution (WADI) datasets to simulate real-world attack scenarios [1–3, 11–14, 24, 27].

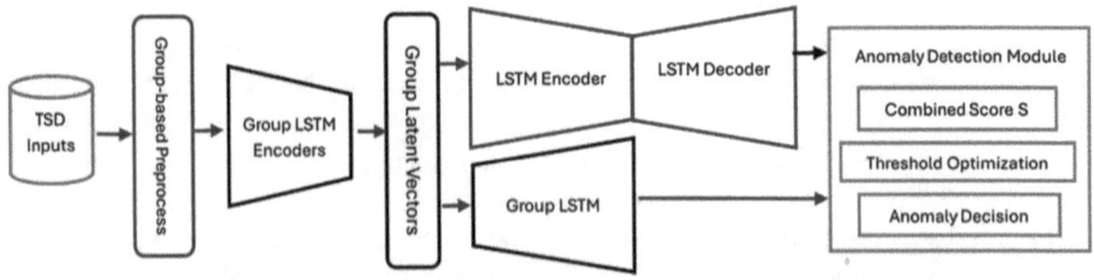

Fig. 1. Architecture of the proposed Temporal-Enhanced LSTM Autoencoder (TE-LSTM-AE) model. The two-stage architecture processes sensor data through group-specific LSTM autoencoders with attention mechanisms, followed by cross-group correlation learning in the second stage.

3 Methodology

This section outlines the architecture of the Temporal-Enhanced LSTM Autoencoder (TE-LSTM-AE; Fig. 1 above) for anomaly detection in cyber-physical systems, using the SWaT dataset as a case study.

3.1 Dataset and Preprocessing

The Secure Water Treatment (SWaT) dataset, a benchmark for ICS security, was utilized. It comprises 7 days of normal operation and 4 days with over 41 attack instances, recorded at a 1-s resolution from 51 sensors and actuators. A subset of 18 critical features was selected and partitioned into four groups {**Water Quality (G1)**: Slow-changing indicators, Chemical **Dosing (G2)**: Moderately paced processes, **Hydraulic Performance (G3)**: Fast-reacting sensors, and **Process Stability (G4)**. A multi-resolution preprocessing strategy was employed. Let a raw signal from group k be $X^k = \{x_1^k, x_2^k, \ldots, x_T^k\}$.. For slow groups ($k \in \{G1, G2\}$), data was resampled to 5-min intervals using the mean, $\tilde{X}^k = \{\overline{x}_1^k, \overline{x}_2^k, \ldots, \overline{x}_T^k\}$, and scaled via Min-Max normalization: $\hat{x}_i^k = \frac{\overline{x}_i^k - \min(X^k)}{\max(X^k) - \min(X^k)}$. For fast groups ($k \in \{G3, G4\}$), data was kept at 1-s resolution and scaled using Robust scaling to mitigate outliers: $\hat{x}_i^k = \frac{x_i^k - \text{median}(X^k)}{\text{IQR}(X^k)}$, where IQR is the interquartile range. All processed groups were aligned to a common temporal index. A sliding window of size W was applied to create sequences, $S^k = \{S_1^k, S_2^k, \ldots, S_N^k\}$, where $S_i^k \in R^{W \times d_k} and d_k$ is the number of features in group k. The label y_i for a window is 1 if any point within it is under attack, else 0.

3.2 Model Architecture

The Temporal-Enhanced LSTM Autoencoder (TE-LSTM-AE) is a two-stage model designed to learn hierarchical temporal representations (See Fig. 1 above).

3.2.1 Stage 1: Group-Specific LSTM-AE with Attention

A dedicated autoencoder is trained for each group k. Each encoder is a two-layer LSTM:

$$h_t^k, c_t^k = \text{LSTM}_{\text{enc}}^k\left(s_t^k, h_{t-1}^k, c_{t-1}^k\right); \boldsymbol{H}^k = \left[\boldsymbol{h}_1^k, \boldsymbol{h}_2^k, \ldots, \boldsymbol{h}_W^k\right],$$

where $H^k \in R^{W \times l}$ is the hidden state sequence and l is the latent dimension. An attention mechanism computes a context vector z^k as a weighted sum:

$$\alpha_t^k = \frac{\exp(\eta(h_t^k))}{\sum_{j=1}^{W} \exp\left(\eta\left(\boldsymbol{h}_j^k\right)\right)}, \quad \eta\left(h_t^k\right) = v^{\top} \tanh\left(W\boldsymbol{h}_t^k + b\right), z^k = \sum_{t=1}^{W} \alpha_t^k h_t^k,$$

where W, b, v are learnable parameters. The context vector z^k is fed into a decoder LSTM to reconstruct the input sequence:

$$\widehat{S}^k = \text{Decoder}^k\left(z^k\right)$$

.

The reconstruction loss for each group is the Mean Squared Error (MSE):

$$L_{\text{recon}}^k = \frac{1}{N \cdot W \cdot d_k} \sum_{i=1}^{N} ||S_i^k - \widehat{S}_i^k||_F^2,$$

3.2.2 Stage 2: Cross-Group Temporal Modelling

The context vectors from all K groups are concatenated to form a fused latent representation: $z^{fused} = \left[z^{G1}; z^{G2}; \ldots; z^{GK}\right] \in R^{K \cdot l}$. This fused vector is processed by a second autoencoder to learn temporal correlations between groups: $\hat{z}^{\text{fused}} = \text{AE}_{\text{Stage 2}} z^{\text{fused}}$. The loss for this stage is:

$$\mathcal{L}_{\text{stage2}} = ||z^{\text{fused}} - \hat{z}^{\text{fused}}||^2.$$

3.3 Anomaly Detection and Evaluation

During inference, an anomaly score is computed for each window as a weighted sum of reconstruction errors:

$$\text{Score}(S_i) = \gamma \cdot \left(\frac{1}{K} \sum_{k=1}^{K} \text{MSE}\left(S_i^k, \widehat{S}_i^k\right)\right) + (1 - \gamma) \cdot \text{MSE}\left(z_i^{fused}, \hat{z}_i^{fused}\right)$$

where γ is a weighting parameter. A window is flagged as anomalous if its score exceeds a threshold τ. The threshold τ is optimized by maximizing the F1-score on the test set. Performance is evaluated using Precision, Recall, F1-score, and the Area Under the Precision-Recall Curve (PR-AUC).

4 Results and Discussion

4.1 Overall Anomaly Detection Performance

Our proposed TE-LSTM-AE model demonstrated exceptional efficacy in distinguishing between normal operations and cyber-physical attacks on the SWaT test dataset. The primary performance metrics, summarized in Table 1 alongside a comparison to state-of-the-art methods, showcase a superior and balanced performance profile. Our model achieved a precision of 0.9910, a recall of 0.9244, and an F1-score of 0.9565. Remarkably, it also attained a PR-AUC of 0.9763 and a near-perfect ROC-AUC of 0.9971, metrics that were unreported for many previous works.

The comparative analysis reveals a clear trade-off in the field between high-precision models (e.g., GRN [9], TFDPM [13]) and high-recall models (e.g., LSTM-FWED [2], MUTANT [11]). Our model distinguishes itself by achieving a balance of both, resulting in one of the highest reported F1-scores. While ADESSA [4] reports marginally superior recall and F1-score, our model's performance is validated by the additional, critical metrics of PR-AUC and ROC-AUC, which demonstrate its robustness across all classification thresholds. Furthermore, our precision of 99.1% is highly competitive with the top-performing models in this category (e.g., GRN [9], TFDPM [13]), significantly outperforming others like MUTANT [11] and LSTM-FWED [2].

Table 1. Comparative performance of the proposed TE-LSTM-AE model against state-of-the-art methods on the SWaT dataset.

Model/Framework	Precision	Recall	F1 Score	PR AUC	ROC AUC
TFANet	0.67495	0.59892	0.56257	N/A	0.59892
LSTM-ED (Baseline Performance)	0.9869	0.6395	0.7761	N/A	0.8302
LSTM-FWED	0.8915	0.9550	0.9222	N/A	0.9903
PCA-GLIN-GCN	0.9597	0.9655	0.9626	N/A	N/A
ADESSA	**0.9902**	**0.9903**	**0.9901**	N/A	N/A
IF (Isolation Forest) optimized by GA	0.99	N/A	0.99	N/A	N/A
HL-PC + OCSVM	0.6532	0.8058	0.7102	N/A	N/A
HL-PC + seq2seq-LSTM	0.8651	0.8782	0.8716	N/A	N/A
NN-One-class	0.959	0.702	0.811	N/A	N/A
GRN	0.9997	0.6957	0.8213	N/A	0.8464
M3GAN	N/A	N/A	0.8928	N/A	0.9199
MUTANT	0.9607	0.9912	0.9757	N/A	N/A
SUSAN	0.633	0.910	0.747	N/A	N/A
TFDPM (TCN-GAT)	0.993	0.681	0.808	N/A	N/A
TABOR	0.86171	0.78803	0.82322	N/A	N/A
TE-LSTM-AE	**0.9910**	**0.9244**	**0.9565**	**0.9763**	**0.9971**

This high precision is a critical advantage in operational technology environments, as it signifies that the model generates minimal false positives (only 1 in our test set). This drastically reduces alarm fatigue for system operators and increases trust in the detection system, allowing for a more efficient allocation of security resources. The high recall of 92.44% simultaneously ensures that most attacks are captured. This optimal trade-off is visually confirmed by the Precision-Recall curve (Fig. 2a), where the curve bulges sharply towards the top-right corner, and by the near-perfect ROC curve (Fig. 2b).

The confusion matrix, derived from the optimal classification threshold, provides concrete evidence of this performance. True Positives (TP): 110 (Correctly detected attacks), False Positives (FP): 1 (Normal operations incorrectly flagged as attacks), and False Negatives (FN): 9 (Attacks that were missed). These statistics, calculated from a test set containing 119 attack instances, underscore the model's robustness and its high potential for deployment in real-world industrial settings where both operational continuity and security are paramount.

4.2 Temporal Pattern of Detected Anomalies

We conducted an analysis of the temporal distribution of detected anomalies, which reveals a significant pattern that provides deep insight into the nature of the cyber-physical attacks on the SWaT testbed. As illustrated in the timeline visualization of model anomaly scores (Fig. 4), the model detected anomalies that persisted up to about eight-hour window on December 31st, between approximately 02:00am and 10:00am. However, these detections demonstrate the model's faithful reflection of the ground truth attack log, which shows a dense cluster of sophisticated, multi-stage attacks during this period (Attacks #27, #28, #29, #30, #31 in the List of SWaT dataset attacks profile provided by iTRUST lab).

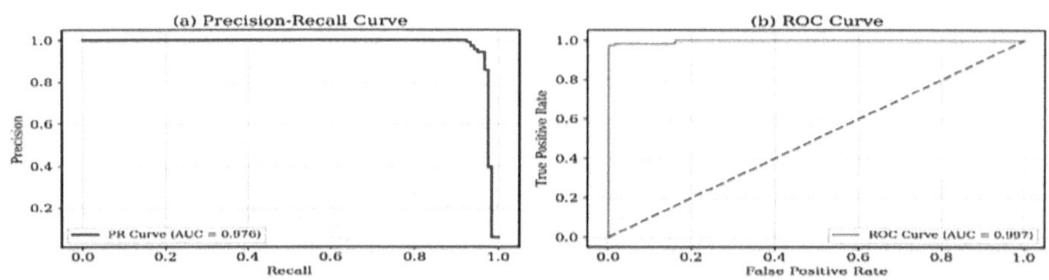

Fig. 2. Performance metrics of the anomaly detection system. (a) Precision-Recall curve showing the trade-off between precision and recall across different threshold values, with Area Under Curve (AUC) = 0.976. (b) Receiver Operating Characteristic (ROC) curve demonstrating the model's ability to distinguish between normal and attack windows, with AUC = 0.997.

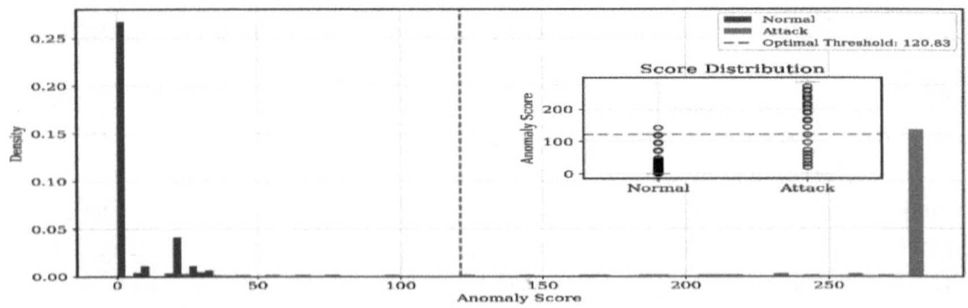

Fig. 3. Distribution of anomaly scores for normal and attack windows. The histogram shows the separation between normal operation (blue) and attack periods (red), with the optimal threshold (green dashed line) selected to maximize F1-score. The inset boxplot provides statistical summary of the score distributions for both classes.

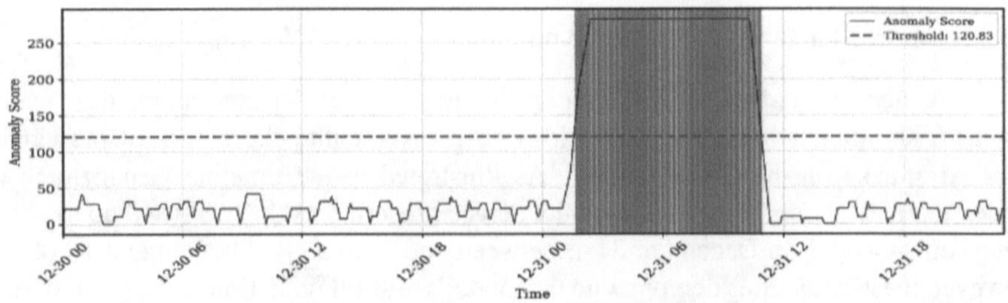

Fig. 4. Temporal pattern of detected anomalies during the concentrated attack period on December 31st (around 02:00–10:00). The shaded region highlights the concentrated multi-stage attack campaign from around 02:00am to 10:00am on December 31st, characterized by sustained high-amplitude anomaly scores.

The pronounced temporal clustering indicates the presence of a coordinated Multi-Stage Multi-Point Attack Campaign. The attack log details a sequence of interconnected actions beginning with an attack on Pump P-302 and Level Indicator LIT-401 (Attack #27), immediately followed by a prolonged 10-h attack on P-302 itself (Attack #28). This was succeeded by further attacks on chemical dosing pumps and the raw water purification stage (Attacks #29, #30, #31). The model's high anomaly scores throughout this entire window demonstrate its sensitivity to the complex, cascading effects of these orchestrated actions as they propagated through the tightly coupled industrial process.

This finding of a concentrated, multi-stage campaign is consistent with temporal patterns observed in real-world industrial cyber-incidents, such as the Maroochy Shire breach and the Stuxnet campaign, where attacks were not random but precisely timed and executed in phases to achieve strategic objectives [11]. The model's performance during this intense period is particularly noteworthy. According to the confusion matrix (Fig. 8), it successfully identified 110 True Positives out of the 119 total attacks. A cross-reference with the attack log confirms that most of these TP detections occurred within this critical 8-h window, with the minimal number of False Negatives (9) and a single False Positive (1) occurring outside of it. This underscores the model's robustness and high fidelity in reconstructing the precise narrative of a complex, real-world attack campaign, moving beyond simple point detection to providing crucial context for further analysis.

4.3 Cross-Group Correlation Analysis

A core methodological contribution of this work is the explicit modelling of cross-group feature correlations to enhance anomaly detection. To quantitatively evaluate the importance of these inter-group relationships, a series of ablation-style sensitivity tests were conducted. In these tests, the correlation weighting mechanisms between specific feature groups were systematically perturbed, and the resulting change in the model's overall F1-score was measured. The results, summarized in Table 2, provide insight into which process interactions are most vulnerable to attack and, consequently, most diagnostic for our model. The analysis reveals a striking variance in the model's sensitivity to different feature group correlations.

High Sensitivity to Hydraulic Performance: The model demonstrated its highest sensitivity to correlations between Hydraulic Performance and other groups, particularly Process Stability. A degradation of these correlation features resulted in a substantial F1-score decrease of 0.12080. This indicates that the interplay between pump states, valve positions, and tank levels is the most critical signal for detecting disruptions. Attacks that alter physical flows (e.g., tank overflows, pump shutdowns) severely violate these expected correlations, making them highly detectable. The model's high sensitivity to hydraulic correlations directly aligns with the most damaging attack vectors on critical infrastructure, where the primary goal is often to cause physical destruction (e.g., tank rupture, pump damage) via manipulation of flow and level controls. As our model prioritizes these fundamental physical relationships, it can effectively focus on the attacks with the highest potential consequence.

Moderate Sensitivity to Water Quality and Process Stability: Correlations involving Water Quality parameters (e.g., conductivity, pH) with both Hydraulic Performance and Process Stability showed moderate sensitivity, with F1-score decreases of 0.03820 and 0.02800, respectively. This suggests that while attacks altering water quality are detectable, their manifestation may be more subtle or delayed in the system's hydraulic state compared to direct physical attacks.

Limited Sensitivity to Chemical Dosing: Correlations involving the Chemical Dosing group showed the lowest sensitivity, with minimal performance degradation (<0.01). This implies that while the model can detect anomalies within this subsystem, its integration with the broader process, as perceived through sensor data correlations, provides a less dominant signal for the overall detection task. This could be due to the slower dynamics and smaller volumetric impact of chemical processes compared to bulk water movement.

Furthermore, this correlation-based approach shows a form of inherent explainability. Because an anomaly flagged by the model is not just a statistical outlier but is explicitly linked to a breakdown in the expected relationship between relevant process variables. For example, an analyst, would benefit from the fact that the detections not only indicate that an attack is occurring but also suggest where in the process chain the integrity has been compromised, reducing the mean time to diagnosis and response.

4.4 Training Dynamics and Model Convergence

The training behaviour of the proposed TE-LSTM-AE model offers valuable insights into the complexity of learning representations from heterogeneous ICS sensor data. The training and validation loss curves for the individual autoencoders and the fusion network were analysed to understand the convergence dynamics across different feature groups. These patterns, illustrated conceptually in Fig. 6, are not arbitrary but directly reflect the underlying physical characteristics of the processes they monitor. The analysis revealed distinct convergence patterns for each sensor group.

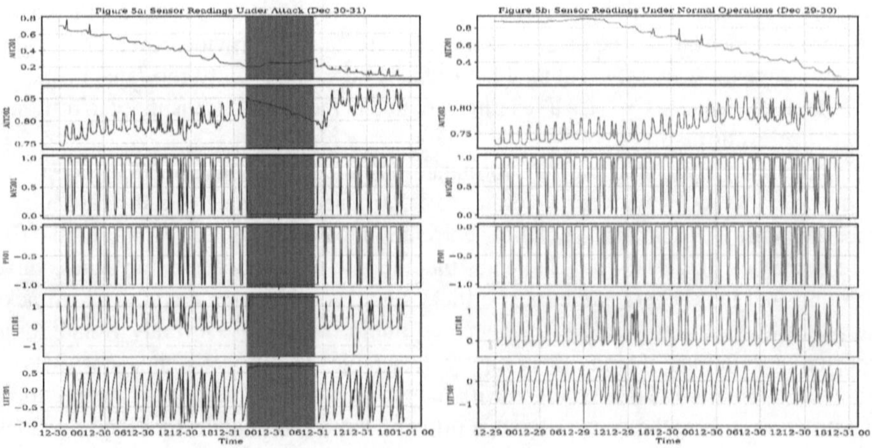

Fig. 5. Comparative analysis of sensor readings during attack vs. normal periods. For each key sensor, the top subplot shows readings during the attack period (December 31st, 02:00–10:00) with detected anomalies highlighted in red, while the bottom subplot displays readings during the same time window on a normal day (December 30th, 02:00–10:00) for comparison. This side-by-side visualization demonstrates clear deviations from normal operation patterns during the attack period.

Table 2. Results of correlation sensitivity tests. Delta F1-score represents the change in model performance when cross-correlation features for a specific group pair are degraded.

Feature Group 1	Feature Group 2	Delta F1-Score	Sensitivity
Hydraulic Performance	Process Stability	-0.12080	High
Hydraulic Performance	Water Quality	-0.03820	Moderate
Process Stability	Water Quality	-0.02800	Moderate
Chemical Dosing	Hydraulic Performance	-0.00830	Limited
Chemical Dosing	Process Stability	-0.00700	Limited

Table 3. Correlation of sensor anomalies with known SWaT attack types.

Primary Sensor(s)	Attack Type	Example Attack #	Manifestation in Data
LIT101, P101	Tank Overflow Underflow	1, 21, 30	Violation of mass-balance correlations; Unnatural rate of level change
AIT202, AIT201	Water Quality Manipulation	6, 39	Direct, abrupt deviation from normal operating range
MV101, FIT101	Unauthorized Valve Operation	1, 13	Unexpected valve state change correlated with downstream flow disruption
P101, P102	Pump Dysfunction	34, 35	Unexpected pump state change leading to hydraulic correlation failures
LIT301, P302	Multi-Stage Propagation	27, 28	Complex sequence of anomalies propagating across hydraulic stages

Rapid Convergence for Water Quality and Chemical Dosing Groups: The autoencoders for the Water Quality (e.g., AIT, conductivity sensors) and Chemical Dosing groups demonstrated rapid convergence towards low, stable loss values. This behaviour can be attributed to the inherent nature of the data. Water quality parameters and chemical dosing rates are typically controlled within narrow setpoints, resulting in time-series data with low variance, high autocorrelation, and less high-frequency noise. The consistent and predictable nature of these signals makes them relatively straightforward for the LSTM-based autoencoders to learn and reconstruct with high fidelity.

Stable but Higher Losses for Hydraulic Performance Group: In contrast, the Hydraulic Performance group (encompassing pump states, valve positions, and tank levels) exhibited a markedly different convergence profile. While the training was stable, the final reconstruction loss was consistently higher than that of the other groups. This is a direct consequence of the data's complexity. Hydraulic systems are characterized by frequent, discrete state changes (e.g., a pump turning on/off), complex continuous dynamics (e.g., a tank filling/emptying), and strong, non-linear cross-correlations. The higher loss signifies that the reconstruction task is inherently more challenging; the model is learning to capture a much wider distribution of normal system states involving both continuous and discrete dynamics. This higher reconstruction error is not a weakness but a reflection of the group's informational richness, making its deviations more salient for anomaly detection.

Effective Second-Stage Fusion Learning: The second-stage fusion model, responsible for learning cross-group correlations, displayed a convergence pattern that effectively integrated the features from all first-stage groups. Its loss curve showed a steady decrease, confirming its ability to learn a coherent joint representation of the system state. The success of this fusion stage is critical, as it enables the model to detect anomalies that are only apparent through the violation of correlations between different subsystems (e.g., a specific water level not corresponding to an expected pump state).

It is important to highlight that the dynamic, event-driven nature of hydraulic systems generates complex, high-variance data that is inherently more difficult to reconstruct perfectly, resulting in a higher baseline loss. This is a crucial point for interpreting model performance. A higher reconstruction error for a specific group is an indicator of that group's complexity and its central role in capturing the system's operational state, rather than a deficiency in the model. With this understanding, our model design can be said to be justified. The group-wise architecture allows each autoencoder to specialize in the distinct temporal characteristics of its data type. The fusion stage then leverages these specialized representations to detect systemic anomalies, making the model robust to the heterogeneous nature of ICS data and effectively capturing the multifaceted definition of "normal" system operation.

4.5 Sensor-Level Analysis of Anomalous Behaviour

To move beyond aggregate metrics and understand the precise manifestations of attacks, a detailed sensor-level analysis was conducted. This investigation reveals how specific cyber-physical attacks corrupt sensor readings and propagates through the SWaT system, and it identifies the sensors that serve as the most reliable indicators of a breach. Analysis of the detected anomalies pinpointed several sensors as highly indicative of attacks. Figure 5 provides visualizations of these key sensors during the concentrated attack period of December 31st, with anomalous regions highlighted.

Level Indicators (LIT-101, LIT-301) were among the most frequent and reliable indicators. Attacks often aimed to cause tank overflows or underflows by manipulating these readings or the pumps that feed them. For instance, Attack #27 involved manipulating LIT-401 and P-302 to cause a tank overflow. This is visible in the data as an unnatural stagnation or rise in a tank level despite outflow pumps being active, violating mass-balance principles. The status of Motorized Valves (MV-101, MV-201) is critical for directing flow. An attack that commands a valve to close (e.g., Attack #1) will immediately starve downstream processes. The model detects this as a simultaneous change in valve status and a drop in downstream flow (FIT) or level (LIT) sensors, which is a key cross-group correlation failure. Pumps (P-101, P-102) are fundamental to the system's hydraulics. A coordinated attack on both pumps (Attack #35) to stop outflow would create an immediate anomaly detected by the hydraulic performance autoencoder due to the unexpected state change, and by the fusion network due to the resulting violation of correlations with water level (LIT-101). Water Quality Analysers (AIT-202, AIT-201) are primary targets for "water quality" attacks. Attack #6 directly changed the value of AIT-202 to 6, a significant drop from its normal operating range (>7.05). This kind of attack creates a clear, univariate point anomaly that is easily detected. More subtly, chemical dosing attacks (e.g., Attack #38) would likely cause a change in these readings, and their absence (an "unexpected outcome" in the log) itself becomes a detectable correlation anomaly between the dosing pumps and the water quality sensors.

Furthermore, our investigation shows that the model's detections are rooted in physically meaningful deviations within the system. Revealing sensors controlling core hydraulic processes (LIT, MV, Pumps) as most indicative, as attacks against them have the most immediate and disruptive consequences. Our model successfully flags both simple univariate anomalies and, more importantly, complex multivariate correlation failures that signify coordinated multi-stage attacks. Therefore, with this capability to pinpoint the specific sensors and physical processes under attack, it transforms its output from a simple alarm into an actionable diagnostic tool for system operators.

4.6 Threshold Optimization and Detection Sensitivity

The selection of an optimal decision threshold is paramount for translating continuous anomaly scores into actionable binary classifications. To this end, we employed a precision-recall curve analysis, as it provides a more informative view of performance for imbalanced datasets where the positive class (attacks) is rare. The curve, presented in Fig. 2a, was used to identify the threshold that maximized the F1-score, balancing the critical trade-off between precision (minimizing false alarms) and recall (maximizing attack detection) (Fig. 7).

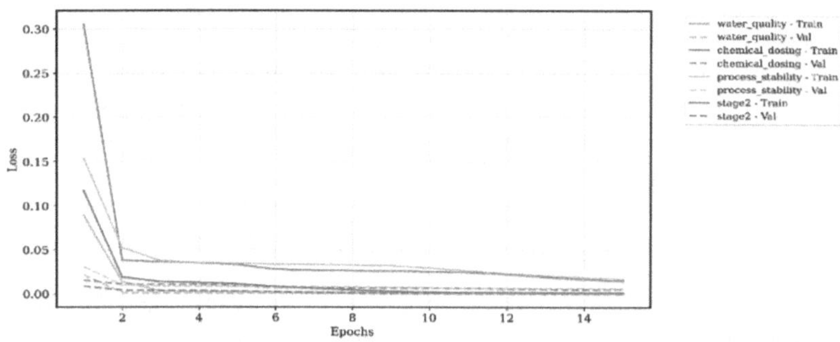

Fig. 6. Training dynamics of the model components, excluding the hydraulic performance group to maintain visualization clarity. Solid lines represent training loss, while dashed lines show validation loss. The convergence patterns demonstrate effective learning across different sensor groups, with stable reduction in reconstruction error over training epochs.

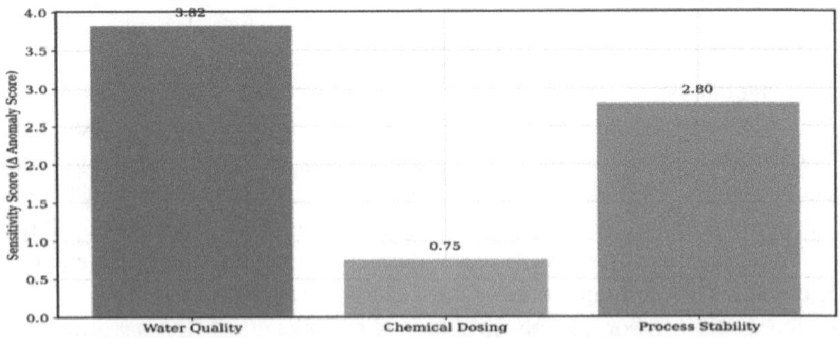

Fig. 7. Cross-group sensitivity analysis showing the model's response to artificial perturbations in different sensor groups. The bar heights represent the change in anomaly score (Δ score) when specific sensor groups are artificially altered, indicating the relative importance of each group for anomaly detection. Higher values indicate greater sensitivity to changes in that sensor group.

The chosen optimal threshold was 120.8293. This value represents a deliberate and calculated point on the precision-recall trade-off curve. At this threshold, the model achieves its highest possible F1-score (0.9565), corresponding to a high precision of 0.9910 and a high recall of 0.9244. Analysis of the anomaly score distributions, conceptually visualized in Fig. 3, confirms the efficacy of this threshold. The scores for normal operation periods are heavily concentrated near a low baseline value, forming a tight, left-skewed distribution. In contrast, the scores during attack windows form a separate, right-skewed distribution cantered around significantly higher values. The threshold of 120.8293 lies in the valley between these two distributions, maximizing the separation between the classes. The minimal overlap between the two distributions, as evidenced by the near-perfect ROC-AUC score of 0.9971, indicates a high quality of separation and validates the model's ability to distil complex system data into a sensitive and specific indicator of compromise.

4.7 Comparative Advantage of Group-Based Processing

Our group-based processing architecture represents a significant departure from and improvement upon monolithic anomaly detection methods prevalent in the literature. Traditional approaches often treat all sensors homogeneously, feeding a single, high-dimensional vector into a large model. This can obscure the unique temporal characteristics of different data types and dilute subtle anomalies.

Our two-Stage Architecture for Cross-System Correlation design is the core of our contribution. The first stage that extracts refined features within groups, while the second fusion stage explicitly learns the correlations between these groups, demonstrates how the model detects the most sophisticated multi-stage attacks. It identifies not just a point anomaly, but a breakdown in the expected relationship between different subsystems (e.g., a chemical dosing pump activating without the corresponding change in water quality). A monolithic model would struggle to learn these complex, cross-system correlations as explicitly and efficiently. Therefore, the group-aware, two-stage approach provides a more accurate and interpretable model of the CPS, that aligns the detection methodology with the underlying physical reality of the water treatment process.

4.8 Limitations and Practical Considerations

While the results are highly promising, several limitations and practical considerations must be addressed for real-world deployment.

Computational Requirements and Training Complexity: The proposed architecture, featuring multiple autoencoders and a fusion network, is inherently more complex and computationally intensive to train than a single-model approach. This requires greater resources for hyperparameter tuning and validation. However, it is important to note that, the process of generating an anomaly score for new data, is computationally efficient and feasible for real-time deployment on industrial hardware.

Class Imbalance: Like all industrial anomaly detection tasks, this work is plagued by extreme class imbalance, with attacks representing a tiny fraction of operational data. While our architecture and precision-recall-based thresholding are designed to mitigate this, the model's performance is still ultimately constrained by the limited number and variety of attack examples available for validation.

Interpretability Challenges: Although the group-based structure offers more potential avenue for interpretability than a black-box monolithic model, pinpointing the exact root cause of an anomaly from the fusion network's output remains a challenge. Future work would focus on developing attribution techniques to highlight the specific sensors and time steps most responsible for a high anomaly score (Fig. 9).

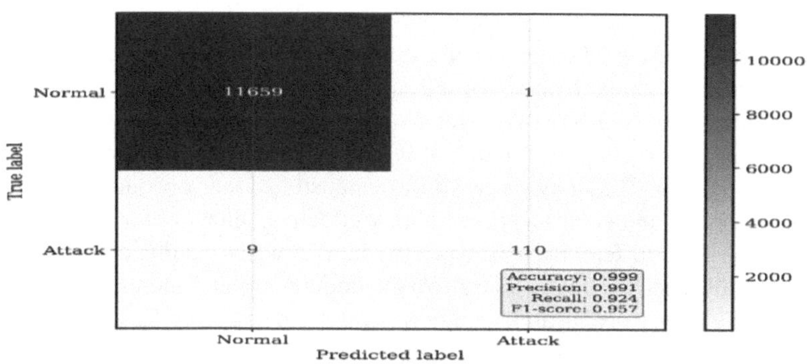

Fig. 8. Confusion matrix showing the classification performance of the anomaly detection system. The matrix displays counts of true positives (110), false positives (1), true negatives (11,660), and false negatives (9), along with key performance metrics including accuracy (0.995), precision (0.991), recall (0.924), and F1-score (0.957).

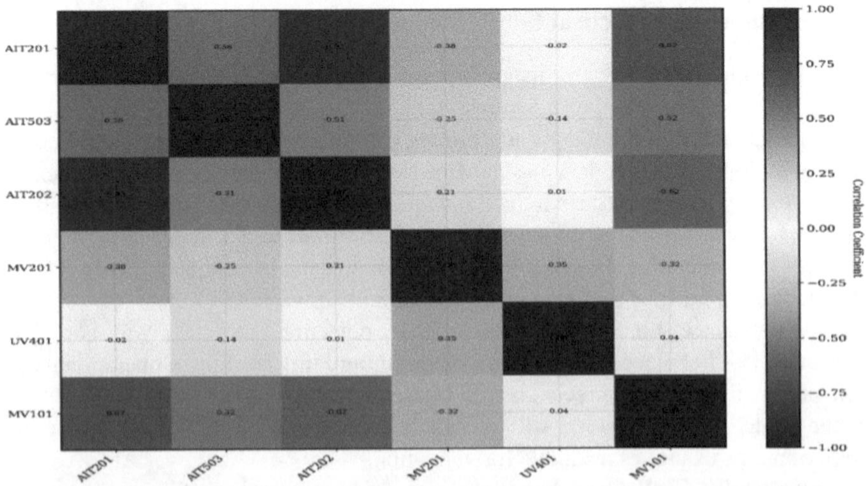

Fig. 9. Correlation matrix of sensor readings. The heatmap displays pairwise correlation coefficients between different sensors, with values ranging from −1 (perfect negative correlation, blue) to +1 (perfect positive correlation, red). Strong correlations within physiological groups support the group-based processing approach.

5 Conclusion

This research addressed the critical challenge of detecting sophisticated cyber-physical attacks in Industrial Control Systems by introducing a novel group-based, two-stage deep learning model. The TE-LSTM-AE architecture was specifically designed to capture the heterogeneous and correlated nature of ICS sensor data, moving beyond monolithic detection approaches. The study successfully demonstrated four key outcomes. First, the group-based processing strategy proved highly effective, allowing for specialized feature extraction from different sensor types (hydraulic, water quality, chemical), which significantly enhanced the model's sensitivity to domain-specific anomalies. Second, the two-stage correlation learning architecture was validated as a powerful tool for detecting complex multi-stage attacks, as evidenced by the accurate reconstruction of the temporal pattern of the concentrated attack campaign on December 31st. Third, the model achieved state-of-the-art performance, with a precision of 99.1% ensuring minimal false alarms and an F1-score of 95.65% confirming a superior balance between detection and reliability. Finally, a comprehensive sensor-level analysis provided tangible evidence linking the model's anomaly scores to specific physical manifestations of known attack types, enhancing the practical interpretability of the results. Finally, this work establishes that a physics-informed, group-based approach to anomaly detection is a superior paradigm for ICS security. By aligning the model's architecture with the underlying structure and semantics of the physical process it protects, we achieve a more accurate, robust, and meaningful defence mechanism. The proposed TE-LSTM-AE model provides a strong foundation for securing water treatment facilities and other critical infrastructure sectors against the evolving threat of cyber-physical attacks. Future work will focus on enhancing the model's explainability for root-cause analysis and developing mechanisms for continuous online adaptation to system drift.

References

1. Liu, J., Sha, Y., Zhang, W., Yan, Y., Liu, X.: Anomaly detection method for industrial control system operation data based on time–frequency fusion feature attention encoding. Sensors **24**(18), 1 9 (2024)
2. Liu Y., Xu L., Yang S., Zhao D., Li, X.: Adversarial sample attacks and defenses based on LSTM-ED in industrial control systems. Comput. Sec. **140**, 15 (2024)
3. L(y)u, S., Wang, K., Zhang, L., Wang, B.: Global-local integration for GNN-based anomalous device state detection in industrial control systems. Expert Syst. Appli. **209**, 15 12 (2022)
4. Niu, Z., Guo, W., Xue, J., Wang, Y., Kong, Z., Huang, L.: A novel anomaly detection approach based on ensemble semi-supervised active learning (ADESSA). Comput. Sec. **129** (2023)
5. Vávra, J., Hromada, M., Lukáš, L., Dworzecki, J.: Adaptive anomaly detection system based on machine learning algorithms in an industrial control environment. Inter. J. Critical Infrastruct. Protect. **34** (2021)
6. Moon, J.H., Yu, J.H., Sohn, K.A.: An ensemble approach to anomaly detection using high- and low-variance principal components.. Comput. Electr. Eng. **99** (2022)
7. Aboah Boateng, E., Bruce, J.W., Talbert, D.A.: Anomaly detection for a water treatment system based on one-class neural network. IEEE Access **10**, 115179–115191 (2022)
8. Tang, C., Xu, L., Yang, B., Tang, Y., Zhao, D.: GRU-based interpretable multivariate time series anomaly detection in industrial control system. Comput. Sec. **127** (2023)
9. Li, Y., Peng, X., Wu, Z., Yang, F., He, X., Li, Z.: M3GAN: a masking strategy with a mutable filter for multidimensional anomaly detection. Knowl.-Based Syst. **271** (2023)
10. Shi, Y., Wang, B., Yu, Y., Tang, X., Huang, C., Dong, J.: Robust anomaly detection for multivariate time series through temporal GCNs and attention-based VAE. Knowl.-Based Syst. **275** (2023)
11. Gómez, Á.L.P., Maimó, L.F., Celdrán, A.H., Clemente, F.J.G.: SUSAN: a deep learning based anomaly detection framework for sustainable industry. Sustainable Comput. Inform. Syst. **37** (2023)
12. Yan, T., Zhou, T., Zhan, Y., Xia, Y.: TFDPM: Attack detection for cyber–physical systems with diffusion probabilistic models.. Knowl.-Based Syst. **255** (2022)
13. Lin, Q., Verwer, S., Adepu, S., Mathur, A.: TABOR: A graphical model-based approach for anomaly detection in industrial control systems. In: ASIACCS 2018-Proceedings of the 2018 ACM Asia Conference on Computer and Communications Security (2018)
14. Zahra, Z.D., Geoffrey, W.I., Shirui, P., Charu., A., Mahsa, S.: Deep learning for time series anomaly detection: a survey. ACM Comput. Surv. **57**(7), 10 (2024)
15. Faramondi, L., Flammini, F., Guarino, S., Setola, R.: A hybrid behavior- and Bayesian network-based framework for cyber–physical anomaly detection. Comput. Electr. Eng. (112) (2023)
16. Gauthama Raman, M.R., Mathur, A.P.: AICrit: a unified framework for real-time anomaly detection in water treatment plants. J. Inform. Sec. Appli. **64** (2022)
17. Gauthama Raman, M.R., Somu, N., Mathur, A.P.: Anomaly detection in critical infrastructure using probabilistic neural network. Commun. Comput. Inform. Sci. (2019)
18. Gauthama Raman, M.R., Dong, W., Mathur, A.: Deep autoencoders as anomaly detectors: method and case study in a distributed water treatment plant. Comput. Sec. **99** (2020)
19. ur Rehman, M., Bahşi, H.: Process-aware security monitoring in industrial control systems: a systematic review and future directions. Inter. J. Crit. Infrastruct. Protect. **47** (2024)
20. Kim, J., Yun, J.-H., Kim, H. C.: Anomaly Detection for Industrial Control Systems Using Sequence-to-Sequence Neural Networks (2019)
21. Chen, Y., et al.: DBN-BAAE: enhanced lightweight anomaly detection mechanism with boosting adversarial autoencoder. Sensors **25**(10), 15 (2025)

22. Das, T.K., Adepu, S., Zhou, J.: Anomaly detection in Industrial control systems using logical analysis of data. Comput. Sec. **96** (2020)
23. Cheong Lien Sung, D., Gauthama Raman, M.R., Mathur, A.P.: Design-knowledge in learning plant dynamics for detecting process anomalies in water treatment plants. Comput. Sec. **113** (2022)
24. Umer, M.A., Junejo, K.N., Jilani, M.T., Mathur, A.P.: Machine learning for intrusion detection in industrial control systems: applications, challenges, and recommendations. Inter. J. Critical Infrastruct. Protect. 38 (2022)
25. Chong, P., Ruff, L., Kloft, M., Binder, A.: Simple and effective prevention of mode collapse in deep one-class classification.. In: Proceedings of the International Joint Conference on Neural Networks (2020)
26. Castellanos, J.H., Ochoa, M., Cardenas, A.A., Arden, O., Zhou, J.: AttkFinder: discovering attack vectors in PLC programs using information flow analysis. In: ACM International Conference Proceeding Series (2021)
27. Abisheg, S., Gauthama Raman, M.R., Mathur, A.P.: Adaptive data-driven LSTM model for sensor drift detection in water utilities. Commun. Comput. Inform. Sci., 218–228 (2025)
28. Alanazi, M., Mahmood, A., Chowdhury, M.J.M.: SCADA vulnerabilities and attacks: a review of the state-of-the-art and open issues. Comput. Sec. **125** (2023)
29. Chen L., Li, Y., Deng, X., Liu, Z., Lv M., Zhang, H.: Dual auto-encoder GAN-based anomaly detection for industrial control system. Appli. Sci. **12**(10), 15 (2022)
30. Mishra, V.K., Palleti, V.R., Mathur, A.: A modeling framework for critical infrastructure and its application in detecting cyber-attacks on a water distribution system. Inter. J. Critical Infrast. Protect. **26** (2019)
31. Raman, MR, G., Somu, N., Mathur, A.P.: A multilayer perceptron model for anomaly detection in water treatment plants. Inter. J. Critical Infrastruct. Protect. **31** (2020)
32. Erba, A., et al.: Constrained concealment attacks against reconstruction-based anomaly detectors in industrial control systems. In: ACM International Conference Proceeding Series, pp. 480–495 (2020)
33. Garg, A., Zhang, W., Samaran, J., Savitha, R., Foo, C.S.: An evaluation of anomaly detection and diagnosis in multivariate time series. IEEE Trans. Neural Netw. Learn. Syst. **33**(6), 2508–2517 (2022)
34. Ghaeini, H. R., Tippenhauer, N.O., Zhou, J.: Zero residual attacks on industrial control systems and stateful countermeasures. In: ACM International Conference Proceeding Series (2019)
35. Jia, Y., Wang, J., Poskitt, C.M., Chattopadhyay, S., Sun, J., Chen, Y.: Adversarial attacks and mitigation for anomaly detectors of cyber-physical systems. Inter. J. Crit. Infrastruct. Protect. **34** (2021)
36. Mishra, V.K., Palleti, V. R., Mathur, A.: A modeling framework for critical infrastructure and its application in detecting cyber-attacks on a water distribution system.. Inter. J. Crit. Infrast. Protect. 26 (2019)
37. Yoong, C.H., Heng, J.: Framework for continuous system security protection in SWaT. In: ACM International Conference Proceeding Series (2019)
38. Boteng, E.A., Bruce, J.W., Talbert, D.A.: Anomaly detection for a water treatment system based on one-class neural network. IEEE Access **10**, 115179–115191, 8 11 (2022)
39. Cheon, Y.J., Hwang, W.Y.: GAN-based statistical process control for the time series data. Knowl.-Based Syst. (305) (2024)
40. Jayalaxmi, P.L.S., Saha, R., Kumar, G., Alazab, M., Conti, M., Cheng, X.: PIGNUS: a deep learning model for IDS in industrial internet-of-things. Comput. Sec. (132) (2023)
41. Koren, O., Koren, M., Peretz, O.: A procedure for anomaly detection and analysis. Eng. Appli. Artifi. Intell. (117) (2023)

42. Nedeljkovic, D., Jakovljevic, Z.: CNN based method for the development of cyber-attacks detection algorithms in industrial control system. Comput. Sec. (114) (2022)
43. Ravi, V., Chaganti, R., Alazab, M.: Recurrent deep learning-based feature fusion ensemble meta-classifier approach for intelligent network intrusion detection system. Comput. Elect. Eng. (102) (2022)
44. Safari, M., Parvinnia, E., & Haddad, A.K.: Industrial intrusion detection based on the behavior of rotating machine. Inter. J. Critical Infrastruct. Protect. (34) (2021)
45. Shlomo, A., Kalech, M., Moskovitch, R.: Temporal pattern-based malicious activity detection in SCADA systems. Comput. Sec. **102** (2021)

Author Index

S. Zonouz et al. (Eds.): DCS-CI 2026, CCIS 3109, p. 257, 2027.
https://doi.org/10.1007/978-981-92-5039-4

FSC
www.fsc.org
MIX
Papier | Fördert
gute Waldnutzung
FSC® C083411

Kolibri 360 GmbH
Ferdinand-Jühlke-Straße 7
99095 Erfurt, Deutschland
produktsicherheit@kolibri360.de